Life

IN AMBER

Life

IN AMBER

GEORGE O. POINAR, JR.

Stanford University Press
Stanford, California
1992

Stanford University Press
Stanford, California

© 1992 by The Board of Trustees
of the Leland Stanford Junior University

CIP data appear at the end of the book

To my wife, Roberta,
who shares my fascination
for life in amber

Preface

As a boy, I used to sit and stare at the illustration of a weevil in amber that decorated the title page of one of my mother's books. The drawing was indelibly imprinted in my mind the moment I saw it, and I decided to find my own weevil in amber someday. I found my first pieces of Baltic amber in 1962 washed up by the North Sea on the western coast of Denmark in Jutland. My search for amber continued and resulted in explorations in Poland, the USSR, Mexico, Morocco, and the Dominican Republic.

My studies of biological inclusions began after I arrived at Berkeley in 1975, when a paleontologist colleague asked me to examine some unknown inclusions in Mexican amber. These turned out to be nematodes (roundworms), a group I was then (and still am) studying. I began searching for roundworms in other amber sources, especially the readily available material from the Dominican Republic. While doing so, I came across many other interesting creatures and wanted to know if they had been found previously in amber. To my astonishment, I discovered that very little had been written, especially in English, on the types of biological inclusions in amber. Sven Larsson's informative book on amber appeared in 1978, but it covered only the Baltic deposits, and I was dealing with amber from the Dominican Republic. Then in 1982 Keilbach published his list and bibliography of animals found in fossilized resins, and Spahr began to publish a series of additions and corrections (1981–present) to Keilbach's list. These publications provided access to the voluminous literature on amber inclusions, which was important because most papers were in foreign journals and many of these were no longer in existence. Despite these useful publications, a synthesis of the biological inclusions in amber was desirable, to include plant and animal remains as well as information on the sources of fossiliferous amber deposits in the world, their location, their history, and, especially, their geological ages. Such a synthesis was the goal of the present work.

Amber preserves fossils in a pristine condition. It also preserves very small and fragile organisms, such as nematodes, rotifers, and even mushrooms, that normally are not preserved under more conventional methods of fossilization. The inclusions are preserved in their normal three-dimensional shapes, complete with minute details of scales, feathers, antennae, and hairs. Biologists can easily compare them with present-day descendants to note what minute evolutionary changes have occurred. Such preservation is not confined to external characters but also extends to cells and tissues. It is quite possible that DNA in some of these cells is still capable of replication.

As our knowledge about the individual plants and animals in particular amber deposits increases, we can begin to reconstruct the paleoecology and paleoenvironment of the original forest. As the jigsaw pieces are fit together, an overall picture is beginning to appear. The fine details continue to be added as new amber fossils are discovered, and these discoveries will continue as long as there are discoverers.

I would like to thank the many people who allowed me to study animal and plant remains (biological inclusions) in their amber collections, including Jake Brodzinsky, Jim Work, Dan McAuley, Didi and Aldo Costa, Manuel Perez, Glen Osborne, Susan Hendrickson, and Heinz Meder. I am also most grateful for the assistance provided me in the identification of amber inclusions by colleagues throughout the world, especially John Doyen for insects and John Strother for plants. I appreciate the support and assistance of Wyatt Durham, above all regarding the Mexican amber deposits, and am also grateful to Peter Rubtzoff for translating pertinent Russian literature, Gerard Thomas for re-shaping amber pieces and printing the photographs, Christina Jordan for preparing the illustrations, Mariko Yasuda for typing the manuscript, and David Phillips for editorial assistance.

Finally, I am especially indebted to my wife, Roberta, who encouraged and supported this endeavor from the beginning.

G. O. P., Jr.
University of California, Berkeley

Contents

Contents

Contents

Tables

Eight pages of color photographs follow p. 216

] xiii [

Life

IN AMBER

CHAPTER ONE

Amber and Its Formation

Human fascination with amber dates back to prehistoric times, when this form of fossilized resin was probably considered to have magical powers and was used for adornment and trade. Amber may well be the first semiprecious gem appreciated by humans. Amber amulets dating from 35,000 to 1,800 B.C. have been found, and amber beads produced by the Magdalenian culture in France and Spain date from 15,000 to 10,000 B.C.

Amber was called *succinum* by the Romans (literally "sap stone"), *elektron* by the Greeks (providing the root of the English word "electricity" because static electricity could be generated on the surface of rubbed amber), and *Bernstein* ("burning stone") by the Germans. The English word "amber" is derived from the Arabic *anbar*, which aside from fossilized resin also means whale; the French made the distinction between the two by speaking about *ambre jaune* (the fossilized resin) and *ambre gris* (the product of the sperm whale).

One of the first serious attempts to describe the properties, occurrence, and uses of amber was made by Pliny (first century A.D.) in his *Natural History* (Book 37, Chapters 11–13). At that time a number of myths suggested sources for amber; for example, amber pieces were considered the solidified tears of the Heliades, the solidified tears of the Meleagrides, or the solidified urine of a lynx. Pliny presented evidence that amber was the discharge of a pine-like tree and, further, that it originated in the north and often contained small insects.

Although Pliny can be given credit for being one of the first persons, if not the first, to note organic inclusions in amber, the quotation of his compatriot Marcus Valerius Martialis (40–104 A.D.) provides a more poetic account of an amber inclusion: "The bee is enclosed, and shines preserved in amber, so that it seems enshrined in its own nectar."

Amber jewelry has always been much sought after and, through trade, amber pieces often ended up in areas quite distant from their source. The discovery of amber thought to have come from the Baltic but found in caches, excavations, and burial grounds throughout Europe and North Africa has led to the reconstruction of the ancient amber trade routes. These routes, the details of which are still somewhat controversial, varied depending upon the amber source and the period, but were active during the Greek and Roman periods. Such distant trade accounts for the finding of Baltic amber beads in Etruscan tombs as well as in excavations in Mycenae, Egypt, and Rome. Amber was also in demand by the Arabic and Oriental cultures, and worn Baltic amber beads still appear today in necklaces from Morocco, Iran, and Tibet.

The lore and study of amber has always been an integral part of the Baltic cultures, and from time immemorial to the present, amber has been collected from the shores of the Baltic Sea. The order of the Teutonic Knights, which occupied the Baltic seacoast in the thirteenth to fifteenth centuries, was formed in 1189 by crusading German knights. They had left the Holy Land, remained for a period in Hungary, and then headed north to Poland and the Baltic area, where they encountered the original Prussians, who at the time controlled the amber trade. Under the guise of Christianization, the Teutonic Knights destroyed these rural people, took their name, regrouped as an efficient military unit, and launched successful military campaigns against Poland, Lithuania, Latvia, and Estonia. They soon monopolized the amber trade and issued orders that all amber found along the shore of the Baltic Sea had to be turned over to them, under penalty of death. After the defeat of the Teutonic Knights by Poland and Lithuania at the battle of Grunwald in 1410, their strict monopoly over the amber market weakened, and other powers emerged to struggle for control.

In the middle of the nineteenth century, an innkeeper named Wilhelm Stantien and a merchant, Moritz Becker, established the first enterprise to dredge amber from the harbors and mine it from the earth. They established a factory near the village of Palmnicken on the west coast of Samland (presently called Yantarny in the Kaliningrad, formerly Königsberg, region of the USSR). The amount of Baltic amber recovered was astonishing; each year between 1875 and 1914 the factory produced between 225,000 and 500,000 tons of raw amber. The higher grades were used for jewelry and carvings, but the inferior grades were melted down for varnish. It is sickening to consider how many thousands of fossils were lost to the melting pots!

Amber was considered talismanic, able to bring supernatural bene-fits and protection to the wearer. Carved pieces and pieces with insects supposedly provided more protection and more magical powers than others. Such pieces were placed in graves, worn as amulets, stationed at the entrance of dwellings, attached to babies' cribs, and even carried into battle.

This believed protective character of amber probably led to its use in the cure of various ailments. Amber worn tightly around the neck was supposed to cure various throat disorders, including goiter. Amber was also used for curing poor eyesight, poor hearing, stomach ailments, asthma, heart disease, convulsions, ulcers, and general infections. It is in-teresting that the first book written on amber, in 1551 by A. Aurifaber (*Succini Historia*, Königsberg), dealt with the medicinal properties of the substance.

Amber has been carved by craftsmen the world over. From the six-teenth to the nineteenth centuries, some of the best amber carvers in the world could be found in two thriving cities, Königsberg and Gdansk, along the Baltic Sea. Sculptures and carvings from the workshops there were carried all over Europe and eventually abroad. Amber caskets, can-dlesticks, statuettes, beakers, cups, boxes, dice, mirrors, chandeliers, cup-boards, desks, game sets, and various pieces of furniture were fashioned by well-established craftsmen.

The really superlative work in amber, however, was an entire room commissioned in 1701 by King Frederick I of Prussia. This chamber con-sisted of a mosaic of multicolored amber depicting various life scenes, each scene focusing on a number of amber figure sculptures. The task took some 10 years to complete, and, in 1717, the room was sent to Leningrad for Peter the Great and Tsarina Elizabeth, and was assembled in their summer residence at Tsarskoe Selo. During the Second World War, the amber room was seized and taken back to Germany.

According to Melchior and Brandenburg (1990), the amber room was moved from Russia via Riga to Königsberg, where it was assembled in room 37 in the Königsberg Castle under the supervision of Gauleiter Koch. When the Russians were advancing on Königsberg, the room was dismantled, packed in large crates, and stored in the basement of the castle. Much intrigue has centered around the current whereabouts of the amber room, which was valued at 200 million marks in 1941. This treasure was and apparently still is the object of searches by German, Pol-ish, and Russian agents. Gauleiter Koch, who knew the whereabouts of the amber room, died in an old Franciscan monastery prison in Barczewo, Poland, in 1986, at the age of 90, apparently without revealing the location.

People have paid with their lives simply for searching for the amber room. One of the latest victims was Georg Stein, a German art dealer, who was helping the Russians in their quest for the treasure. He was found stabbed to death in a forest near Munich several years ago. Just as in the thirteenth century, the price of searching for amber can be paid in blood.

We may never know what happened to this famous room, but many historic amber carvings can be seen today in British and European museums. One especially interesting museum containing amber shrines and altars is located in the reconstructed castle of the Teutonic Knights in Malbork, Poland. Amber carving has continued to the present, and one can purchase Baltic, Dominican, and Mexican sculptures made by local craftsmen, although few of these can match the artistic and technical creations of the sixteenth to eighteenth centuries.

The preservative qualities of plant resins, including amber, were known by the ancients. Egyptians used resins to embalm their dead, and Greeks used them to preserve their wines. These resins also preserved, often in a remarkable state, all kinds of organisms that made contact with the sticky substance. Although we know much more today about the way life forms became entombed in amber, the quotation by Alexander Pope (1688–1744) still applies in some instances: "Pretty in amber to observe the forms of hairs, or straws, or dirt or grubs, or worms! The things, we know, are neither rich nor rare, but wonder how the devil they got there."

The present work surveys all life forms, from microbes to vertebrates, that have been reported from amber deposits throughout the world. These begin with the earliest known fossiliferous amber, dating from the Carboniferous Period, some 300 million years ago, to the more recent prolific deposits of the Tertiary. The book also treats the formation of amber from both resin and copal, as well as the location, geological history, and early exploration of the major world amber deposits.

The present work serves several purposes. By covering all life in amber (down to the generic level) it provides a guide to those interested in identifying organisms found in amber, at the same time showing what types have been reported by past workers. Practical information on how to determine fake amber and copal containing present-day forms of life is also presented.

This work can serve as a beginning for tracing the geological history of a particular group of animals or plants. Researchers may discover species related to the groups they are now studying, and because amber fossils are preserved so completely, they can be intimately compared with related living species. Finally, amber fossils demonstrate the past distribution of plants and animals, and this information can be used to reconstruct paleoenvironments in various regions of the world.

The present work also brings together the scattered, varied, multilingual literature that is inaccessible to so many. In so doing, it serves as a compendium on fossil life in all of the world's amber deposits.

Copal and Amber: Definitions and Characteristics

Amber is an amorphous polymeric glass with mechanical, dielectric, and thermal features common to other synthetic polymeric glasses (Wert and Weller, 1988). Amber, however, is a natural fossilized resin that shows a conchoidal fracture, but lacks cleavage and crystalline structure. It oxidizes on exposure to the air, and older amber is often brittle and crumbly. And because all amber can break and shatter when dropped or struck, care is required when handling it or excavating it. The chemistry and overall terpenoid hydrocarbon distributions of amber show a pattern resembling that of modern plant resins (Gough and Mills, 1972; Mills et al., 1984/85; Urbanski et al., 1976; Nicoletti, 1975; Beck, 1986; Grimalt et al., 1987; Simoneit et al., 1986).

Amber originates from resin, commonly recognized as the tacky, odoriferous "pitch" that sticks to us when we brush up against a pine or another resin-producing tree. Resins are not restricted to the conifers, however, and occur also in a wide range of flowering plants (angiosperms). Botanists recognize resins as complex mixtures of terpenoid compounds, acids, and alcohols secreted from plant parenchyma cells. Resins are insoluble in water, which distinguishes them from gums, which slowly dissolve (gum arabic is similar in appearance to some types of amber). Resins also differ in this respect from "sap," the watery fluid containing dissolved minerals that is transported from the roots to the leaves by the xylem tissues of plants.

When first deposited, most resins are mixed with essential oils (named after the French word *essence*) and are called oleoresins. The function of the essential oils, which are composed of complicated mixtures of terpine derivatives, is not well understood. These oils, often called volatile oils because they volatilize at normal environmental temperatures, are gradually lost as the resin ages.

Through the aging processes of oxidation and polymerization, the resin becomes harder and forms a semifossilized product known as copal. The word "copal" is derived from the Aztec *copalli*, which means resin (Santamaria, 1978), but unfortunately it has been used to denote types of trees and their general exudations (gums and oils), as well as hardened resin. In Mexico, 12 separate tree species of the genus *Bursera* are known by the term "copal," and another 19 species in the same genus can be included when the term is combined with a modifier (Martinez, 1979). All of the above-mentioned *Bursera* species produce an aromatic resin that is used as incense. Copal is also used to describe many of the hardened varnish resins originating from a range of tropical trees. Some of the more common types are Manila and Kauri copal, from coniferous *Agathis* trees, and Zanzibar, Congo, South American, or Colombian copal, from leguminous trees.

In the present work, copal is defined as recently deposited resin that can be distinguished from amber by its physical properties. In this definition, "recently deposited" extends from when resin hardens and cannot be molded by hand (if moldable, it is still resin) up until 3 to 4 million years, by which time, and under the right conditions, it acquires the unyielding properties of amber. Copal may be found on the ground, in the earth, and occasionally still attached to the tree that produced it.

The best criteria for judging whether a piece of fossilized resin is copal or amber are to be found in its physical characteristics. Without sophisticated scientific equipment, the most reliable criteria are hardness, melting point, and solubility. Other useful features include specific gravity, appearance under UV light, and burning characteristics (see Table 1). As pointed out previously (Poinar, 1982A), it is unwise to use a single test to distinguish between amber and copal (or a synthetic substitute). The best way to determine the hardness and odor and obtain a rough estimate of the melting point is by watching, under magnification of 10–20 times, a hot needle as it enters the resin (this is the hotpoint technique). In copal, the hot needle enters easily and immediately melts the surrounding resin. At the same time, a puff of white smoke and usually a sweet, lemony, resinous odor is produced. With amber, the needle must be pushed into the matrix, and the surrounding area does not become liquid but changes to a granular or gelatinous form. The emitted smoke is usually dark and the odor slightly acrid, with a burnt resinous tinge.

A second important test examines surface solubility when a drop of acetone is applied. This is best done on an unpolished but cleaned surface; a polish or an artificial coating may affect the results. The best pro-

Table 1. Characteristics of amber and copal

Characteristic	Amber	Copal
Melting point	200–380°C	under 150°C
Hardness (Moh's scale)	2–3	1–2
Solubility with acetone drops (on solid surface)	not soluble	soluble (surface becomes sticky)
Refractive index	1.5–1.6	1.5–1.6
Specific gravity	1.04–1.10	1.03–1.08
Appearance under UV light (on freshly broken unpolished surfaces)	usually a distinct bluish color	a faint light sheen, at most
Reaction when burnt	steady flame with black smoke	sputtering flame with whitish smoke
Odor from hot pointing	mostly acrid, burnt-resinous	mostly sweet, lemon-resinous

cedure is to apply three successive drops of acetone on the same spot, allowing one to dry before applying the next. After the third drop has evaporated, press your finger or an instrument against the area. With copal, the surface will be tacky or sticky because the acetone has partially dissolved the resin. With amber, there will be no change, although the area will sometimes feel cool to the touch as a result of acetone evaporation. Other solvents that can be substituted for acetone are ethanol (95 percent) and ether. An indication of hardness can be had by polishing the material in question. The surface of copal normally melts from the heat caused by friction against the polishing wheel, whereas the surface of amber remains solid.

The refractive index (a measure of the amount a light ray is bent as it enters or leaves a given substance) of amber is usually given as 1.54, although I have found a range of 1.5 to 1.6 among ambers from different geographical areas (see also Dahms [1906] for a discussion of the refractive indices of fossil resins). Most copals have a refractive index similar to that of amber so this test is not helpful in distinguishing between them. The specific gravity (the ratio of the density of a substance to that of water) of amber varies from 1.04 to 1.10. Some white varieties of Baltic amber are so light, owing to multitudes of minute air bubbles, that they will float in fresh water. Even a bore hole in a piece of amber may contain enough air to affect the specific gravity reading. The specific gravity values of copals presented in Table 1 are from the author's tests on a variety of copal samples from throughout the world. As a rule, the specific gravity of a resin increases with age. There is, however, considerable overlap between the values for copal and amber in the region of 1.04 to 1.08, so this feature is not especially useful for distinguishing between them. A simple method for determining the specific gravity of an amber sample is to place it in a series of solutions containing various amounts of dissolved table salt (purified if possible). For example, a specific gravity of 1.10 is achieved by dissolving 15.4 grams of salt in 100 milliliters of water. All amber (and copal) should float in a solution with such a high specific gravity, whereas many plastics will sink.

If resinous material has enclosed insects, they can be compared with present-day counterparts. Copals will contain contemporary (extant) insects or occasionally extinct species (Hills, 1957). Amber normally contains insect species that are now extinct. Specific identification, of course, can only be made by specialists with a knowledge of the insect groups in question.

A word might be said here concerning the presence of insects in copal that is passed as amber. As explained by Troost in 1821,

> I have seen large collections of amber, but found only one variety of insects in the true. The greatest part of the specimens in the cabinets, labeled amber, with insects, is not amber but copal—I myself have assisted, in Holland, one of my friends in selecting from copal, found at different druggists, a large collection with insects, which was cut and polished. This collection, after the death of the owner, was sold as one of true amber with insects, which the most practised eye was not able to distinguish.

If someone is already going to the trouble of imitating amber, then the enclosure of a "bug" or something else in the material would further increase the value. The rarer and more unusual the inclusion, the greater the price. Lizards are a favorite for imitations because they genuinely occur in amber and bring a high price. Copal is easy to use as an embedding medium since it is inexpensive, is easy to obtain, has many of the physical properties of amber, and can be melted to receive the inclusion. Also used commonly as an amber substitute are pieces of synthetic plastic or a combination of natural and synthetic resins.

In some circumstances, the imitator has undergone considerable research in order to make the "perfect" fake. One scientifically executed imitation fabricated in Germany had a basic matrix consisting of recent copal that had been chemically treated, leaving it clear and transparent. The pieces were then threaded on a string and sprayed with shellac. The final product possessed a specific gravity and refractive index close enough to those of amber to be approved by the American and German Gem Institutes. The material was sold commercially in the United States in the form of jewelry (necklaces, etc.) and many pieces contained insects. Also difficult to detect are pieces of authentic amber into which insects have been placed. The hole is usually filled with melted copal and the artificial cavity can sometimes only be detected under perfect optical conditions.

Amber ranges in color from light yellow to deep brownish and in light transmission from complete opacity to nearly complete transparency. Some amber appears black due to a high level of organic debris, but no natural amber is completely solid black. Melted amber with added carbon black produces a black ambroid material that was popular around the turn of the last century. This is similar in appearance to material sold in parts of Poland as "black amber," which is really jet, or fossilized wood. From the early days of Baltic amber exploration, workers recognized and named several different types of amber recovered from the sea. The most desirable in the trade was clear or transparent. The color of clear amber varies from almost colorless (water clear or yellow clear) to reddish yellow (red clear) or reddish brown. The great majority of amber from Mexico and the Dominican Republic is clear golden and, for this reason, the biological inclusions are generally easily visible.

A large amount of Baltic amber is cloudy or milky in appearance. This cloudiness, which is attributed to the presence of microscopic air bubbles (possibly originally derived from moisture droplets), varies in quality and quantity, and has its own special terminology. Bastard amber, the most common cloudy type, was extensively used for jewelry. The "clouds" or air bubbles in bastard amber are not uniformly distributed but are arranged in "streams" or flows that result in patterns collectively known as the "Baltic swirl." Cloudy amber is of little interest to fossil-hunters because it normally is inclusion poor and, when present, the inclusions are masked from view. In the past, cloudiness or turbidity was removed by clarifying the amber. Clarification was performed by immersing the material in oil (rapeseed oil was often used), slowly raising the temperature to the boiling point, and then allowing it to cool. The hot oil penetrated the fine cracks and pores of the amber and replaced the air, thereby causing cloudy amber to become transparent. This treatment was not always fully successful because many of the inner bubbles, which are from 8 to 20 micrometers in diameter, did not become filled if the piece was too large. The oil had a refractive index of 1.48, which was close enough to that of Baltic amber (1.53–1.56) to produce transparency. Ear-

lier commercial operations involving these clarifying operations sometimes added various dyes to the oil, thus imparting an artificial color
(often red) to the amber.

The air bubbles and water droplets enclosed in amber have been the
subject of recent debate. Berner and Landis (1987, 1988) and Horibe and
Craig (1987) crushed amber in a sealed container and analyzed the composition of the gases enclosed in the bubbles. They supposed that amber
was a sealant and that the composition of enclosed bubbles would reflect
either the atmospheric composition or plant gases produced when the
amber was formed. As suggested by the earlier clearing and coloring
treatments of amber, however, Hopfenberg et al. (1988) demonstrated that
amber is not a sealant. Instead, it has the ability to absorb, retain, and
release gases according to the laws of kinetics that apply to organic gases.

When amber is heated, minute disk-like fractures appear, and grow
in size until they become quite noticeable. Although considered un-
desirable at one time and evidence that the amber had been treated, these
so called "sun spangles" or "fish scales" are now common in amber jewelry and can be standardly produced by heating the amber to critical temperatures in dry sand. Present-day amber coming from Poland and the
USSR often contains these fractures because the iridescent colors that ac-
company them are considered to enhance the beauty of the amber. Some
dealers still insist they are natural biological inclusions.

The minute air bubbles are even more extensive (reaching 900,000
per square millimeter) in osseous, or bone, Baltic amber. As the name
implies, this type of amber is usually opaque white and has the general
appearance of chalk or bone. It graduates into a softer variety that has
been called frothy amber. Both bone and frothy amber are so light that
they may float in water, in spite of the iron pyrite crystals that are frequently embedded in these types.

As noted before, amber of the cloudy type is restricted almost entirely to the Baltic deposits and the characteristic cloudiness probably
stems from the original coniferous resin. Milky deposits or areas of
cloudiness often surround insect inclusions in Baltic amber, and these deposits are similar to those observed when incompletely dehydrated insects are mounted in balsam. The water escaping from the insects forms
an emulsion in the resin and turns it cloudy. Thus, the cloudiness adjacent to Baltic amber insects is probably the result of moisture escaping
from the entrapped insects. The water has long disappeared from these
droplets and upon examination now they appear to be air bubbles
(Mierzejewski, 1978). It is interesting that such cloudiness is rarely ob-

served surrounding insects in Mexican and Dominican amber, suggesting a difference in water solubility within the resins originating from leguminous trees.

Aside from the normally yellow, yellow-orange, or orange-brown color of amber, some deposits occur naturally blue or green. Such colors may be associated with turbidity in Baltic amber when the light rays are scattered by striking air bubbles of the proper size at a particular angle. Clarifying the amber causes it to lose both the turbidity and color. Blue or green amber from the Dominican Republic is mostly clear and insects are rarely found in such amber. The insect inclusions that are found in Dominican blue amber tend to be "washed out" or partially dissolved, as if they had been subjected to an acid treatment. When examined by transmitted light, Dominican blue amber appears a normal reddish brown, but when viewed with reflected light, a blue "sheen" is evident. Sometimes this sheen appears to be on the surface, and sometimes it radiates from the center of the piece. It is possible that the blue amber was produced from the roots of the tree and the blue color represents some component from the soil that entered the hardening resin, or it could have resulted from (volcanic?) heat that contacted the amber. Curiously, the blue color is not always stable: after 1 or 2 years, or even sooner, the blue in some Dominican amber completely disappears (even when kept in the dark) or becomes green. Some workers claim that the blue will return if the piece is repolished but this is not always the case.

Green amber occurs in the Baltic, Mexico, and the Dominican Republic. In Baltic amber, the green color is again associated with turbidity within the piece. In Mexico, the green color appears under reflected light and is associated with a surface phenomenon. In the Dominican Republic, the green color radiates from the matrix and is frequently found in amber closely associated with lignitic coal. Green amber usually has few recognizable inclusions and the color may gradually fade to yellow brown; a similar change also occurs in Baltic green amber as a result of exposure to the atmosphere. Light colored amber will gradually darken with age and the term "antique amber" is sometimes applied to aged Baltic cloudy amber, which is commonly found in necklaces made during the Victorian period.

The outer surface of amber that has been exposed to the air for some time gradually begins to oxidize and decompose. Decomposition usually begins with a network of fine hairline surface cracks, followed later by a crumbling of the outer layer. This "weathering" process may be associated with a reddening of the surface and subsurface areas. Even if the weathered crust is removed, the red color may persist, imparting a red color to the entire piece. This is how natural red amber is normally formed.

The deterioration of amber is indeed disturbing to museum curators responsible for the preservation of amber objects, and various methods have been employed to postpone or prevent this process. One method is to keep the specimens in containers that are as airtight as possible. Other methods include storing the pieces in water, mineral oil, or glycerin to reduce contact with the atmosphere. Small specimens can be embedded in a mounting medium such as Canadian balsam, clarite, or a synthetic plastic. Placing an air-resistant wax or resin coating on the amber surface is also employed.

Amber can be easily ground, sanded, and polished by methods currently available for plastics. But it is not sectile (capable of being cut into slices or shavings). When rubbed on some surfaces, amber becomes negatively charged and this character was used by the ancients to recognize amber; however, the property is shared with many modern plastics. When rubbed vigorously with the hand, amber will impart a characteristic odor, but this odor can be masked by commercial polishes or coatings now placed on many finished amber products. A variety of amber fakes occur on the market, and care should be taken to avoid confusing a man-made substitute for the real product (see Poinar 1982A for tests to determine amber forgeries).

Formation of Copal and Amber

How amber is formed has intrigued man for some time. By using the correct temperature and maintaining a constant pressure, would it be possible to create amber from tree resin? Attempts have been made but the final product was not amber. Apparently, we do not know all of the physical and chemical factors involved in amber formation. We can, however, discuss some of the probable ones.

First, the starting resin must be resistant to decay. Many trees produce resin, but in the majority of cases this deposit is broken down by physical and biological processes. Exposure to sunlight, rain, and temperature extremes tends to disintegrate resin, and the process is assisted by microorganisms such as bacteria and fungi. For resin to survive long enough to become amber, it must be resistant to such forces or be produced under conditions that exclude them. Because such conditions are highly unlikely, it is improbable that Baltic amber was formed from trees similar to present-day pines, spruce, or fir. Resins from these trees do not persist and have never been documented as occurring in fossil form of any great age. It should be noted that a single tree species may produce several types of resins, each differing chemically, and each having a different rate of decay. The Kauri gum tree or Kauri pine (*Agathis australis*), for example, produces at least five widely different resins from the heartwood, bark, and leaves (Thomas, 1969). The source of the resin that produced the famous Kauri "gum" was the resin canals of the bark. The special chemical composition of the bark resins may have made them less vulnerable to decomposition.

Any type of protection that would limit oxidation and exposure to the elements would favor preservation and amber formation. For example, resin that is rapidly covered by vegetation and soil would have reduced exposure to sunlight (UV rays and temperature) and air (rapid oxidation causes deterioration of the surface). Such covered resin would have a better chance of becoming fossilized than the same material left exposed to the elements.

The first step in the fossilization of resins is polymerization, a process whereby small molecules (monomers) combine chemically to produce a large network of molecules, or polymers. Polymerization makes the resin hard enough to be fractured (no longer pliable) and upon reaching such a state, the material is called copal, which, as mentioned earlier, is a subfossil resin (Stach et al., 1975) (Table 1). Copal is usually found on or in the ground under the resin-producing tree, but it can also be found in the earth far from the trees of origin. After a few million years under the right conditions, copal becomes amber. By this time the essential oils have become greatly reduced in concentration, and oxidation and polymerization have produced a fossilized resin with a hardness of between 2.0 and 3.0 on the Mohs scale, a specific gravity of 1.04–1.10, and a melting point of 200–380°C (see Table 1). However, the rate of transformation from resin to amber can vary depending on the physical and biochemical conditions that were present. All known amber-bearing beds are or have been associated with marine deposits, indicating an inundation with salt water at some time in their past. If seawater is a catalyst for the amberization process, then resin that was covered by the sea would form amber at a faster rate than resin of the same age not exposed to seawater. In addition to seawater, two other factors affecting the rate of amberization are probably temperature and pressure. Without the proper conditions, beds with a geological age of 3–4 million years (Ma) may contain a fossilized resin having the properties of copal and not amber. Such may be the case for Tanzanian "amber," which is said to be Pliocene in age (1.6–5 Ma), but still shows physical characters typical of copal.

The processes that turn wood into coal may closely parallel those that turn resin into amber, and many amber deposits occur in association with low grades of coal. Geologists, in noting the presence of amber in coal veins, have traditionally called it "resinite." This "resinite" or "ter-pene resinite" may occur as a microscopic (occluded) form derived from intercellular deposits of the wood, bark, and leaves, or as macroscopic (non-occluded) lumps derived from exudations (Selvig, 1945) (Poinar, 1991B). Although coals are known in rocks dating from the Precambrian to the Recent, large-scale accumulations became possible only after es-tablishment of the land flora toward the end of the Silurian (Cooper and Murchison, 1969). During the earth's history, there have been two great coal-forming periods, the first lasting from the Carboniferous to the Per-mian, some 120 million years in duration, and the second in the Tertiary Period, 20–60 Ma. Smaller deposits were also formed in many parts of the world during the Mesozoic Era. It is significant that the Baltic, Mexi-can, and Dominican amber deposits all correspond to the second or Ter-tiary coal-forming period and that the Alaskan, Canadian, Middle East, and Taimyr amber deposits correspond to the Mesozoic deposits. There are no significant amber deposits from the first coal-forming period. This is probably due to the absence of plants capable of producing large amounts of preservable resins and to the rigorous physical conditions (pressure and temperature) that any resin would have been subjected to during formation of the bituminous and anthracitic coals typical of the Carboniferous Period. Even so, spores, pollen, and portions of coniferous flowers have been described from minute pieces of Carboniferous amber (Smith, 1896). These fossils constitute the earliest record of life in amber.

Coalification is a continuing process, starting with decaying vegeta-tion, which then becomes peat, then lignite, then bituminous, and finally anthracitic coals. Throughout this process there is a darkening in color and an increase in lustre, accompanied by a general rise in carbon content and a decrease in moisture, volatile matter, and oxygen (Cooper and Murchison, 1969). This is roughly what also happens to resin during am-berization. Unfortunately, scientists know very little about the processes involved in turning wood into coal and have simply broken coalification into two stages. The first stage is relatively short and occurs during peat formation when microorganisms break down the original wood. This stage does not apply to amber formation because there is no evidence that amber-producing resins were directly modified by microbial activity. The second or geochemical stage in coal formation, however, involving chemical and physical changes resulting from temperature and pressure, occurs also during amberization. Forces that would tend to modify or destroy aging resin in the ground would be oxidation, weathering, heat (thermochemical reactions), shear forces (piezochemical reactions), and possibly radiation (radiochemical reactions) (Gray and Boucot, 1975). These are the same forces that alter organic microfossils in sedimentary rocks. Thermal destruction and shear are probably the most widespread, destructive forces affecting world amber deposits.

As previously mentioned, marine deposits often occur in amber beds, and such deposits may also occur in coal deposits. During amberization, the seawater may alter the amount of oxygen, maintain a lower temperature, or provide salts that catalyze the polymerization process.

Amberization is a continuous process and is progressing in many parts of the world today. The Kauri pine, *Agathis australis*, exists today in northern New Zealand, where it produces resin (Thomas, 1969). Historical accounts of the recovery of Kauri copal, formed some 30,000–40,000 years ago and now far removed from living trees, is well documented (Reed, 1972). Amber, too, occurs in New Zealand, produced by Kauri trees some 36–40 Ma. Young *Agathis* amber from the Pliocene (3–4 Ma) containing invertebrates and older *Agathis* amber have been reported in Australian brown coal deposits dating from the Oligocene (25–40 Ma) (Hills, 1957), and related trees are still growing in their natural habitat in Queensland, Australia. In Mexico and South America, resin from living leguminous trees of the genus *Hymenaea* has been recovered from the trunk of trees and the ground under trees, and *Hymenaea* copal has been found while cutting out road beds in Colombia. Amber from *Hymenaea* trees occurs in Mexico and the Dominican Republic.

In summary, we can suppose that the conditions required for amberization, which parallel in part those required for coalification, include a decay-resistant resin that is protected from direct contact with the elements under conditions allowing for a progressive oxidation and polymerization of that resin. Exposure to high temperatures (over 80°C), high pressures, and air would be detrimental to resin preservation, and would result in crumbling and fragmentation, which is one reason older amber is so fragile.

CHAPTER TWO

The World's Amber Deposits: Location, Age, and Source

Amber occurs in deposits throughout the world. The oldest fossiliferous amber was recovered from deposits of the Carboniferous Period (360–286 Ma) of the Paleozoic Era (590–248 Ma), but such material occurs in small fragments and has not been well studied (Smith, 1896). All substantial, investigated amber deposits date from the Mesozoic Era (248–65 Ma) and the Cenozoic Era (<65 Ma). Only those deposits containing fossiliferous amber (they occur in all of the major amber areas; see Fig. 1, Table 2) will be discussed here. These amber remains can be divided into two main categories on the basis of age, the Cenozoic Era deposits (Fig. 2) and the Mesozoic Era deposits (Fig. 3). Those in the former category all developed during the Tertiary Period (65–2 Ma), and deposits younger than the Tertiary fall into the copal category. Most remains from the Mesozoic developed during the Cretaceous Period (144–65 Ma). The amber deposits covered in the present work are listed in Table 2.

Tertiary Amber

The Tertiary is the period that saw the rise of the mammals, including the primates, and the dominance of flowering plants, birds, and insects. Most of the life forms preserved in Tertiary amber have close relatives in contemporary fauna and flora.

Baltic Amber

Certainly the most well-known amber in the world is Baltic amber, which occurs naturally in Poland, the Soviet Union, Germany, Lithuania, Latvia, Estonia, Denmark, Sweden, Great Britain, and Holland. Archaeological excavations have revealed amber beads in Bronze Age tombs in

Yugoslavia and Great Britain, and Stone Age amber artifacts have been recovered in northern Germany.

The most concentrated deposits of fossiliferous Baltic amber came and still do come from the Samland Peninsula, a roughly square area of land approximately 40 kilometers from east to west and 32 kilometers from north to south. The area is presently in the Kalinin district of the Russian Republic, and the administrative center is the seaport Kaliningrad (formerly Königsberg).

The now historic name, Samland, was derived from a province in Prussia that existed in the thirteenth century when the Teutonic Knights occupied the area. In the records of the Teutonic Knights from 1224 to 1246, the area was referred to as Samblandia or Sambia, which included the northern region of Slavonia and Nadrovia (Gusovius, 1966). The landmass of present-day Samland is much higher than the surrounding sloping beaches and is covered with patches of forests. It is in this area that the famous "blue earth" with its amber deposits occurs (Ley, 1951).

The soil profile of this region is composed of a series of overlying sands and clays (Fig. 4). The "blue earth" layer was some 28–30 meters beneath the surface and located in Tertiary deposits. It was about 5 meters beneath sea level, which explains how amber could be picked up on the Baltic Sea coast after a storm—the wave action loosened amber pieces from the exposed blue earth layer and washed them ashore. "Blue earth" is actually a misleading term for this stratum because it is really greyish-green when dry and black when wet. The green color comes from deposits of the greenish ferric-iron silicate mineral glauconite, which is characteristically formed on submarine elevations some 30 to 1,000 meters below sea level. Glauconite, generally formed as a precipitate from seawater, contains potassium and yields radiometric ages that measure the time of sedimentation. Thus, sedimentary samples can be analyzed for radiogenic argon, and a rough age of the stratum can be obtained.

Once the precise location of amber on the Samland Peninsula had been discovered, in the 1800's, mining operations began, and it is staggering to realize that over half a million kilograms of amber has been retrieved from the ground during the past century (Fig. 5).

The age of amber cannot yet be determined by direct analysis. In most cases, dating is achieved by analyzing the strata mineralogically or by examining fossils that occur in the amber strata. The latter method consists of searching for index fossils that identify and date the stratum. Such index fossils include swimming or floating marine organisms such

Figure 1. Well-known deposits of fossiliferous amber. 1, Baltic; 2, Dominican Republic; 3, Mexican; 4, Chinese; 5, Romanian; 6, Burmese; 7, Sicilian; 8, Canadian; 9, Alaskan; 10, Middle East (Lebanese); 11, Siberian (Taimyr); 12, New Jersey.

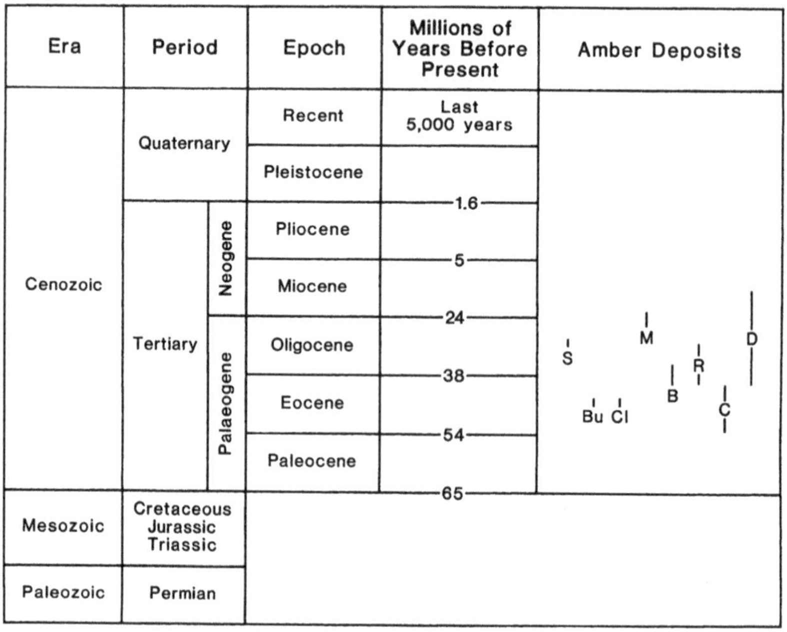

Era	Period		Epoch	Millions of Years Before Present	Amber Deposits
Cenozoic	Quaternary		Recent	Last 5,000 years	
			Pleistocene		
	Tertiary	Neogene	Pliocene	—1.6—	
				—5—	
			Miocene		
		Palaeogene	Oligocene	—24—	S M R D
				—38—	B
			Eocene		Bu Cl C
			Paleocene	—54—	
Mesozoic	Cretaceous Jurassic Triassic			—65—	
Paleozoic	Permian				

Figure 2. Geological time scale with Cenozoic Era amber deposits (B, Baltic; BU, Burmese; C, Chinese; Cl, Claiborne; D, Dominican; M, Mexican; R, Romanian; S, Sicilian).

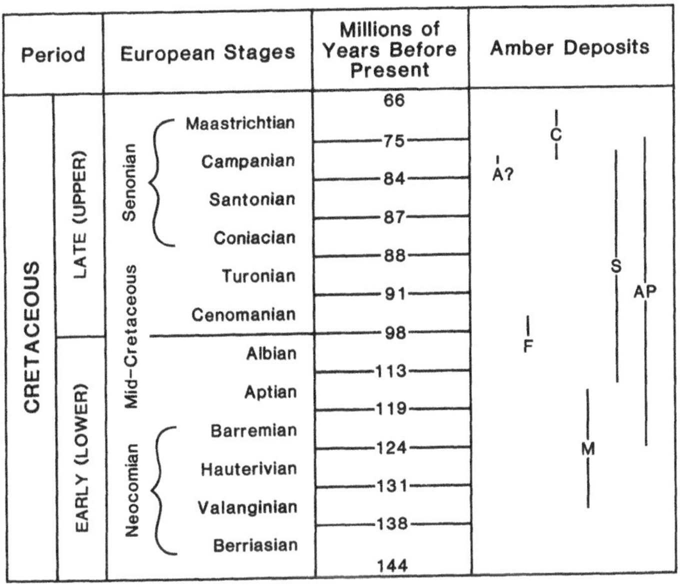

Period			European Stages		Millions of Years Before Present	Amber Deposits
CRETACEOUS	LATE (UPPER)	Senonian	Maastrichtian		66	
			Campanian		—75—	C
			Santonian		—84—	A?
			Coniacian		—87—	
		Mid-Cretaceous	Turonian		—88—	
			Cenomanian		—91—	S AP
	EARLY (LOWER)	Neocomian	Albian		—98—	F
			Aptian		—113—	
			Barremian		—119—	M
			Hauterivian		—124—	
			Valanginian		—131—	
			Berriasian		—138—	
					144	

Figure 3. Geological time scale with Cretaceous amber deposits (A, Alaskan; AP, Atlantic Coastal Plain; C, Canadian; F, French; M, Middle East; S, Siberian).

as graptolites, ammonites, or foraminifera that evolved rapidly, were widely distributed for a known time, and then disappeared. Their presence in a stratum corresponds to a particular period or epoch in the earth's past. Different types of shell-producing protozoa belonging to the foraminifera are usually used to identify the ages of amber-bearing strata, because these organisms were abundant during the Mesozoic and Cenozoic eras.

On the basis of these index fossils, much of the Baltic amber from the blue earth area of Samland (northern Poland and USSR) is contained in a mineral complex containing nannoplankton (microscopic, unicellular organisms) of zone 21. This zone (a specific band of earth materials) is categorized as Early Oligocene on the basis of stratigraphy, but amber in sediments of zones NP19 and 20, corresponding to the Late Eocene, and in sediments of zone NP16, corresponding to the Mid Eocene, were also found (Odrzywolska-Bienkowa et al., 1981; Kosmowska-Ceranowicz, 1987; Kosmowska-Ceranowicz and Müller, 1985). Although small amounts of Baltic amber have been recorded from the Early Eocene (55 Ma) in North Jutland (Larsson, 1978), the age of most Baltic amber varies from Early Eocene (55 Ma) to Early Oligocene (35 Ma). The majority of the

Table 2. Amber deposits containing biological inclusions

Type of amber	Geographical area	Approximate age (millions of years)	Approximate number of fossiliferous pieces recorded	References
Tertiary deposits				
Baltic	Most from the "Samland" Peninsula along the Baltic Sea, but other deposits in countries bordering on the Baltic and North seas	35–40 (some in North Jutland 50)	200,000	Ley, 1951 Kosmowska-Ceranowicz, 1987
Burmese	Burma	45?	<400	Cockerell, 1922
Claiborne	Arkansas	45	700	Saunders et al., 1974
Dominican	Dominican Republic	15–40	20,000	Baroni Urbani and Saunders, 1980 Lambert et al., 1985
Chinese	Fu Shun, China	40–53	2,000	Hong, 1981
Indonesian	Sumatra	10–20	<20	Durham, 1956
Mexican	Chiapas, Mexico	22–26	3,000	Langenheim, 1966
Philippine	Luzon	5–10	<20	Durham, 1956
Romanian	Romania	30–40 (some 70)	500	Protescu, 1937
Sicilian	Sicily	30	50	Emery, 1890
Cretaceous deposits				
Alaskan	Arctic Coastal Plain of Alaska	80?	<300	Langenheim et al., 1960
Atlantic Coastal Plain	Northeastern United States	75–124	<300	Langenheim and Beck, 1968
Canadian	Cedar Lake, Manitoba	70–80	5,500	McAlpine and Martin, 1969
French	Northwestern France	95–100	<200	Schlüter, 1978A
Middle East	Lebanon, Israel, Jordan	120–130	4,000	Bandel and Vavra, 1981 Nissenbaum, 1975 Schlee and Dietrich, 1970
Siberian	Eastern Taimyr (Soviet arctic)	78–115	4,000	Zherichin and Sukacheva, 1973

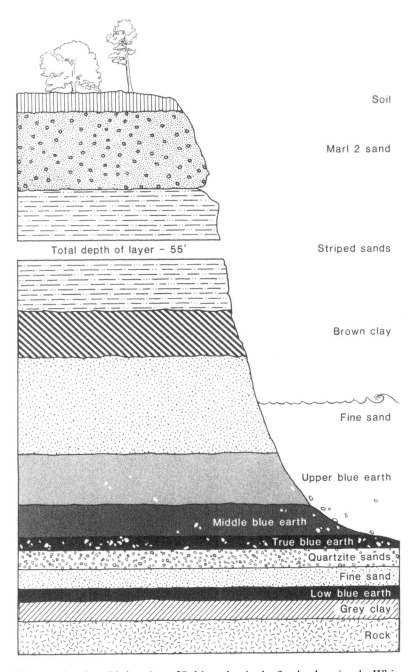

Soil

Marl 2 sand

Total depth of layer – 55′

Striped sands

Brown clay

Fine sand

Upper blue earth

Middle blue earth

True blue earth

Quartzite sands

Fine sand

Low blue earth

Grey clay

Rock

Figure 4. Stratigraphic location of Baltic amber in the Samland peninsula. White specks indicate location of amber, most of which is concentrated in the true blue earth layer (after Ley, 1951, and others).

Figure 5. Collecting Baltic amber. A. Ancient net method. B. Modern suction pump method.

fossiliferous material comes from the blue earth zone and was probably deposited during the period dating from the Late Eocene to the Early Oligocene, making it roughly 35 to 40 million years old.

While commenting on Baltic amber in his *Natural History*, Pliny (1st century A.D.) emphatically dismissed earlier mythological views on the origin of amber and boldly stated that it represented a hardened secretion from trees. His statement has withstood the test of time and it was some hundreds of years later that scientists began to discuss the types of trees that produced the resin precursors of amber. In earlier times, Baltic amber was the predominantly available form and thus the object of the first scientific experiments. Agricola (1546) first obtained succinic acid from Baltic amber and, as a result of his and subsequent studies, the presence of succinic acid was considered to be a special property of Baltic amber, which was called succinite. The idea arose that only succinite could be properly called amber: all other fossil resins from different parts of the world were thought to lack succinic acid, and so were called retinites. This idea was shattered in 1935 when La Baume reported that amber from Portugal, Sicily, and Romania also contained succinic acid. Later, Rottländer (1970) concluded that succinic acid, in the strict sense, does not occur naturally in amber, but is formed only during the alkaline hydrolysis of amber (as an alkaline salt) or during dry distillation (as its anhydride). He presented arguments to show that succinic acid (as a salt or anhydride) occurs as a natural oxidation product of amber and, therefore, is indicative of an aging process. Because many studies of the early amber trade routes identified Baltic amber as the "source" on the basis of succinic acid content, Rottländer feels that these proposed routes could be erroneous: the amber could have come from any location. Further discussion on this topic can be found in the 1980, 1981, 1984, 1985, and 1986 issues of *Acta Praehistorica et Archaeologica* (Berlin).

Other types of amber in addition to Baltic amber have been reported from northern Europe, and all of these have been analyzed by ^{13}C Nuclear Magnetic Resonance Spectroscopy (Lambert and Frye, 1982; Lambert et al., 1988; Beck et al., 1986). These results show that material referred to as beckerite belongs to the Baltic amber group and is a contaminated form of Baltic amber. Simetite from Sicily, rumanite from Romania, schraufite from Bukovina (Romania), walchowite from Czechoslovakia, and delatynite from Delatyn (USSR) all fall into a second category characterized by weak or absent exomethylene resonances. A material called statienite produced a spectrum more reminiscent of a coal than amber, and it may represent resin-impregnated wood. Gedanite, named after the

Latin *Gedanum* for the city of Gdansk, is a soft resin found in association with Baltic amber (succinite). Two different materials seem to be called gedanite because analyses of several samples yielded one set of samples with spectra identical to Baltic amber and a second set with quite different spectra (Beck et al., 1986; Lambert et al., 1988).

The type of tree that produced Baltic amber was first described by Aycke (1835) as a species of *Pinus*, but Göppert (1836) described it as an extinct genus and species, *Pinites succinifer*. Since then, the Baltic amber-producing tree has been given several names, depending on the investigator (see Table 3). Of all the early studies that considered the tree a member of the pine family, the most noteworthy are the studies by Conwentz (1890) and Schubert (1961). The former author studied the woody remains in Baltic amber, and although he redescribed the amber-bearing tree as *Pinus succinifera* (Göppert) Conwentz, he admitted that the material had characteristics of both *Pinus* (pines) and *Picea* (spruces). Schubert (1961) conducted a detailed study on the bark found in Baltic amber and discovered calcium oxalate crystals in the parenchyma cells, a characteristic he claimed would remove it from the genus *Picea* and definitely place it in the genus *Pinus*. The amber tree is most often cited as *Pinus succinifera* (Conwentz) emended Schubert 1961, although the correct name should be *Pinus succinifera* (Göppert) Conwentz, since Göppert first named the species.

Although referring to *Pinus succinifera* as the source tree of Baltic amber has become somewhat of a tradition, more modern methods of analysis have raised some annoying questions. In 1958, Hummel showed that infrared spectroscopy could be performed on tree exudates, and in 1964, Beck and others reported on their investigation of the infrared spectra of whole amber (in the solid state) dispersed in potassium bromide pellets (Beck et al., 1964; Savkevich and Shakhs 1964A, 1964B). These authors investigated amber from a number of different sources, and of 69 spectra of Baltic amber, all but one had an easily recognizable absorption band of medium intensity at 1,160–1,150 cm^{-1} (8.6–8.7 micrometers) that was preceded by a more or less flat shoulder ("Baltic shoulder") of nearly 0.5 micrometers width. Non-Baltic amber lacked this shoulder.

Infrared spectroscopy has been a successful technique in comparing fossil and Recent resins because the polymerization that occurred during fossilization preserved the simple functional groups of Recent resin (with the exception of carbon-carbon double bonds). Following the original studies of Beck et al. (1964), the spectra of various Recent and fossil resins were compared (Langenheim and Beck, 1965, 1968). There are, however, some potential problems in dealing with infrared spectra of amber. First, a single tree can produce resins of different composition (Thomas, 1969) and because an infrared spectrum reflects the nature of the original resin, different resins from the same tree could give different spectra, as suggested by Savkevich and Shakhs (1964A). Second, these authors showed also that infrared spectra from the same type of amber vary according to the degree of oxidation (aging) the samples have received (Savkevich and Shakhs, 1964B). Third and more recently, Mustoe (1985) presented evidence indicating that the burial temperature to which amber has been exposed may significantly affect infrared spectra. Notwithstanding the above considerations, the infrared method has provided us with interesting insights into the original plant producers of amber.

It was surprising to discover that the infrared spectra of Baltic amber were not similar to those of any modern pine resin, but were most similar to resin of the araucarian tree *Agathis australis* that grows in New Zealand (Langenheim, 1969). Quite similar infrared spectra between the copal of *Agathis australis* and Baltic amber were also presented by Thomas (1969), although he noted significant differences in the fingerprint region and in the carbonyl absorption near 1,700 cm^{-1}. After the hydrolysis of Baltic amber, which removes succinic acid from the insoluble portion, the resulting infrared spectrum was almost superimposable on that of the polymer from *Agathis australis* (Gough and Mills, 1972; Mills et al., 1984/85). Gough and Mills pointed out that most pine resins are characterized by a high abietane (derived from abietic acid) content and do not seem to polymerize or to survive for long periods, whereas resins with a high labdane (derived from agathic acids) content, such as in most *Agathis*, begin to polymerize as soon as they leave the tree and survive for long periods in the soil. Recent analyses of *Agathis* copal and Baltic amber using pyrolysis mass spectrometry have supported the results obtained with infrared spectroscopy (Poinar and Haverkamp, 1985). This analysis also demonstrated some heterogeneity among samples of Baltic amber, which was also found in thin-layer chromatographic studies of Baltic amber samples (Kucharska and Kwiatkowski, 1979). This heterogeneity raises the question of whether a single tree species was responsible for the amber production throughout the Baltic.

Thus, while analytical methods of analysis indicate that an *Agathis*-like araucarian tree was the Baltic amber resin producer, earlier morphological studies indicated a *Pinus*-like conifer. Although Rottländer (1970, 1974, 1980/81, 1984/85) presented chemical arguments showing

Table 3. Plants considered as the source of different amber deposits

Amber deposit	Type of examination	Proposed plant species or genus	Plant family	Reference
Baltic	Morphological-anatomical	*Pinus* sp.	Pinaceae	Aycke, 1835
Baltic	Morphological-anatomical	*Pinites succinifer* Göppert	Pinaceae	Göppert, 1836
Baltic	Morphological-anatomical	*Abies bituminosa* Haczewski	Pinaceae	Haczewski, 1838
Baltic	Morphological-anatomical	*Pinites succinifer* Göppert and Berendt	Pinaceae	Göppert and Berendt, 1845
Baltic	Morphological-anatomical	*Taxoxylum electrochyton* Menge	Pinaceae	Menge, 1858
Baltic	Morphological-anatomical	*Pityoxylon succiniferum* Krasu	Pinaceae	Schimper, 1870–72
Baltic	Morphological-anatomical	*Picea succinifera* Conwentz	Pinaceae	Conwentz, 1886A
Baltic	Morphological-anatomical	*Pinus succinifera* (Göppert) Conwentz	Pinaceae	Conwentz, 1890
Baltic	Morphological-anatomical	*Pinus succinifera* (Conwentz) emd. Schubert (= *P. succinifera* (Göppert) Conwentz)	Pinaceae	Schubert, 1961
Baltic	Chemical	*Pinus*	Pinaceae	Rortländer, 1970
Baltic	Infrared spectra	*Agathis*	Araucariaceae	Langenheim, 1969
Baltic	Infrared spectra	*Agathis*	Araucariaceae	Thomas, 1969
Baltic	Infrared spectra and chemical analysis	*Agathis*	Araucariaceae	Gough and Mills, 1972
Baltic	Resin analysis	*Pinus halepensis* Miller	Pinaceae	Mosini and Samperi, 1985
Baltic	Pyrolysis mass spectrometry	*Agathis*	Araucariaceae	Poinar and Haverkamp, 1985
Mexican	Infrared spectra, inclusions	*Hymenaea*	Leguminoseae	Langenheim and Beck, 1968
Dominican	Infrared spectra, inclusions, pyrolysis mass spectrometry, nuclear magnetic resonance spectroscopy	*Hymenaea*	Leguminoseae	Langenheim and Beck, 1968 Poinar and Haverkamp, 1985 Lambert et al., 1985

Location	Type of evidence	Taxon	Family/Order	Reference
Dominican	Morphological Plant fossils in amber beds	*Hymenaea protera* Poinar	Leguminoseae	Poinar, 1991D
Alaskan		*Sequoiadendron, Metasequoia, Taxodium*	Taxodiaceae	Langenheim et al., 1960; Langenheim, 1969
Alaskan	Pyrolysis mass spectrometry	*Agathis*-like	Araucariaceae	Poinar and Haverkamp, 1985
Alaskan	Nuclear magnetic resonance spectroscopy	*Agathis*	Araucariaceae	Lambert et al., 1989
Canadian	Infrared spectra, pyrolysis mass spectrometry	*Agathis*	Araucariaceae	Langenheim and Beck, 1965, 1968; Poinar and Haverkamp, 1985
Canadian	Nuclear magnetic resonance spectroscopy	*Agathis*	Araucariaceae	Lambert et al., 1989
Jordanian (Middle East)	Infrared spectra, mass spectrometry, thin layer chromatography	*Agathis*	Araucariaceae	Bandel and Vavra, 1981
Lebanese (Middle East)	Infrared spectra, mass spectrometry, thin layer chromatography	Araucarian	Araucariaceae	Bandel and Vavra, 1981
Romanian	Associated fossils	*Abies*	Pinaceae	Protescu, 1937
Romanian	Plant tissue in amber, silicified wood, fragments in amber strata	*Sequoiaoxylon gypsaceum* Göppert	Taxodiaceae	Ghiurca, 1988
Atlantic Coastal Plain (New Jersey)	Wood anatomy	*Cupressinoxylon bibbinsi* Knowlton	Coniferales	Knowlton, 1896
Atlantic Coastal Plain (New Jersey)	Infrared spectra	*Liquidambar*	Hamamelidaceae	Langenheim, 1969
Atlantic Coastal Plain (New Jersey)	Infrared spectra	*Sequoiadendron, Metasequoia*	Taxodiaceae	Langenheim in Wilson et al., 1967
Atlantic Coastal Plain (New Jersey)	Nuclear magnetic resonance spectroscopy	*Agathis*	Araucariaceae	Lambert et al., 1989
Arkansas	Infrared spectra	*Shorea*	Dipterocarpaceae	Saunders et al., 1974

how Baltic amber could have been produced from pine resin, much of the argument is hypothetical. Furthermore, Rottländer's experiments regarding the artificial production of amber could not be duplicated by Kucharska and Kwiatkowski (1978), who felt that Baltic amber was probably the secretion of an *Agathis*-like araucarian. In New Zealand today, there is "living" evidence that *Agathis* trees can produce copious amounts of preservable resin and that this resin can harden in the ground; in contrast, there is no evidence of pine resins becoming fossilized. On the other hand, there are numerous "*Pinus*" fossils in and associated with Baltic amber, whereas no araucarian remains have been reported. This discrepancy is indeed strange and raises the question of whether araucarian remains were in the amber but not recognized. Unfortunately, many of the original plant fossils in Baltic amber have been lost. In his book on Baltic amber, however, Bachofen-Echt (1949: 28) shows a photograph of a flower cone of *Pinites*. The similarity of this cone and the staminate cones or strobili of *Agathis* (Araucariaceae) is striking. The Araucariaceae date back to the Mesozoic Era, when the family was abundant in both the Northern and Southern Hemispheres. The family began to decline in the early Tertiary, however, and today the Araucariaceae is represented by only two genera, *Agathis* and *Araucaria*, both of which are restricted to the Southern Hemisphere. The Pinaceae made their appearance during the Cretaceous Period, although certain pine-like ancestral plants have been recovered from the Middle Jurassic. The family today is generally, with one exception, restricted to the Northern Hemisphere (Stewart, 1983).

Another possible explanation of why Baltic amber shows an *Agathis*-like spectrum but contains only fossils that were thought to belong to the Pinaceae is that the proposed tree, *Pinus succinifera*, was a now-extinct form that shared characters with present-day pines and araucarians. This tree could have been morphologically similar to pines but chemically similar to araucarians. In discussing this possibility, Larsson (1978) points out that there exist today a few pines, including the northwestern North American sugar pine, *Pinus lambertiana* Dougl., in which the main resin acids are the labdane and not the abietane type, and whose infrared spectrum reveals the "Baltic shoulder" characteristic of Baltic amber. We can hope that further analytical methods such as nuclear magnetic resonance, mass spectroscopy, X-ray diffraction, neutron activation analysis, and fission-track dating will provide answers to this age-old question. At the present time, I feel that the evidence supports the view that Baltic amber was, for the most part, produced by an *Agathis* or *Agathis*-related tree of the family Araucariaceae.

Because the Baltic amber deposits are so large, it has often been asked how so much resin could have been secreted to produce such a large quantity of amber. Conwentz (1890) presented the hypothesis that the resin production was abnormally high because the trees were attacked in mass by a parasite that produced a disease called "succinosis." This hypothetical parasite could have been a bark beetle or possibly a fungus similar to those that have wiped out the chestnut or elm trees in various parts of the world. There seems to be no direct evidence to support this hypothesis, however. Instead, Schubert (1961) presented a theory that the sudden increase in resin production was due to a change in climate that resulted from the transgression of the Tertiary Sea. This continual production of resin might have weakened the trees and made them susceptible to attack by parasites, which may have resulted in a still heavier supply of resin.

Another possibility is that the trees produced only their normal amount of resin, which slowly built up over a period of millions of years. Wyatt Durham (personal communication) has estimated that the resin production of the trees that resulted in the Mexican amber deposits continued for 4 million years; a substantial amount of resin could have been produced in that duration of time. Over 100 thousand metric tons of fossilized Kauri resin was removed from New Zealand soil between 1853 and 1970 (Reed, 1972) and it has been estimated that resin production by *Agathis australis* in the worked areas continued for only thousands, not millions, of years. Thus, it is conceivable and probable that the observed amounts of amber could have come from a succession of normal, healthy trees over a long time.

Dominican Amber

From the standpoint of availability, abundance, and number of fossil inclusions, the amber deposits in the Dominican Republic are second only to the Baltic. In fact, because of the present-day relative scarcity of Baltic amber inclusions, Dominican amber fossils are easier to acquire and have become the predominant amber sold in many parts of the world. As of June 5, 1987, however, a decree issued by the Santo Domingo government proclaims that taking amber fossils out of the country is prohibited without official permission issued by the National Museum of Natural History (Balaguer, 1987).

Historical records of amber in the Dominican Republic date back to the journals of Christopher Columbus (Hale, 1891). The first disclosure that the amber was fossiliferous (contained organic inclusions) was made by Lengweiler in 1939, followed by Renato Zoppis in 1949. Earlier geological reports had confirmed the presence of amber in the country and it had been found in Indian tombs in the form of ear ornaments, but no one showed much interest. Despite these earlier descriptions, it was the report by Sanderson and Farr (1960), discussed below, that sparked additional exploration and scientific investigation of the amber inclusions.

Hispaniola is one of the larger islands in the Greater Antilles Archipelago of the Caribbean Sea. The Dominican Republic comprises the eastern two-thirds of the island, Haiti the other third. The island is composed of five distinct upland ranges running along a northwest to southeast axis. The most extensive range is the Cordillera Central, which includes Pico Duarte (elevation 3,175 meters), the highest mountain in the West Indies. No amber has been recovered from this range or the two smaller ranges located to the southwest of the Cordillera Central. A minor range in the northeastern portion of the country (the Cordillera Oriental) contains the amber mines La Medita and Ya Nigua near El Valle (Fig. 6, Table 4). A large percentage of Dominican amber, however, comes from mines located in the Cordillera Septentrional, a range to the north and roughly parallel to the Cordillera Central. The original mine sites were located in this area (Sanderson and Farr, 1960). These mines, all of which are in the general area between Santiago and Puerto Plata, include Palo Quamado, Palo Alto, Juan de Nina, La Toca, Las Cacaos, Los Aguitos, La Cumbre, Los Higos, La Bucara, Pescado Bobo, El Naranjo, Las Auyamas, El Arroyo, Aquacate, Carlos Diaz, and Villa Trina (Fig. 7, Table 4). Two additional sources of amber include the sites Comatillo and Sierra de Agua, which are located at Bayaguana in the lowlands and near Cotui in the valley between the Cordilleras Central and Septentrional. Both of these sites produce a lighter, softer form of amber than is mined in the highlands, although recently some hard amber has come from Bayaguana.

After conducting a preliminary investigation, Sanderson and Farr (1960) concluded that the amber in both Cordilleras sites (Oriental and Septentrional) occurred in Tertiary sandstone beds. For the mine at Palo Alto, they described an upper leached area that was up to 15 meters in depth, followed by a soft clay shale (0.5–2.0 meters thick), then a hard silty shale (2.0–2.5 meters thick), and finally a gray sandstone of un-

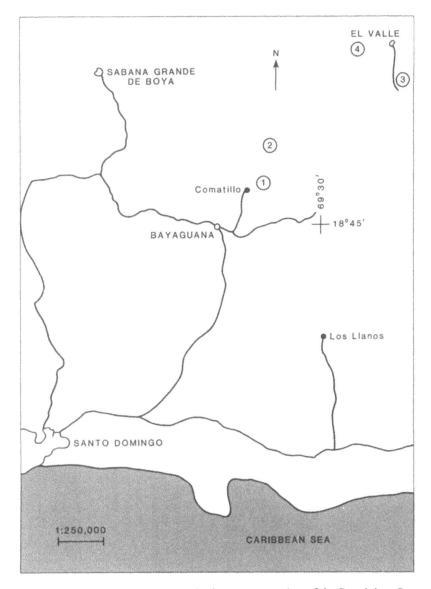

Figure 6. Location of amber mines in the eastern portion of the Dominican Republic: Comatillo (1) and Sierra de Agua (2) in the Bayaguana area, and La Medita (3) and Ya Nigua (4) in the El Valle area.

Table 4. Characteristics of amber from various mines in the Dominican Republic

Mine[a,b]	Location	Amber features (personal communications from Rubio and many other amber workers and dealers)	Age estimated from nuclear magnetic resonance data (Lambert et al., 1985)	Age of surrounding sedimentary rock on basis of nannofossils[c] (Baroni Urbani and Saunders, 1980; Cepek and Eberle, personal communication)
La Toca (including La Peña mine)	Cordillera Septentrional	Hard, clear, yellow to red. La Peña mine noted for its vertebrate fossils.	30–40 Ma	30–45 Ma
La Cumbre	Cordillera Septentrional	Hard, yellow to red. Few invertebrates.	N.A.	N.A.
Las Cacaos (Tamboril)	Cordillera Septentrional	Hard, clear, yellow to red, but also known for blue amber.	30–40 Ma	N.A.
Palo Quemado	Cordillera Septentrional	First of the Dominican amber mines opened about 50 years ago; hard, mostly red to yellow amber, but also some blue.	N.A.	12–23 Ma
Los Higos	Cordillera Septentrional	Hard, clear, yellow to red amber, also some green. Few insects (many decomposed).	N.A.	N.A.
La Bucara	Cordillera Septentrional	Hard, clear, yellow (in rock) to red (in soil), but also some blue amber.	N.A.	20–40 Ma
Palo Alto	Cordillera Septentrional	Clear, yellow to red amber. Many of the early invertebrate fossils came from these deposits.	20–30 Ma	23–30 Ma
Los Aguitos	Cordillera Septentrional	Similar to Palo Alto. Some invertebrate fossils.	20–30 Ma	N.A.
Pescado Bobo	Cordillera Septentrional	Similar to Las Cacaos, but also some dark blue amber. Few invertebrates.	N.A.	N.A.
El Naranjo	Cordillera Septentrional	Clear, yellow amber, sometimes with bluish tinge, some with metal deposits. Few fossils.	N.A.	N.A.

Name	Location	Description		
Juan de Nina	Cordillera Septentrional	Pale yellow, fairly soft amber; contains metal associated with organic material.	N.A.	N.A.
Las Auyamas	Cordillera Septentrional	Yellow amber. Some fossils.	N.A.	N.A.
El Arroyo	Cordillera Septentrional	Yellow and yellow-blue amber. Very few fossils.	N.A.	N.A.
Aguacate	Cordillera Septentrional	Yellow to red amber. Some fossils.	N.A.	N.A.
Carlos Diaz	Cordillera Septentrional	Yellow to red amber. Some vertebrates.	N.A.	N.A.
Villa Trina	Cordillera Septentrional	Yellow to red amber.	N.A.	N.A.
El Valle (La Medita) (Ya Nigua)	Cordillera Oriental	Mostly hard, brittle, yellow to red amber; also a type of blue (light) amber. Many pseudoscorpions and some vertebrates.	20–30 Ma	N.A.
Bayaguana (Sierra de Agua)	Foothills in eastern part of country	Hard and brittle pale yellow to red amber. Few fossils.	N.A.	N.A.
Bayaguana (Comatillo)	Foothills in eastern part of country	Yellow-grayish, fairly soft (some hard) amber. Filled with decayed wood and organic matter.	15–17 Ma for soft material	N.A.
Cotui	In valley between Cordillera Septentrional and Cordillera Central	Clear amber, sometimes with yellow tinge, relatively soft.	15–17 Ma (some material may be younger)	N.A.

NOTE: N.A. = not analyzed.

[a]The term "mine" can refer to a slide area, exposed hillside, or one or more excavation sites (up to 200 in some mine areas, although only a few may be worked) that consist of variously sized tunnels into the hillside to intersect the veins containing the amber.

[b]For additional information on mine characteristics, see Martinez and Schlee (1984).

[c]Because the amber was redeposited in a sedimentary rock formation after the amber had already been formed, the use of index fossils provides a minimum age of the amber.

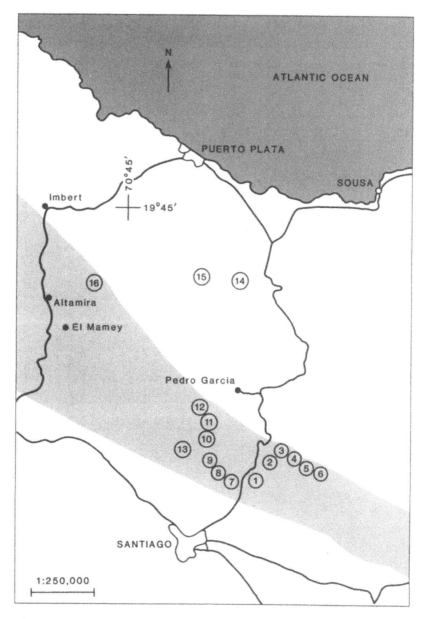

Figure 7. Location of amber mines in the Cordillera Septentrional of the Dominican Republic: 1, Palo Quemado; 2, Las Cacaos; 3, La Toca; 4, La Cumbre; 5, Carlos Diaz; 6, Villa Trina; 7, Los Higos; 8, La Búcara; 9, Aguacate; 10, Palo Alto; 11, Las Auyamas; 12, Los Aguitos; 13, El Arroyo; 14, Juan de Nina; 15, El Naranjo; and 16, Pescado Bobo. Lower shaded area represents the "El Mamey" formation, which yields a Late Eocene age of approximately 40 million years.

known thickness. This latter layer, described as fine grained, micaceous, carbonaceous, laminated graywacke, contained the amber. Recent studies by Champetier et al. (1980) on the profile of the mining area at Palo Alto differ somewhat from Sanderson and Farr's original report. Champetier et al. describe amber at Palo Alto as occurring in narrow sequences of lignite or carbonaceous clay alternating with layers of sandstone. In non-disturbed series at Palo Alto, amber appears in the form of small fragments in the carbonaceous face and in the form of nodules in highly lignited (= carbonaceous) clays. In disturbed series, the amber occurs in fine sandstone of adjacent layers. At the mine near El Valle, the situation is similar. The base is composed of a layer of sandstone, which varies from 4 to 12 meters in width, depending on the terrain. Pieces of driftwood and flattened carbonaceous material occur in this sandstone. As one ascends the series, the sandstone disappears, and beds of silt, silty clay, and clay appear. Amber is found in these heavily carbonaceous (lignited) upper layers of clay (Fig. 8).

Considerable amounts of amber remain in the mountains of the Dominican Republic and possibly in Haiti. Although no extensive collections have yet been made in the latter country, minute amounts of amber were noted in a core sample from a lignite deposit near Maissade in the Central Plateau of Haiti (Lemoine, cited in Sanderson and Farr, 1960).

The exact age of the Dominican amber is still being investigated. Recent studies indicate that the age of Dominican amber varies considerably depending on the region and depth of the sampled deposit. Sanderson and Farr (1960) mentioned that Brouwer believed it to be Oligocene in age but that further investigation was needed. In a letter to Sanderson, W. Durham (November 1961) stated that the matrix from a sample of Dominican amber contained several kinds of marine foraminifera. One was *Miogypsina*, which occurs only from the Late Oligocene and Mid to Early Miocene, thereby placing the amber-bearing beds at the Oligo-Miocene boundary (25–30 Ma). The source of this matrix was not stated, however, and the amber had clearly been redeposited.

Recent studies indicate that the base of the Cordillera Septentrional rests on folded metamorphic rocks resulting from plate margins being consumed (Miyashiro, 1972; Eberle et al., 1980). Shallow-water rock-fragment sediments from the late Eocene cover the basement disconformably and were folded before a late Tertiary (Neogene) marine sedimentation took place. After the deposition of limestones, there were extensive and intense lateral and vertical tectonic movements (Eberle et al., 1980). This

Figure 8. Dominican amber mines. A. Searching for the amber vein. B. Entrance to the famous La Toca mine. C. Mine entrances on the slope of a steep hill. D. Jim Work, a professional amber dealer known for locating scientifically valuable amber fossils (especially vertebrates).

would indicate that the Cordillera Septentrional is mostly covered by sedimentary rocks of Tertiary age. Most of the amber mines in this mountain range occur in or on the border of the Altamira faces in the "El Mamey" formation, which is a shale-sandstone formation interspersed with a conglomerate composed of well-rounded pebbles. Organic matter and extensive coal seams are common and the amber occurs in the lignite-rich sandstone beds or in lignite seams. The shales and sandstones at El

Mamey yielded a Late Eocene age (40 Ma) on the basis of coccolite age determinations (Eberle et al., 1980).

On the basis of the foraminiferal fauna, Baroni Urbani and Saunders (1980) produced an age corresponding to the lower part of the Early Miocene (20–25 Ma) for a sample from the Palo Alto mine in the Cordillera Septentrional. They point out, however, that the amber had been washed into the marine environment so that this age should be regarded as a minimum.

Amber samples from different mines in the Dominican Republic vary in color and hardness, suggesting a variation in age. A study using nuclear magnetic resonance spectroscopy has revealed a difference in the exo-methylene group among ambers from different regions. The softest and presumably youngest ambers had the most pronounced exo-methylene groups (Lambert et al., 1985), and Recent resin produced a very high exo-methylene resonance. By extrapolating from the relative dates obtained with this technique, a range of tentative absolute dates of Dominican amber were obtained. These range from 15–17 Ma (Mid Miocene) for the softer Cotui and Bayaguana ambers in the eastern portion of the country to 33–40 Ma (at the Oligocene-Eocene border) for the harder La Toca ambers. This scale was based on an Early Miocene age (20–25 Ma) for amber from the Palo Alto mine on the basis of results by Baroni Urbani and Saunders (1980).

On the basis of floral and vegetative structures found in amber, the Dominican amber-producing tree has been described as *Hymenaea protera* Poinar (1991D). This fossil species of leguminous tree most closely resembles the present-day *H. verrucosa* Gaertner, which occurs in East Africa and the adjacent islands (Color Plate 1). Although Dominican amber was originally assumed to have been produced from pines (Langenheim, 1964), studies with infrared spectrometry (Schlee and Glöchner, 1978), nuclear magnetic resonance (Cunningham et al., 1983; Lambert et al., 1985), and pyrolysis mass spectrometry (Poinar and Haverkamp, 1985) have shown the similarity of Dominican amber with resin from present-day species of *Hymenaea*, especially that of *H. verrucosa* (Hueber and Langenheim, 1986).

Mexican (Chiapas) Amber

The first discovery of Mexican (Chiapas) amber is difficult to determine, although the use of amber for Indian jewelry was noted in the writings of early European explorers (Blom, 1959). The geographical location of amber in the Chiapas highlands was first reported by Kunz

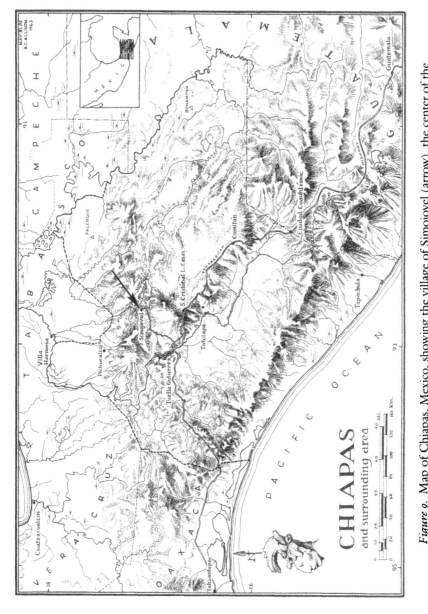

Figure 9. Map of Chiapas, Mexico, showing the village of Simojovel (arrow), the center of the amber area. (Courtesy of Richard C. Allison and May Blos.)

(1889). Buddhue (1935) appears to have first reported that this amber was fossiliferous, although he stated that the amber came from Baja California; a subsequent report (Langenheim et al., 1965), however, revealed that his material originated from a jewelry store in Mexico City and probably was from Chiapas. Buddhue (1935) called his material "bacalite" and, although it probably was not from Baja California, the term has been retained for amber originating from Baja California (Langenheim et al., 1965). Amber from Lower or Baja California was first noted in 1962 but, to date, no fossil inclusions have been reported.

The typical "Mexican amber," which constitutes the great majority of amber from Mexico, originates in Chiapas. This is the southernmost Mexican state, bordered on the south by Guatemala, on the west by Oaxaca, on the northwest by Veracruz, and on the north by Tabasco. Chiapas occupies some 73,000 square kilometers in the Pacificio Sur zone (Fig. 9), and is relatively isolated from the rest of the country. It has coastal plains on both the north and south borders, mountain ranges rising from the inner portions of the coastal plain, and a central depression. Most of the amber deposits occur in the northern mountain ranges, commonly referred to as the Chiapas highlands (Figs. 10–12). The climate in this area is tropical-subtropical but approaches temperate on some of the mountain tops. The area is inhabited mostly by Indians, mainly Tzeltales and Tzotziles, and some of these served as guides for early exploration of the area by University of California scientists searching for amber (Fig. 13). More recently, amber has been recovered near the village of Totolapa, also in Chiapas, where it is mined from the banks of an intermittent stream (Bryant, 1983). When characterized by ^{13}C nuclear magnetic resonance spectra, amber from both sites gave identical spectra, indicating a single paleobotanical source (Lambert et al., 1989).

Most amber "mines" occur in the Simojovel area, roughly from Tapilula in the west to Yajalon in the east and Los Cruses in the south, and are in a sequence of primarily marine calcareous sandstones and silt with beds of lignite (Fig. 14). The amber can be found directly in the lignitic beds or some distance from them. In both the Simojovel and Ixtapa-Soyalo areas, the amber occurs in association with the Balumtun Sandstone of the Early Miocene and the La Quinta formation of the Late Oligocene (Wyatt Durham, personal communication). The latter formation has been assigned to the planktonic foraminiferal zones represented by *Globigerina ciperoensis* and *Globorotalia kuglieri* (Frost and Langenheim, 1974). In the Cenozoic Planktonic Foraminiferal Zonal Sequence, this interval is within zones N3 and N4, which has been assigned radio-

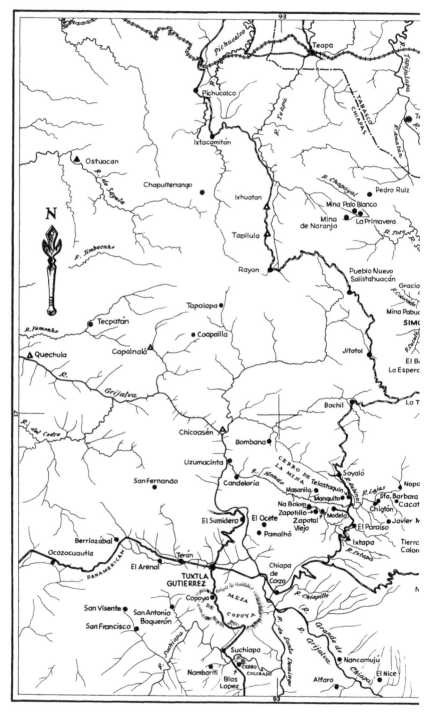

Figure 10. Detailed map of the Simojovel area showing the location of various amber mines (black triangles). (Courtesy of Richard C. Allison and May Blos.)

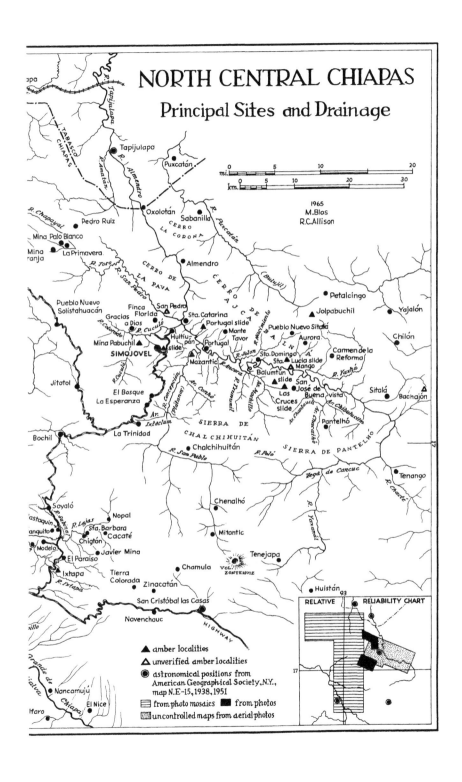

NORTH CENTRAL CHIAPAS
Principal Sites and Drainage

1965
M.Blos
R.C.Allison

▲ amber localities
△ unverified amber localities
◉ astronomical positions from American Geographical Society, N.Y., map N.E-15, 1938, 1951
▤ from photo mosaics ■ from photos
▨ uncontrolled maps from aerial photos

RELATIVE RELIABILITY CHART

Figure 11. Amber area in Chiapas, Mexico. A. Entrance to an amber mine (photo taken in 1983 by G. Poinar). B. Local residents digging through rock slide debris for amber at the Jolpabuchil amber outcrop (photo taken in 1956 by W. Durham).

Figure 12. Miner collecting Mexican amber from an exposed amber vein in the Simojovel area (photo by Gertrude Blom).

metric ages ranging from 22.5 to 26 Ma (Early Miocene–Late Oligocene) (Berggren and Van Couvering, 1974). The presence of amber in these adjacent ranges means that, on the basis of present-day chronologies, some 4 million years elapsed between the beginning and end of the amber-producing period (Wyatt Durham, personal communication).

Infrared spectra have given convincing evidence that Mexican (Chiapas) amber was derived from an ancestral *Hymenaea* tree. Langenheim and Beck (1965) examined resin from the following resin-producing tree genera that occur in Chiapas today: *Amyris, Myroxylon, Guaicacum, Pistacia, Bursera, Protium, Hymenaea, Pinus, Picea,* and *Taxodium.* The resin from each genus provided a characteristic, reproducible spectrum, and when compared with the spectrum of amber from Chiapas, a close

Figure 13. Tzeltal Indian girl wearing amber beads (photo taken in Santa Lucia in 1954 by Gertrude Blom).

similarity with the resin from *Hymenaea courbaril* was obvious. This result, coupled with the discovery of sepals and a leaf of *Hymenaea* included in the amber, support the conclusion that Mexican amber was produced by an ancestral *Hymenaea* tree (Langenheim, 1966). *Hymenaea* resin is known to accumulate in large quantities in soil under present-day conditions (Langenheim and Beck, 1965).

Obtaining fossiliferous amber for scientific study from Mexico and the Dominican Republic was facilitated through the activities of the Bloms and Brodzinskys, respectively, who lived in the region, knew the

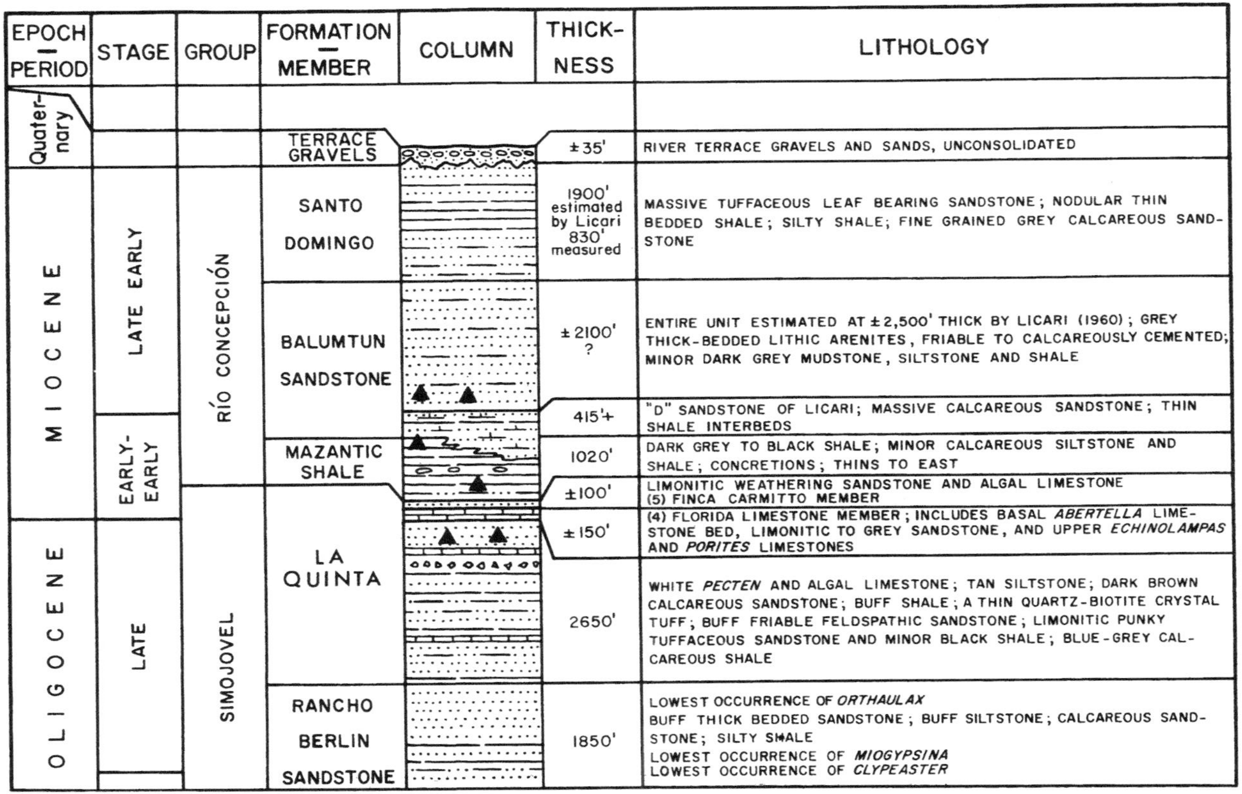

EPOCH / PERIOD	STAGE	GROUP	FORMATION / MEMBER	COLUMN	THICKNESS	LITHOLOGY
Quaternary			Terrace Gravels		±35'	RIVER TERRACE GRAVELS AND SANDS, UNCONSOLIDATED
MIOCENE	LATE EARLY	RÍO CONCEPCIÓN	Santo Domingo		1900' estimated by Licari 830' measured	MASSIVE TUFFACEOUS LEAF BEARING SANDSTONE; NODULAR THIN BEDDED SHALE; SILTY SHALE; FINE GRAINED GREY CALCAREOUS SANDSTONE
	EARLY- EARLY		Balumtun Sandstone		±2100' ?	ENTIRE UNIT ESTIMATED AT ±2,500' THICK BY LICARI (1960); GREY THICK-BEDDED LITHIC ARENITES, FRIABLE TO CALCAREOUSLY CEMENTED; MINOR DARK GREY MUDSTONE, SILTSTONE AND SHALE
					415'+	"D" SANDSTONE OF LICARI; MASSIVE CALCAREOUS SANDSTONE; THIN SHALE INTERBEDS
OLIGOCENE	LATE	SIMOJOVEL	Mazantic Shale		1020'	DARK GREY TO BLACK SHALE; MINOR CALCAREOUS SILTSTONE AND SHALE; CONCRETIONS; THINS TO EAST
					±100'	LIMONITIC WEATHERING SANDSTONE AND ALGAL LIMESTONE (5) FINCA CARMITTO MEMBER
					±150'	(4) FLORIDA LIMESTONE MEMBER; INCLUDES BASAL *ABERTELLA* LIMESTONE BED, LIMONITIC TO GREY SANDSTONE, AND UPPER *ECHINOLAMPAS* AND *PORITES* LIMESTONES
			La Quinta		2650'	WHITE *PECTEN* AND ALGAL LIMESTONE; TAN SILTSTONE; DARK BROWN CALCAREOUS SANDSTONE; BUFF SHALE; A THIN QUARTZ-BIOTITE CRYSTAL TUFF; BUFF FRIABLE FELDSPATHIC SANDSTONE; LIMONITIC PUNKY TUFFACEOUS SANDSTONE AND MINOR BLACK SHALE; BLUE-GREY CALCAREOUS SHALE
			Rancho Berlin Sandstone		1850'	LOWEST OCCURRENCE OF *ORTHAULAX* BUFF THICK BEDDED SANDSTONE; BUFF SILTSTONE; CALCAREOUS SANDSTONE; SILTY SHALE. LOWEST OCCURRENCE OF *MIOGYPSINA*. LOWEST OCCURRENCE OF *CLYPEASTER*

Figure 14. Geological formation (generalized composite columnar section) of the Simojovel area, Chiapas, Mexico. Black triangles indicate formations containing amber deposits. (Courtesy of Richard C. Allison.)

Figure 15. People who have greatly aided scientists in the study of New World fossiliferous amber. A. Francis and Gertrude Blom in San Cristóbal de las Cases, Chiapas, Mexico (photo taken in 1956 by W. Durham). B. Jake and Marianella Brodzinsky, Santo Domingo, Dominican Republic (photo taken in 1986 by G. Poinar).

local miners and amber sites, and assisted scientists in collecting expeditions (Fig. 15). Their efforts are greatly appreciated.

Chinese (Fu Shun) Amber

Fossiliferous amber dating from the Eocene Epoch (40−53 Ma) was reported near the city of Fu Shun in Liaoning province (Hong et al., 1974). The amber occurred in coal beds of the Guchenzgi formation of the Eocene Fu Shun group in association with the remains of some 50 plant species of gymnosperm and angiosperm origin. The amber beds extend from just beneath the surface to some distance underground (200 meters) and occurred near a prehistoric settlement. The coal seams that contained the amber were interspersed with beds of shale and sandstone. On the basis of the amber inclusions, some of which are deposited in the Peking Geological Museum, it was concluded that the area had a subtropical climate during the Eocene.

Romanian Amber

Romanian amber, or Rumanite, has an ancient history, similar to that of Baltic amber. Amber from the Carpathians is said to have been used in neolithic times by the Dacians and later by the Romans (Protescu, 1937; Williamson, 1932). The first historical account of its occurrence was made by Raicevich in 1788. Romanian amber, which is difficult to obtain at present, was known for its beautiful and varied colors, especially deep red, deep blue, and greenish blue, which are uncommon natural colors for amber.

Although amber has been found by geologists in over 360 outcrops in Romania (e.g., near the Monastery of Alunis on the Danube, near Craiova and Olanesei, in tar deposits of Putna, in the petrol wells of Ocmta, and by the banks of the rivers Dimbovita, Colti, Roscouel, Venetisul, Frasinul, and Corbul), most of the studied material containing organic inclusions originates from the locality of Colti, in the Sibicui valley of the Buzau District. One major collecting locality is located about 13 kilometers northwest of Patarlage.

Romanian amber has been recovered from two distinct geological formations, Cretaceous and Tertiary (Eocene-Oligocene), depending on the geographical location. The material found in Colti dates from the Early Oligocene (30–35 Ma). The amber is found in strata of sandstone alternating with beds of lignite, and associated shells indicate a period when the area was covered by a shallow sea (Protescu, 1937). Silicified wood fragments among the deposits have been identified as *Sequoioxylon gypsaceum* Göppert and *Taxodioxylon gypsaceum* Göppert. Colti amber samples contained material of *S. gypsaceum*, and Ghiurca (1988) considered that this tree was probably responsible for the Tertiary amber deposits. Amber-bearing strata also contained high pollen levels of *Cupuliferoidaepollenites liblarensis* Thomson and *Ulmipollenites undulosus* Wolf.

Burmese Amber

Although Burmese amber has been mined for centuries, its occurrence in the Hukong Valley in upper Burma was first recorded in 1892 by Noetling, who described the amber as occurring in a soft blue clay that was superficially discolored brown. The amber was limited to the upper part of the clay and was considered to be secondarily deposited there during the Miocene. Later, Chhibber (1934) revisited the area, described the mines and mining operations, and characterized the amber as coming from deposits of Eocene age. This was in agreement with the conclusions of Stuart (1923), who collected the nummulite foraminifer *Nummulites biarritzensis* from a piece of chalky limestone found in the amber-bearing beds. The foraminifera of the "Nummulite Limestone" are characteristic of the uppermost zone of the Lower Khirthar corresponding with the Lutetian, which represents the earlier part of the Mid Eocene (45 Ma) (Cockerell, 1922). On the basis of his investigations, Cockerell (1917A), who studied and described 42 arthropods, including primitive types, from Burmese amber, considered that the amber may be Late Cretaceous. If the amber was secondarily deposited during the Eocene, it is possible that the amber was originally deposited in the Cretaceous. Further evidence is lacking, however, and Burmese amber is generally considered to be Eocene in age. The plant source of Burmese amber is not known. An account of the physical and chemical properties of the amber was presented by Helm (1893), who proposed the name Burmite for this deposit. A listing of the families of arthropods discovered in Burmese amber was provided by Cockerell (1920B).

According to Chhibber (1934), numerous (over 800) amber mines, most just shallow pits about one meter square and up to 14 meters deep, were dug into the ground and worked by the local peasants. Amber was gathered from the clay deposits and coal seams. Localities that contained "mines" were the Nangtoimow Hills near the village of Shingban and in the vicinity of the villages of Pangmamaw, Khanjamow, Ninkundup, and Wayrtmaw.

Sicilian Amber

Amber from the island of Sicily, or simetite, has an extensive history. Buffum (1898) mentions early trading of Sicilian amber with the Phoenicians. According to all accounts, Sicilian amber has more beautiful color patterns than other ambers and can vary from deep red to blue or green. These characters have been attributed to its proximity to Mt. Etna and a possible volcanic influence (Williamson, 1932). The name, simetite, is derived from the river Simeto, near Catania, where it was first discovered. The amber is found in the central part of the island, especially near streams where it has been washed out of the clay-filled soil. Sicilian amber was reported to have been formed during the Mid Miocene, which would give it an age of about 11–17 Ma (Emery, 1890); later studies, however, indicate an Oligocene age, similar to that of fossiliferous amber from Mercato Saraceno in the Apennines of northern Italy (Skalski and

Veggiani, 1988). The tree source is unknown and few fossil insects have been reported from the amber.

Other Fossiliferous Tertiary Amber

Fossiliferous amber has been recovered from commercially exploited refractory clay in the central Arkansas Coastal Plain (Saunders et al., 1974). The amber-bearing beds occur near Malvern, Hot Spring County, Arkansas, within a sequence of sand, clay, and lignite. The amber occurs in the lignitic and lignitic-clay beds as isolated irregular lumps ranging in diameter from 3 to 8 centimeters. Most of the amber is fragile and the outer surface splinters easily. This amber is called Claiborne because it occurs in lower Claiborne strata of the Mid Eocene. It gives an infrared spectrum similar to present-day *Shorea* resin, suggesting that a member of the Dipterocarpaceae was the botanical source. Over 700 arthropod inclusions representing eight insect orders have been identified from the amber.

Small quantities of insect-containing amber dating from the Miocene and Miocene-Pliocene boundary were reported from marine sedimentary formations in Java and Sumatra, Indonesia, and from Luzon Island in the Philippines by Durham (1956). The amber was collected from exposed rocks along stream beds.

In 1982, Barthel and Hetzer reported on fossiliferous amber from the area around Bitterfeld, a city some 60 kilometers southeast of Magdeburg in the province of Saxony in Germany. This deposit of what is called Saxonian or Bitterfeld amber was discovered in Early Miocene layers in Aquitanian brown coal beds and was considered 12 million years younger than the deposits of Baltic amber. On the basis of infrared spectrometry, Barthel and Hetzer (1982) suggested that the resin producing the Bitterfeld amber came from the Tertiary conifer *Cupressospermum saxonicum* Mai, which was widespread in central Europe during the Tertiary. The spectra of Bitterfeld and Baltic amber are also similar in some respects, and some investigators claim that the Bitterfeld deposits are redeposits of Baltic amber. About 48 species of invertebrates and plants have been identified in the Bitterfeld deposits.

Various deposits of amber ranging in age from 6–9 Ma (Pliocene) to 225–231 Ma (Triassic) have been reported from Austria. A few fossils have been reported in amber from the Gablitz deposits, which are considered to be of the Eocene Epoch (Vávra, 1984).

Amber also occurs in at least eight different localities in Switzerland.

Most have been dated from the Late Paleocene (55 Ma) but one deposit stems from the Late Triassic (200 Ma). Representatives of the insect families Psychodidae (Diptera), Sciaridae (Diptera), Ceratopogonidae (Diptera), Proctotrupidae (Hymenoptera), and Chalcididae (Hymenoptera), as well as a portion of a leaf, have been identified from the Tertiary deposits (Soom, 1984).

Amber from the state of Washington in North America has been recovered from the Tiger Mountain Formation near Issaquah. This formation constitutes part of the Puget Group, and according to paleobotanical correlation with other formations from the West Coast, its age is judged to be Mid Eocene or 45 Ma (Mustoe, 1985). A few unreported fossils have been found in these deposits.

Another recent find of amber comes from southeastern Nigeria. This material, called Amekit, occurs in Eocene clay-sandstone deposits of the Ameki formation near Umuahia (Sawkiewicz and Arua, 1988). No organic inclusions have yet been reported in this material.

Cretaceous Amber

The deposits of Mesozoic fossiliferous amber were mostly formed during the Cretaceous Period, ranging from 144 to 65 Ma (Fig. 3). These deposits overlap with the period of the dinosaurs and ammonites. Amber from this period is especially interesting because it was formed during the development of the angiosperms and contains extinct insects that differ enough from present-day forms to be placed in separate families.

Canadian (Cedar Lake) Amber

Fossilized resins have been found in various regions of Canada. These localities are cited by McAlpine and Martin (1969), who state that, with the exception of the Cedar Lake deposits, Canadian amber has been little collected or studied. Thus, the term "Canadian amber" generally refers to amber collected in the vicinity of Cedar Lake, Manitoba, the major source of fossil-bearing amber reported from Canada (Fig. 16).

Although collected earlier by the local Indians, Cedar Lake amber was first officially noted in 1893 by Tyrrell. This geologist reported that the amber deposits extended to a depth of almost one meter for a distance of nearly 2 kilometers along the lake shore and extended back 25 to 36 meters from the shore line; he estimated that this area contained over 600 metric tons of amber. The Cedar Lake amber was named Chemawinite by Harrington (1891), after an Indian tribe living in the area (Fig. 17A).

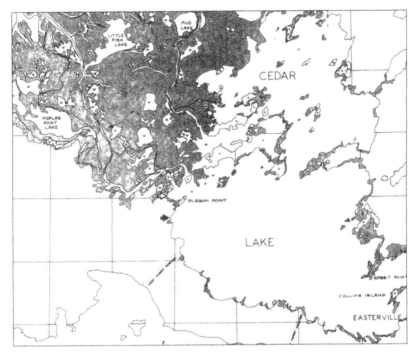

Figure 16. Location of Canadian amber (arrow) at Cedar Lake, Manitoba, Canada.

Later, apparently not realizing the amber had been already described, Klebs (1897) called it Cedarite. The former name has generally been used when describing these deposits.

The first discovery that this amber was fossiliferous was made by Walker in 1934. On the basis of reports by J. B. Tyrrell and O. J. Klotz of the Geological Survey of Canada, W. C. King, then manager of the Hudson's Bay Company, began a commercial operation and somewhat less than a ton of amber was collected between 1895 and 1937. Most of the pieces were small and much of it was sold to varnish manufacturers.

The amber had been secondarily deposited along the beach at Cedar Lake, somewhat south of the mouth of the Saskatchewan River, where the beach area is a low shelving shore. The amber is washed ashore by wave action and can be found mingled with shells, coal fragments, sand, and an assortment of organic debris. Unfortunately, a dam was constructed at the foot of Cedar Lake, and with the rising water, the amber shore and surrounding area were completely inundated. Even so, wave action has washed some of the amber up on the newly formed "higher"

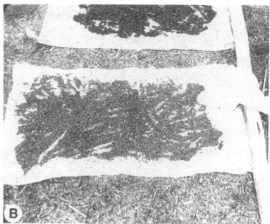

Figure 17. Canadian amber deposits. A. Cree Indian collecting Canadian amber along the shore of Cedar Lake (photo taken in 1941 by William M. Legg). B. Canadian amber drying on sacks after being removed from debris by the flotation method (photo courtesy of Frank M. Carpenter).

beach area, where it can still be collected. In the 1930's, Frank Carpenter collected several million pieces of Cedar Lake amber by sifting through large amounts of debris and filtering out the amber pieces by using a water-flotation method (Fig. 17B). This unexamined amber is now housed at the Museum of Comparative Zoology at Harvard University. The Cedar Lake deposits are thought to have been carried into the lake by the easterly flowing Saskatchewan River, which runs through lignitic beds in Alberta and south Saskatchewan.

Amber resembling the Cedar Lake deposits was found in coal deposits near Medicine Hat, Alberta. These coal deposits were determined to belong to the Foremost Formation (Belly River Series) and are overlaid by Upper Cretaceous bentonites (soft clay-like rocks) that indicated an age of 75–78 Ma by radiometric dating (Folinsbee et al., 1964). This would correspond to the minimum age of the Cedar Lake amber deposits, which were laid down before the close of the Campanian Epoch of the Late Cretaceous Period (McAlpine and Martin, 1969). Thus, the amber is generally believed to be about 70–80 million years old.

Canadian (Cedar Lake) amber produced infrared spectra similar to that of resin from *Agathis australis*, the New Zealand Kauri pine, although one sample produced a spectrum similar to that of Baltic amber (Langenheim and Beck, 1965, 1968). A similarity between Canadian amber and the resin of *Agathis australis* was also demonstrated with pyroly-sis mass spectrometry (Poinar and Haverkamp, 1985). Analysis by ^{13}C nu-clear magnetic resonance spectroscopy showed Canadian amber to be quite similar to Atlantic Coastal Plain amber as well as Washington and Alaskan amber (Lambert et al., 1989). Unfortunately, no botanical inclu-sions have been described from Canadian amber and further studies are necessary to locate and examine the bedrock flora, which occurs some distance away from the amber deposits.

Alaskan Amber

Dall (1870) was one of the first to mention Alaskan or Arctic amber, noting its occurrence in lignitic beds on the Alaskan peninsula, in the al-luvium of the Yukon delta, and near coal deposits on Fox Island. Usinger and Smith (1957) cite reports from the Eskimos about an amber bay and amber deposits in the Aleutian Island chain, especially in the mountains of Unalaska and in Umnak. According to these authors, in order to ob-tain amber from Umnak Island, the natives spread a walrus skin between two boats at the base of a cliff and pulled down debris and amber onto the skin.

In 1955, Usinger and Smith from the University of California at Berkeley spent from July 12 to August 9 exploring for amber in the Alas-kan Arctic north of the Brooks Range. Localities where amber had been reported were Ninuluk Bluff near the Killik Bend of the Colville River about 80 kilometers upstream from Umiat, Kay Creek (a tributary of the Ikpikpuk River), and Maybe Creek, all on the arctic slope between Point Barrows and the Brooks Range. Natives reported finding amber on the beach at Smith Bay and at an exposed cliff 16 kilometers up the Ketic River. The amber occurred in coal and harder lignite in the last of the tilted beds downstream on the south side of the Colville River at Ninuluk Bluff. The amber was brittle and shattered or powdered when uncovered from small pockets in the shales, which were near but not in the coal beds. At Umiat, Usinger and Smith recovered pea-size amber pockets on the gravel banks and found a larger, usable piece of amber in the beach drift (Fig. 18).

After proceeding up the Kuk River, the mouth of which empties into the Arctic Ocean near Wainwright, Usinger and Smith discovered more amber. Some distance up the Kuk (at "Coal Mine No. 3") they found two pieces of amber and then gathered several hundred pieces (up to 4 centimeters in diameter) on a beach downstream from the mouth of the Omalik River. The amber occurred in beach drifts (composed of

Figure 18. Alaskan Amber. A. Robert Usinger collecting pieces of Alaskan amber washed out of veins in the banks of the Kuk River. B. Piece of Alaskan amber (arrow) in drift near the mouth of the Omalik River. (Photos taken in 1955 by R. Smith.)

small coal and driftwood) for kilometers along the upper reaches of the Kuk River, but the best collecting was at the mouth of the Omalik and across the Kuk at the mouth of the Pugnik River. Abundant amber also occurred at Kuk Inlet, near the junction of the Avalik, Ketic, and Kaolak rivers. Usinger and Smith continued up the Ketic River for about 16 kilometers and discovered exposed cliffs on the right bank composed of gray mud and shale. Thin coal veins running in the clay contained pieces of amber similar to those found in the beach drift. Similar amber-bearing exposed bluffs were found up and down the rivers. Caribou and ground squirrel activity loosened the amber, which was then washed down into the rivers and deposited in beach drift.

Further exploration was reported by Langenheim et al. (1960) who found amber in most of the outcrops along the Kaolak and Ketic rivers. The amber had accumulated on the ground under and downslope from the coal outcrops. Most of the amber was found in coal beds, and although some occurred in shale, none was found in sandstone. The authors concluded that amber occurred only in association with coal or carbonaceous sediments and was originally deposited in combination with woody debris. They described the amber as ranging in color from light golden yellow to deep red and almost black. The clear amber was nearly crack free but the opaque material was mostly granular. The small size of most pieces (a few millimeters in diameter) and their shape (teardrop masses, subcylindrical or irregular mammillary blobs) caused the authors to conclude that the original resin was exposed to the atmosphere. This conclusion is supported by the presence of insects in some pieces. The discovery of amber in coal having a woody texture indicates that some resin was fossilized in situ in pockets within the tree limbs.

The Alaskan amber was found in layers of Cretaceous rocks occurring between the Brooks Range and the northern coastline of Alaska. This area comprises the Arctic Coastal Plain and Arctic Foothills (Langenheim et al., 1960). More than 75 percent of the Cretaceous rock in the Kuk drainage basin is composed of gray to black shale. The remainder is made up of relatively thin layers of sandstone, coal, bentonite, and concretionary beds. The shale is noncalcareous and contains scattered fragments of coaly material and amber. The coal and amber were probably originally deposited in thin coal beds or carbonaceous shale layers, and became mixed with the shale in slumping; some amber, however, was scattered though the clay rock as originally deposited. The coal associated with the amber was described as black with a vitreous or dull luster, finely laminated, and formed of alternating layers up to 2 millimeters thick. This type of coal is placed in the categories of argillaceous (clay-like) coal and carbonaceous shale.

Langenheim et al. (1960) obtained paleontologic data on the age of the amber by examining the megafossil and microfossil floras. An abundance of angiosperm megafossils, mainly foliage, was found, suggesting a Late Cretaceous age, but an abundance of gymnosperm microfossils was also present, indicating Early Cretaceous. The authors concluded that the deposits could be either Early or Late Cretaceous. Later authors have cited the deposits as Late Cretaceous (Langenheim, 1969).

Langenheim et al. (1960) considered the amber to have been produced by members of the Taxodiaceae. This conclusion, however, was not substantiated by comparing the infrared spectra of Alaskan amber and present-day resins of the taxodiaceous genera *Sequoiadendron*, *Metasequoia*, or *Taxodium* (Langenheim, 1969). Results of pyrolysis mass spectrometry showed a similarity of Alaskan amber to recent *Agathis* resin, suggesting an araucarian origin (Poinar and Haverkamp, 1985); this conclusion is further supported by recent analysis by ^{13}C nuclear magnetic resonance spectroscopy (Lambert et al., 1989). The discovery of *Podozamites* leaf impressions in the amber-bearing layers (Langenheim et al., 1960) is interesting because some *Podozamites* leaves are very similar to those of present-day *Agathis* and certain species could well belong to the Araucariaceae (Stewart, 1983).

Middle East Amber (Lebanon, Israel, and Jordan)

Mountainous areas in the Middle East comprising the countries Lebanon, Israel, and Jordan have produced fossiliferous amber dating from the Early Cretaceous (Fig. 19). This amber has the distinction of being the oldest extensive fossiliferous amber known and was deposited some 135–120 Ma. The abundance of Lebanese amber was first noted by the German paleontologist Oscar Fraas (1878) over 100 years ago, when he was working at the Museum of Natural History in Stuttgart. Most investigations of the material that Fraas brought back to Stuttgart were conducted by his successors, W. Hennig, D. Schlee, and H. G. Dietrich of Tübingen. A description of Lebanese amber was reported by Schlee and Dietrich (1970) and Schlee and Glöckner (1978).

The Lebanese amber beds are located in the district of Jezzine, about 32 kilometers east of Saida, in "Gres de Bose" beds of the Early Cretaceous Period (probably Hauterivian Stage). Schlee and Dietrich (1970) describe the amber as occurring in primary deposits of the Neocomian division of the Early Cretaceous as well as in secondary deposits, also in

Figure 19. Location of Middle East amber (stars indicate localities where amber has been recovered).

the Neocomian division and the younger Aptian Stage. Amber was also collected by Aftim Acra in the vicinity of Dar al-Baidha, between Beirut and Damascus (Munro, 1981).

The same amber deposits found in Lebanon are presumed to extend southward into Israel (Nissenbaum, 1975), making the Middle East amber belt almost as extensive as the Baltic deposits. In Israel, amber has been recovered from the southern slopes of Mt. Hermon, from the eastern escarpment of the Naftali Mountains, near Qiryat Shermona in northern Israel, and from drillings in the Kokhov and Barboor areas in southern Israel (Fig. 19). The amber is found in rock matrix in the form of droplets and nodules ranging in size from a few millimeters to several centimeters in diameter. The color is similar to that reported for Lebanese amber, ranging from translucent yellow to faintly translucent dark brown to honey colored. Amber from all localities in Israel occurs in the Early Cretaceous strata of possibly the Hauterivian or Valanginian stages, which corresponds to the same age as Lebanese amber (about 135 Ma). Israeli amber is located in sequences of sandstones and silty shales and is frequently associated with lignites, fossil wood, and highly carbonized plant remains. The chemical and carbon isotopic compositions of Israeli amber are similar to those of Baltic amber, although the infrared spectra of amber from the two sources varies somewhat.

Fossiliferous amber from the Early Cretaceous was reported in Jordan, too (Bandel and Vavra, 1981). This amber was found in the Kurnub Series in the area of Wadi Zerka, north of Amman. Also discovered in the amber-bearing strata were fossils of *Agathis*-like plants, and results of infrared spectroscopy indicated an araucarian origin of not only the Jordanian but also the Lebanese material. A similarity was noted, too, between Lebanese and Austrian amber and the latter was attributed to an araucarian producer also (Bandel and Vavra, 1981).

Siberian (Taimyr) Amber

There are five major amber deposits within the geographical boundaries of the Soviet Union (Savkevich, 1974). One includes a large portion of the Baltic amber deposits, which is incorporated in the Baltic-Dneprovsk area, and extends into the Kaliningrad district or peninsula (Sambia), Lithuania, Latvia, Byelorussia SSR, and beyond the Baltic in the Pravoberezhaya portion of the Ukraine. The second deposit is in the Carpathian area, which includes Ciscarpathia, the Skibov zone of the Soviet Carpathians within the basins of the upper course of the Prut River and the Dniester River as far as Khotin. Amber here occurs in coastal-marine sandstone of Late Eocene age. The third area is called Transcaucasia and extends along the northeastern slope of the Lesser Caucasus from the Akstafa River to the Araks River. It is estimated that some 576 tons of amber, possibly of Cretaceous age, occur in this region alone. The fourth, or Far East, area covers the territory of the Primor'ye district and Sakhalin as well as the eastern shore of Kamchatka. Most of this amber is made into jewelry, but fossiliferous amber has been recovered on the eastern coast of southern Sakhalin and can be found on the beach near the villages of Vzmor'ye, Firsovo, and Starodubskoye.

The fifth deposit, one that is known to be highly fossiliferous, is the Taimyr deposits, which are grouped with the "arctic" deposits. These deposits occur over a wide geographic area in the Soviet arctic bounded on the west by the Mezen Gulf of the White Sea and in the east by the Verkhoyansk marginal depression (Fig. 20). Taimyr amber was first investigated in 1970 by members of the Paleontological Institute of the USSR Academy of Sciences who organized an expedition to the Khatanga Basin in eastern Taimyr. During this and subsequent expeditions to the Khatanga and Agapa basins, about 4,000 insect inclusions were collected, making this the largest known collection of studied Cretaceous amber inclusions. The Taimyr deposits, which include some 10 different sites, are between 80 and 105 Ma. The Khatanga Basin amber occurs in marine deposits that range from the Barremian Stage to the Santonian Stage of the Cretaceous. The amber deposits located at Yantardak (Amber Mountain) are also from the Santonian Stage (Rodendorf and Zherichin, 1975). Further descriptions of the amber-bearing strata and biological inclusions recovered are provided by Zherichin and Sukacheva (1973), and a list of arthropod families collected from the Yantardak (major) site of the Taimyr deposits is presented in Table 5. The botanical origin of the Taimyr amber is unknown.

Atlantic Coastal Plain Amber

Small outcrops of Upper Cretaceous beds containing amber occur along the North American Atlantic Coastal Plain. Amber from these beds has been found in Maryland, Martha's Vineyard, New Jersey, New York, Delaware, and South Carolina (Langenheim and Beck, 1968). A single piece of amber collected, almost by chance, from the Magothy exposure at beach cliffs, Cliffwood, New Jersey, was found to contain the first Mesozoic ant, later described by Wilson et al. (1967). More recently,

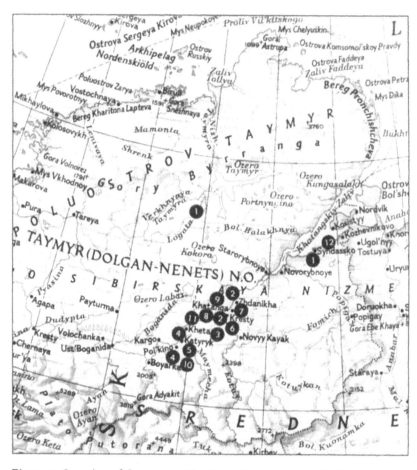

Figure 20. Location of Cretaceous Siberian amber deposits in the Taimyr area of the USSR. 1, Ogenv formation (Albian age); 2, Begichev formation (Albian-Cenomanian age); 3, Turon formation (Coniacian age); 4, Khet formation (Coniacian-Santonian age); 5, Yantardak site (Santonian age); 6, Kresty site (undetermined age); 7, Zhdanikha site (undetermined age); 8, Sokolovskii site (undetermined age); 9, Gubina Mountain site (undetermined age); 10, Romanikha site (undetermined age); 11, Isaevskii site (undetermined age); 12, Khatang Guba site (undetermined age). (Based on data in Zherichin and Sukacheva, 1973.)

Table 5. Arthropods recorded from Cretaceous Siberian amber
at the Yantardak site

Arachnida (69)	Heteroptera (5)
Acari (19)	Largidae (1)
Araneae (47)	Microphysidae (1)
Oribatids (4)	Homoptera (138)
Camisiidae (1)	Aleurodoidea (1)
Oribatidae (1)	Aphidoidea (124)
	Coccoidea (7)
Insecta (about 2,098)	Fulgoroidea (6)
Blattoidea (22)	Hymenoptera (271)
Coleoptera (19)	Bethylidae (18)
Cerophytidae (2)	Braconidae (21)
Helodidae (1)	Ceraphronidae (1)
Lathridiidae (1)	Chalcidoidea (25)
Passandridae (1)	Cleptidae (= Chrysididae) (2)
Scraptiidae (2)	Cynipidae (1)
Staphylinidae (2)	Formicidae (2)
Collembola (2)	Mymaridae (12)
Arthropleona (1)	Scelionidae (94)
Symphypleona (1)	Serphitidae (72)
Diptera (1,418)	Sphecidae (1)
Bibionidae (1)	Tiphiidae (1)
Bombyliidae (8)	Isoptera (1)
Cecidomyiidae (7)	Lepidoptera (4)
Ceratopogonidae (305)	Micropterigidae
Chironomidae (390)	(Mnesarchaeiinae) (1)
Dolichopodidae (1)	Neuroptera (1)
Empididae (8)	Coniopterygidae (1)
Limoniidae (4)	Plecoptera (1)
Mycetophilidae	Psocoptera (16)
(= Fungivoridae) (7)	Amphientomidae (1)
Platypezidae (6)	Lophioneuridae (6)
Psychodidae (1)	Psocidae (1)
Rhagionidae (1)	Thysanoptera (9)
Scatopsidae (28)	Ceratothripidae (2)
Sciadoceridae (19)	Pygothripidae (1)
Ephemeroptera (71)	Thysanura (4)
Isonychiidae (1)	Lepismatidae (1)
Leptophlebiidae (7)	?Machilidae (1)
	Trichoptera (45)
	Sericostomatidae (4)

NOTE: Assembled from data in Zherichin and Sukacheva (1973); numbers in parentheses following each group represent number of individuals recovered. Because some amber-borne arthropods could be identified only to the level of order, the family totals in this table do not always add up to the order totals.

these deposits have revealed the oldest fossil bee (Michener and Grimaldi, 1988) and the oldest ceratopogonid from North America (Grogan and Szadziewski, 1988).

The three major formations of exposed black clay where fossils have been found in New Jersey are the Merchantville formation (Campanian to Santonian stages, 75–87 Ma), the Magothy formation (Coniacian to Turonian stages, 87–91 Ma), and the Raritan formation (Turonian to Cenomanian stages, 88–98 Ma). The oldest amber deposits of the Atlantic Coastal Plain occur in the Patuxent formation of the Patomac Group (Barremian Stage, 119–124 Ma) (Langenheim and Beck, 1968).

Most of these deposits seem, on the basis of chemical analyses, to have originated from araucarians (Langenheim, 1969; Lambert et al., 1989). Analysis of some samples, however, indicated *Liquidambar* (Hamamelidaceae) as a resin source (Langenheim, 1969), and still other samples indicated a possible taxodiaceous source (*Sequoiadendron* or *Metasequoia*) (Langenheim, in Wilson et al., 1967). Thus, the resin-producing forest was either mixed or a succession of resin-producing plants appeared during the Cretaceous in this region. Knowlton (1896) earlier had named the amber-producing tree *Cupressinoxylon bibbinsi* after conducting anatomical studies on wood samples. Further information on these deposits is provided by Grimaldi et al. (1989).

Other Fossiliferous Cretaceous Amber

Cretaceous amber has been described from at least two localities in Japan. Kuji amber is from the Tamagawa Formation of the Kuji Group of the Late Cretaceous and Choshi amber is from the Kimigahama Formation of the Choshi Group of the Early Cretaceous. No report of biological inclusions in either of these amber deposits could be obtained. Mizunami amber, also from Japan, is considered to be from the Pleistocene Epoch and has been dated in the thousands rather than millions of years (Schlee, 1984).

Fossiliferous amber from the Late Cretaceous (Cenomanian Stage) of the Paris and Aquitan Basin in northwestern France has been reported by Schlüter (1978A). This amber is especially interesting because many of the entrapped insects had been partially pyritized, which preserves fine details and allows them to be revealed by high resolution radiographs. Arthropod remains in the allochthonous (produced elsewhere) amber were found in clay and silt deposits near Bezonnais, Durtal, and Fouras. The amber is cloudy, and because it deteriorates immediately upon contact with air, all examinations were conducted under water. A few plant remains (a bud, hairs) and some 40 arthropods representing 11 orders (5 new genera and 6 new species) were recovered. On the basis of infrared spectroscopy, trees of the family Araucariaceae have been considered to be the amber source.

Copal Deposits

There are various deposits of semifossilized resin throughout the world, and many of these include insects and other organic remains. These semifossilized deposits are considered as copal, and if they survive some additional millions of years they may develop the physical characteristics of amber. They represent resin that has been in the earth less than 3 to 4 million years (see pp. 6–8). A single geographical region may have copal deposits that range from recent to many thousands of years old, depending on where the material was obtained (e.g., the Kauri deposits in New Zealand).

Most copal can be readily distinguished from amber by its softness and lighter color (Table 1). Because of the relative ease of melting copal and embedding modern insects or other life forms in various-sized pieces, it is sometimes difficult to determine if the inclusions are natural or man-made. Copal deposits from which fossils have been described occur in South America (resin from *Hymenaea*); in Mizunami, Japan; in East Africa (Madagascar, Kenyan, or Tanzanian copal); Allendale, Victoria (Australia), northern New Zealand and the South Pacific region (resin from *Agathis*); in West Africa (resin from *Copaifera*); and in Israel and the Middle East (resin from *Pistacia*). Few of these records are included here, because the inclusions are all extant species and the present work includes amber inclusions which deal almost exclusively with now-extinct species.

Until recently, all fossil resins from Africa fell into the copal category. However, amber that has been named amekit was found in Eocene clay sandstone deposits of the Ameki Formation (facies D) in southeastern Nigeria near Umuahia (Sawkiewicz and Arua, 1988). To date, no organic inclusions have been reported from this amber.

CHAPTER THREE

Major Collections of
Fossiliferous Amber

Amber has always been admired and valued for its beauty as a gem. The value of amber, however, depends on whether it is being appraised for public or scientific purposes. The public desires a piece of beautiful color, polish, and form, whereas the biologist looks for an interesting inclusion that can be clearly seen and examined. The greatest commercial value of amber, for both the public and scientists, comes from the inclusions it contains, although large pieces of polished amber (especially Dominican blue) are rare and valuable without any inclusions. From the biological standpoint, the value of fossiliferous amber depends on the following:

1. The rarity of the inclusion. Most ants and fungus gnats are fairly common in amber, and such common inclusions are not as valuable as such rare ones as a flea or tick. Well-preserved vertebrates (lizards, frogs) and unusual arthropods (scorpions) have values in the thousands of dollars, whereas a well-preserved ant might fetch only 15 dollars.

2. The state of the inclusion's preservation. The value, from both a scientific and aesthetic point of view, is increased when a specimen is well preserved (in life-like form). Frequently, a specimen will be partially disintegrated from microbial activity, obstructed by milky deposits or mold, or "washed out" from other causes. Fossils that are complete and can be identified are more valuable than those that are incomplete and cannot.

3. The clarity of the amber. Cloudy amber usually obstructs the examination of inclusions, and air bubbles and other deposits including organic debris may block the fossil from clear view. Sometimes the obstruction can be ground away but this always involves the risk of damaging the specimen.

4. The position of the inclusion. A specimen positioned along the longitudinal axis of an oval piece of amber is more easily seen than one positioned perpendicular to the axis. (Sometimes the amber can be re-

shaped if fractures are not extensive.) Some insects are rolled up ventrally or wrapped around themselves, making identification difficult.

5. The size of the inclusion. Anything large enough to be distinctly seen without the aid of a hand lens will be appealing to the public, but most insects enclosed in amber are small (less than 5 mm long). From the scientific standpoint, size is important only in its relation to the rarity of the inclusion. Because large insects in amber are generally uncommon, species of large insects may be rare; species of small insects can be just as rare, however.

Unfortunately for biologists, fossiliferous amber does have popular appeal, and many rare fossils in amber bring high prices from private collectors, thus eliminating them from study by scientists. For this reason, the acquisition of fossiliferous amber by various institutions should be encouraged. Most natural history museums contain small collections of fossiliferous amber, and some of the larger public collections are listed in Table 6.

The assembly of many collections, biological and otherwise, deposited in museums throughout the world, often stems from the energy and finances of nonprofessionals, whose enthusiasm and finances often allow them to gradually amass large collections. Almost all of the significant collections of fossiliferous amber that have been deposited in museums throughout the world and studied by experts have been amassed by amateurs, many of whom were or are involved commercially in the sale of amber.

Undoubtedly the largest assemblage of amber was the famous Stantien and Becker collection of Baltic amber originally held at the Königsberg University Geological Institute Museum in Samland (Klebs, 1910). This collection included some 120,000 animal and plant fossils that had been gathered during the extensive amber mining operations by Stantien and Becker. This venture started in 1860 when Wilhelm Stantien, an innkeeper in Memel, formed a partnership with the merchant, Moritz Becker (Ley, 1951). Their first activities included dredging Frisches Bay and extracting amber from the sand; later they began regular mining operations in Samland. Over the years, their collection of Baltic amber provided the great majority of specimens used throughout the world for scientific study. It was feared that during the Second World War the entire collection had been destroyed by bombing (Wenzel, 1953); however, it was later learned that before the bombing, the collection had been divided and deposited in various localities, with the result that at least a portion of it was saved. Many private collections of Baltic amber were

Table 6. Some public institutions holding collections of fossiliferous amber, with sources and approximate numbers of pieces

Institution	Source	Number of pieces
British Museum of Natural History, London	Baltic	25,000
	Burmese	300
	Sicilian	50
Museum of the Earth, Warsaw	Baltic	25,000
Zoological Institute, Leningrad	Baltic	25,000
Museum of Comparative Zoology, Harvard University, Cambridge	Baltic	16,000
	Canadian	5,000
Palaeontology Museum, Humboldt University, Berlin	Baltic	20,000
Institute for Geology and Paleontology, Göttingen	Baltic	11,000
Geological Institute, Moscow	Baltic	5,000
	Siberian	4,000
Zoological Museum, Copenhagen	Baltic	7,600
Natural History Museum, Stuttgart	Baltic	2,500
	Dominican	4,600
Smithsonian Institution, Washington, D.C.	Dominican	5,500
Geological Institute, Hamburg	Baltic	4,000
Florida State Collection of Arthropods, Florida Department of Agriculture, Gainesville	Dominican	3,500
Field Museum of Natural History, Chicago	Baltic	2,600
American Museum of Natural History, New York	Baltic	500
	Dominican	2,000
University of California, Berkeley	Mexican	3,000
	Alaskan	200
Geological Museum, Beijing	Chinese	2,000
State Museum, Ludwigsburg, Germany	Baltic	2,000
National Museum of Natural History, Paris	Baltic	2,000
Biosystematics Research Institute, Ottawa, Ontario, Canada	Canadian	300

either sold or lost during the war. The Bachofen-Echt collection was apparently vandalized and the remainder sold piecemeal to jewelers (Wenzel, 1953).

The major North American collections of Baltic amber, which are located at Harvard University and the Natural History Museum in Chicago, owe their presence, in part, to a gentleman named William A. Haren, who lived in St. Louis. Haren had a private collection of amber he had obtained from Germany. At his death, the largest portion of his collection was purchased by the Museum of Comparative Zoology at Harvard and the remainder by A. F. Kohlman of Racine, Wisconsin. When Kohlman died, his collection was purchased by a Racine high school science teacher, F. E. Trinklein. Trinklein then sold the collection to the Chicago natural history museum, where it is now maintained as the Kohlman collection. It is interesting to note that when the Kohlman collection of approximately 2,632 fossiliferous pieces was purchased in 1953, the Field Museum paid only $500, or approximately 19 cents apiece. Needless to say, the price of fossiliferous amber has risen considerably since then, and it is now possible to pay $500 for a single large piece of fossiliferous Baltic amber!

The Museum of Comparative Zoology at Harvard also acquired the Herman Hagan collection of approximately 8,000 pieces of fossiliferous Baltic amber. Herman August Hagen was born in Königsberg in 1817 and received an education in medicine at the then famous University of Königsberg. After practicing as a physician for 50 years, he was invited by Louis Agassiz to become director of the entomology section of the then Zoological Museum at Harvard. He had published extensively on various insect groups and continued to do so. Having been born in the middle of the most famous amber region of the world, and into a family with a strong amber tradition (both his father and grandfather published papers on amber inclusions), it was natural for Hagen to continue the family tradition.

Another famous amber collection is the Brodzinsky Lopez-Penha collection of Dominican amber, which was recently purchased by the Smithsonian Institution. This collection, consisting of approximately 5,500 fossiliferous pieces, was slowly acquired by Jake Brodzinsky and his wife (Lopez-Penha) during their retirement years in the Dominican Republic. Having spent his working life following non-biological pursuits, Brodzinsky started an amber business, purchased a microscope, and learned to recognize the orders and many families of insects found in the amber (see Fig. 15B). In this way, he was able to assist many specialists. The Brodzinsky Lopez-Penha collection contains a wide range of biological inclusions that were selected by Brodzinsky for their excellent preservation and scientific value.

Collecting amber is challenge enough, but polishing and preparing pieces for examination can be tedious and involve a certain amount of care. A piece may be brittle or contain internal fractures that result in breakage during the shaping and polishing processes. Too much sanding may remove the specimen or a portion of it, and constant examination is necessary. Finally, a clear polishing is important, so that scratches do not interfere with the viewing. All this takes time and equipment possessed by few scientists.

CHAPTER FOUR

Biological Inclusions in Amber

A range of life forms, from microbes to vertebrates and higher plants have been identified in amber. This chapter examines them all, in roughly phylogenetic sequence. The information presented here will form the basis for discussing, in the concluding chapter, the biological implications of amber for studies of evolution, extinction, and biogeography.

Kingdom Monera (Bacteria)

Owing to the difficulty of examination under the light microscope, few studies have been performed on microorganisms in amber.

Bacteria in Baltic amber were first studied by Blunck (1929). The bacterial cells were obtained by partially dissolving the amber in turpentine oil, then sieving and centrifuging the dissolved portion. Examination of the pellet with the light microscope at a magnification of 800 times revealed pollen, fungus spores, and bacteria. Among the latter were micrococci (1 micrometer in diameter), short rods (2–3 micrometers in length), longer rods (5–8 micrometers in length), and spiral forms. On the basis of these shapes, Blunck established the taxa *Micrococcus elektroni*, *Bacillus elektroni*, *Longibacillus elektroni*, and *Spirillum elektroni*, respectively. Clusters of spherical bacterial cells (*Succinococcus*) in Baltic amber are figured by Katinas (1983).

While sectioning Mexican amber for electron microscope observations, Poinar and Hess (unpublished information) found bacterial cells attached to the pseudocoel of a nematode. These cells were approximately 2 micrometers long and 0.6 micrometers wide (Fig. 21), and may have been anaerobes that developed in the nematode's body after entrapment in the amber. Studies involving attempts to revive

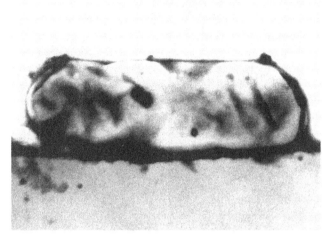

Figure 21. Electron micrograph of a bacterial cell in Mexican amber (photo by R. Hess-Poinar).

bacteria and other cells in amber are discussed in a later section of this work.

Kingdom Protista (Algae and Slime Molds)

The report by Kirchner (1950) on the presence of the alga *Discophyton electroneion* (Cyanophyceae) in Baltic amber requires verification. No other algae in amber have been reported.

The description of the slime mold *Stemonites splendens* Rost. forma *succinifera* Domke (1952) from Baltic amber and a plasmodium from Dominican amber (Color Plate 8) are the sole reports of myxomycetes from amber.

Kingdom Fungi

The oldest organisms found in amber have been reported from Carboniferous amber in Scotland. From it, Smith (1896) described a new order and new genera and species of fungi, pollen, and various portions of coniferous flowers. This amber, called Middletonite, is associated with coal deposits in northern England (near Leeds) and southern Scotland

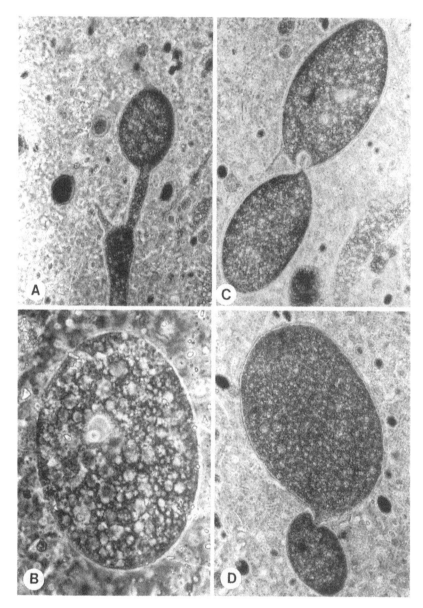

Figure 22. Fungi reported from Middle East amber. A. Zoosporangium and vesicles of *Peronosporites pythius*. B. Gametangial bodies of *P. pythius* (each body about 450 μm long). C. Sporangium of *P. pythius*. D. Gametangial interactions of *P. pythius* (longest length of larger body is 700 μm). (Photos courtesy of A. Nissenbaum.)

(Kilmarnock district). The next oldest fossiliferous amber is from the Late Triassic (Carnian stage) of Germany (Color Plate 6).

Various phycomycetes have been described from Middle Eastern (Israeli) Cretaceous amber (Ting and Nissenbaum, 1986). These unicellular organisms, described in the genera *Phycomycitis, Peronosporites,* and *Blastocladitis,* are thought to represent primitive fungi and range from 300 to 700 micrometers in largest diameter. Some are thin-walled (Fig. 22C) and contain structures reminiscent of spores borne on conidiophores (Fig. 22A). Also present and still more intriguing are stages suggesting the emergence of germ plasm from resting bodies (Fig. 22D) and the transfer of cytoplasmic elements between individuals (Fig. 22B).

Saprophytic fungi are common in amber and are mostly represented by a mycelium growing on the remains of arthropods and plants. Legg (1942) reported fungal mycelia in Canadian amber and representatives of the genera *Acremonium, Brachycladium, Cladosporium, Fungites, Fusidium, Gonatobotrys, Oidium, Mucorites, Penicillium, Pezizites, Polyporus, Ramularia, Schizosacharomycetes, Sporotrichites, Stilbum, Streptotrix, Torula, Tramaites,* and *Veionella* have been described from Baltic amber (Czeczott, 1961; Conwentz, 1890; Caspary and Klebs, 1907; Katinas, 1983; Berkeley, 1848; Göppert, 1853). In addition, Grüss (1931), examining Baltic amber, described species of yeastlike fungi (Moniliales) in the genus *Anthomycetes,* as well as species of *Melanosphaerites, Arachnomycelium,* and *Cladosporium.* Pampaloni (1902) described a *Monilites* from Sicilian amber, and *Geotrichites glaesarius* (Stubblefield et al., 1985) and an *Aspergillus* sp. (Thomas and Poinar, 1988) (Fig. 23) have been described from Dominican amber. A representative of the Dermatiaceae described from Austrian amber was dated as between the Paleocene and Mid Eocene (Bachmayer, 1962). Insect pathogenic fungi in amber include an entomophthoralean on a termite (Poinar and Thomas, 1982) and a *Beauveria* sp. on an ant (Poinar and Thomas, 1984). Nematophagous fungi have been described from Mexican amber where they were infecting populations of *Oligaphelenchoides atrebora* (Jansson and Poinar, 1986) (Fig. 24). The earliest known fossil gilled mushroom ("Agaricaceae") is *Coprinites dominicana* (Coprinaceae) described from Dominican amber (Poinar and Singer, 1990) (Color Plate 7). The latter find also represents the only known fossil "mushroom" (Agaricales) from the tropics and is important to considerations of the evolutionary development of the Basidiomycetes.

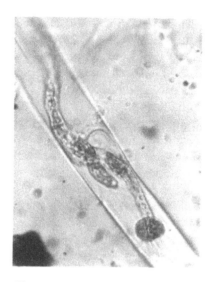

Figure 23. An *Aspergillus* fungus grow-ing on a fossil insect in Dominican amber.

Figure 24. A nematophagous fungus growing inside a nematode (*Oliga-phelenchoides atrebora* Poinar) in Mexi-can amber (Museum of Paleontology collection, University of California, Berkeley).

Kingdom Plantae (Plants)

Bryophyta (Mosses and Liverworts)

Fossil bryophytes are rare and the best-preserved representatives are found in amber. Portions of moss were noted by Legg (1942) in Cana-dian amber but were never described. As noted by Czeczott (1961) and Mägdefrau (1957), who presented synopses of bryophytes in Baltic am-ber, descriptions of species in the following genera of mosses have been published: *Catharinea, Dicranites, Dicranum, Grimmia, Muscites, Phas-cum, Pogonatum, Polytrichum, Rhytideadelphus, Trichostomum,* and *Weisia.* No descriptions of fossil moss species from other amber sources have ap-peared, although representatives of the genera *Bryopteris* and *Octoble-pharum* have been noted in Dominican amber (Hueber and Langenheim, 1986).

Grolle (1988) summarized the hepatics or liverworts in amber and recorded species of the genera *Bazzania, Cephaloziella, Cheilolejeunea,*

Frullania, Jungermannia, Lejeunea, Nipponolejeunea, Notoscyphus, Radula, Spruceanthus, and *Trocholejeunea* from Baltic amber and *Calypogeia, Cephaloziella, Cheilolejeunea, Frullania, Radula,* and *Trocholejeunea* from Saxonian or Bitterfeld amber. Grolle (1988) also cited species of *Bazzania, Bryopteris, Cyclolejeunea, Leucolejeunea, Prionelejeunea,* and *Radula,* as well as representatives of *Drepanolejeunea, Frullania, Lejeunea,* and *Stictolejeunea,* from Dominican amber. A single species of hepatic, *Lejeunea palaeomexicana* Grolle, is known from Mexican amber.

Pteridophyta (Ferns)

Fern remains in amber are rare. Czeczott (1961) reports only two species from Baltic amber, namely *Pecopteris humboldtiana* Göpp. and Beren. and *Alethopteris serrata* Casp. More recently, Gómez (1982) described an epiphytic fern, *Grammitis succinea* (Grammitidaceae), from Dominican amber.

Gymnosperms and Angiosperms (Higher Plants)

Most analytical studies on Cretaceous amber indicate that the deposits originated from gymnosperms, which is understandable because the only other possible group that could have produced large amounts of resin, the angiosperms, do not appear in the fossil record until the Early Cretaceous (Stewart, 1983). According to Hickey and Doyle (1977) the oldest true angiosperms are represented by the remains of small pinnately veined simple leaves from the Early Cretaceous in Siberia. These remains are followed by a rapid diversification of both monocot and dicot leaf types in the subsequent deposits of the Cretaceous. By the Late Cretaceous, angiosperm leaves predominate in floras uncovered in Alaska, western Canada, the United States, Europe, and Asia (Stewart, 1983). This rapid radiation and diversification of the angiosperms extended into the Tertiary and was especially notable during the Eocene Epoch. Arnold (1947) estimated that a little less than half of the present-day 300 families of angiosperms have a fossil record. This record, mostly leaf compressions or impressions, is based mainly on Tertiary deposits.

Unfortunately, botanical remains in Cretaceous amber have been little studied. Bits of decayed woody material and fragments of bark and leaves were reported in Canadian amber (Legg, 1942), and shreds of plant debris are widespread in Alaskan amber (Langenheim et al., 1960), but no identifications were made. Plant remains occur also in Lebanese amber, but identifications have not been made. Plant species that have been proposed as the source of fossiliferous amber deposits are listed in Table 3.

The most extensive studies on flora in amber have been conducted on Baltic amber. Between the years of 1830 and 1937, up to 700 species of spore- and seed-producing plants have been described from either impressions or inclusions in Baltic amber. Most of the botanical investigations have been summarized in four general works, prepared by Conwentz (1886B), Göppert and Berendt (1845), Göppert and Menge (1883), and Caspary and Klebs (1907). Two classical studies on the Baltic amber tree by Conwentz (1890) and Schubert (1961) illustrated the leaves, the cones, and the fine detail of wood remains in amber.

Of the 750 separate plants described from Baltic amber over the past two and a half centuries, only 216 are valid species according to Czeczott (1961). Included among the 216 species are 63 cryptograms (5 species of bacteria, 1 myxomycete, 18 fungi, 2 lichens, 18 liverworts, 17 mosses, and 2 ferns), 52 gymnosperms, and 101 angiosperms.

An examination of the gymnosperms in Baltic amber reveals many present-day genera (Fig. 25, Table 7). It is also noteworthy that the present-day species most closely related to the fossil gymnosperms occur in North America, China, Japan, and Africa. Many modern-day genera of angiosperms are also present (Table 8). Approximately 67 percent of the angiosperms in Baltic amber have been identified from flowers, fruits, or seeds; the remainder have been described from leaves and twigs.

A great diversity of angiosperm families is represented in Baltic amber. Some represent temperate groups while others are representative of Mediterranean, subtropical, and even tropical climates. The distribution of closely related contemporary species of angiosperms is intriguing, as it was with the gymnosperms. For example, the fossil angiosperm *Drimysophyllum* (Magnoliaceae) most closely resembles the contemporary *Drimys* of the Winteraceae; this genus now occurs in the Malay Archipelago, New Caledonia, New Zealand, and Central and South America. The nearest present-day locality for *Clethra*, found in Baltic amber, is Madeira Island. Other tropical or subtropical families represented in Baltic amber are the Palmae, Lauraceae, Dilleniaceae, Myrsinaceae, Ternstroemiaceae, Commelinaceae, Araceae, and Connaraceae. According to an analysis by Czeczott (1961) about 23 percent of the families are tropical, 12 percent are restricted today to temperate regions, and the remainder are cosmopolitan or have an anomalous or discontinuous distribution.

Figure 25. Twig of a member of the cypress family (Cupressaceae) in Baltic amber.

Table 7. Families and genera of gymnosperms reported from Baltic amber

Family	Genera
Cycadaceae	*Zamiphyllum*
Podocarpaceae	*Podocarpites*
Pinaceae	*Pinus, Piceites, Larex, Abies*
Taxodiaceae	*Glyptostrobus, Sciadopitys, Sequoia*
Cupressaceae	*Widdringtonites, Thuites, Libocedrus, Chamaecyparis, Cupressites, Cupressinanthus, Juniperus*

SOURCE: Modified from Czeczott, 1961.

Table 8. Families and genera of angiosperms reported from Baltic amber

Family	Genera
Monocots	
Araceae	*Acoropsis*
Commelinaceae	*Commelinacites*
Gramineae	*Graminophyllum*
Liliaceae	*Smilax*
Palmae	*Bembergia, Palmophyllum, Phoenix*
Dicots	
Aceraceae	*Acer*
Apocynaceae	*Apocynophyllum*
Aquifoliaceae	*Ilex*
Campanulaceae	*Carpolithus*
Caprifoliaceae	*Sambucus*
Celastraceae	*Celastrinanthium*
Cistaceae	*Cistinocarpum*
Clethraceae	*Clethra*
Connaraceae	*Connaracanthium*
Dilleniaceae	*Hibbertia*
Ericaceae	*Andromeda, Ericiphyllum, Orphanidesites*
Euphorbiaceae	*Antidesma*
Fagaceae	*Castanea, Dryophyllum, Fagus, Quercus*
Geraniaceae	*Erodium, Geranium*
Hamamelidaceae	*Hamamelidanthium*
Lauraceae	*Cinnamomum, Trianthera*
Leguminoseae	*Dalbergia, Leguminosites*
Linaceae	*Linum*
Loranthaceae	*Enantioblastos, Loranthacites, Patzea*
Magnoliaceae	*Drimysophyllum, Magnolilepis*
Myricaceae	*Myriciphyllum, Myrica*
Myrsinaceae	*Berendtia, Myrsinopsis*
Olacaceae	*Ximenia*
Oleaceae	*Oleiphyllum*
Oxalidaceae	*Oxalidites*
Pittosporaceae	*Billardierites*
Polygonaceae	*Polygonum*
Proteaceae	*Dryandra, Lomatites, Persoonia*
Rhamnaceae	*Rhamnus*
Rosaceae	*Mengea*
Rubiaceae	*Sendelia*
Salicaceae	*Saliciphyllum*
Santalaceae	*Osyris, Thesianthium*
Saxifragaceae	*Adenanthemum, Deutzia, Stephanostemon*
Ternstroemiaceae	*Pentaphylax, Stuartia*
Thymeleaceae	*Eudaphniphyllum*
Umbelliferae	*Chaerophyllum*
Urticaceae	*Forskohleanthium*

SOURCE: Modified from Czeczott, 1961.

The mixture of plants that today occur in different geographical parts of the world has led to several theories regarding the climatic conditions of the Baltic amber forest. Schubert (1953) and earlier workers imagined that the ecology of the amber forest was similar to that of southern Florida, where isolated "hammocks" of tropical plants occur among forests of subtropical and temperate forms. Here also can be found three groups representing the variety of the Baltic amber forest: palms, pines, and oaks. Ander (1941), on the other hand, felt that the amber forest was denser and the climate moister than that presently in Florida. He imagined the area as partly mountainous, with the tropical groups occurring on south-facing slopes. Heer (1859) had originally explained the mixture of floral types by the immensity of the forest area, which he supposed stretched over an area from present-day Scandinavia to Germany and Poland. The amber-bearing trees lived in a temperate to a subtropical climate, reaching the former by possibly extending up the sides of mountains in the northern portion. Because the production of Baltic amber probably extended over several million years, Wheeler (1915) supposed that the climate could have changed over this period, perhaps producing a shift of vegetation that in turn was reflected in the amber inclusions.

Plant hairs are commonly found in Baltic amber where they are often referred to as oak hairs because they resemble those found today on oak bud scales. A recent analysis indicated that there are five major trichome types in Baltic amber, and all occur today among representatives of the Fagaceae (Jay H. Jones, personal communication) (Figs. 26, 27).

Plants reported from other amber deposits are relatively few compared with the rich flora of Baltic amber. Of special interest was the discovery of a *Myrica* (Myricaceae) fruit in Austrian amber in deposits ranging from the Paleocene to the Mid Eocene (Bachmayer, 1968). Two other plants have been described from Mexican amber (Miranda, 1963). One was a species of *Acacia* (Leguminosae), identified on the basis of leaflets and pinna of a bipinnate compound leaf; the shape of the leaflets is similar to those of the present-day *Acacia angustissima* (Mill.) Kuntze and *Acacia milleriana* Standl. *Acacia* is a genus of world-wide distribution in tropical and subtropical areas. The other species described from Mexican amber was *Tapirira durhamii* Miranda, a member of the Anacardiaceae. This genus occurs in humid zones of tropical America, including the area where Mexican amber is mined. In fact, the fossil species somewhat resembles the local *Tapirira mexicana*, which aside from existing in humid tropical coastal areas, can also be found in higher mountainous, somewhat temperate regions (Miranda, 1963). On the basis of this and other

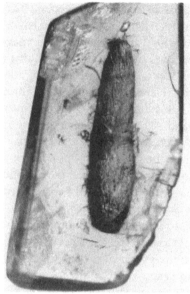
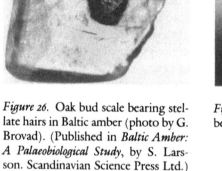
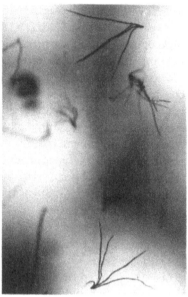

Figure 26. Oak bud scale bearing stellate hairs in Baltic amber (photo by G. Brovad). (Published in *Baltic Amber: A Palaeobiological Study*, by S. Larsson. Scandinavian Science Press Ltd.)

Figure 27. Oak bud hairs in Baltic amber (Poinar collection).

fossils from Miocene deposits, Miranda concludes that "few evolutionary changes have occurred in the genus *Tapirira* during the last 30 million years." The fossil *Tapirira* flowers possess a short stem that is characteristic of *T. guianensis* and other South American species. This greater resemblance to a South American species than to a Mexican species suggests that the composition of the past flora has changed somewhat during the Tertiary, and climatic factors are suggested as a possible cause.

Also enclosed in Mexican amber were a possible oak leaf, stellate oaklike trichomes, part of a flower of a legume (other than *Hymenaea*), a branch belonging to a member of the Jungermaniales (a liverwort), numerous seeds, spores, and pollen grains, and the leaf, sepal, and pollen remains of the tree legume *Hymenaea* sp. (Langenheim, 1966, 1973). The *Hymenaea* leaf and sepals were identified by Miranda just before his death, and possessed characters belonging to the present-day species *H. courbaril* and *H. intermedia.* Furthermore, the infrared spectrum of Mexican (Chiapas) amber showed a close similarity with that of resin from these two tree species. The spectra were also similar to Miocene amber obtained from Para, Brazil and to Colombian copal from Pleistocene or earlier deposits in Girón and Medellín.

Amber from the Dominican Republic also contains floral and leaf remains of trees of the genus *Hymenaea* (Schlee, 1980, 1984; Poinar, 1985A; Hueber and Langenheim, 1986) (Color Plate 1), and recent studies have indicated that the amber was produced, at least in part, by trees of the extinct species *H. protera* Poinar (1991D). Morphologically, *Hymenaea protera* most closely resembles the present-day *H. verrucosa*, which occurs in East Africa and adjacent islands. The latter species is placed in the more "primitive" *Trachylobium* section of the genus and is restricted in distribution to lowland evergreen and semi-deciduous forests. The remaining species of *Hymenaea* are all New World, ranging from central Mexico through Central and South America and the West Indies (Lee and Langenheim, 1975).

The similarity of *H. protera* in Dominican amber to *H. verrucosa*, now restricted to East Africa, raises some interesting questions about the origin and geographical distribution of the genus *Hymenaea*. Hueber and Langenheim (1986) consider the petals and ovaries of *H. verrucosa* and *H. oblongifolia* as relatively primitive, and suggest that this portion of the genus *Hymenaea* (section *Trachylobium*) had an African origin. They propose that fruits from this stock drifted from West Africa to the east coast of Brazil and the Antilles by way of the south equatorial current. The one remaining East African species, *H. verrucosa*, is then considered a relict of an earlier Pan-African distribution of this and other species.

The present-day African–South American distribution of *Hymenaea* could be explained more plausibly by plate tectonics. The genus could have originated and evolved at a period when the two continents formed a common landmass (Late Cretaceous). Climatic changes in Africa associated with the movement of the African plate eradicated many species and left *H. verrucosa* as an East African relict. The present co-existence of the related resin-producing tree genus *Copaifera* with 30 neotropical and four African species (Langenheim, 1973) supports this hypothesis. Most of the neotropical species of *Copaifera* are found in the eastern part of South America (the majority in Brazil); in Africa, three species occur in the western part and one occurs in Angola and northern Rhodesia.

Hymenaea, Copaifera, and possibly other Caesalpinioideae probably originated on the joint South American-African landmass sometime in the Early Cretaceous. When South America finally became separated from Africa, now considered to be Late Cretaceous (Cox and Moore, 1985), common species were left on the separate continents, where they underwent their own speciation based on genetic change and environmental selection.

The natural history of most *Hymenaea* species has not been studied. Some information, however, is available on *H. courbaril* (the guapinol, locust, algarroba, Jatoba, or stinking toe), which may be closely related to the species that produced Mexican amber. *Hymenaea courbaril* is a polymorphic species, and six varieties have been described, mainly on the basis of morphological differences. The species ranges from Mexico through Central and South America (except Argentina, Uruguay, and Chile) and on to the Antilles. Trees of this species, which grow to a height of 55 meters and may have a trunk diameter of almost two meters, occur in a wide variety of habitats, including moist evergreen forests, savanna areas, and relatively dry forests.

Although *H. courbaril* is considered an evergreen tree, it is deciduous and has a synchronized leaf drop, usually in late December to mid-January, with new leaves appearing within two weeks. Each tree produces flowers for about 45 days, but only several of the 50–150 buds open each night. The petals (2–4 centimeters across) and stamen filaments (2 centimeters long) are white, and pollination is mainly effected by nectarivorous bats (*Glossophaga* sp.), which receive pollen on their head and chest fur while obtaining nectar from the flowers (bees also have been reported as pollinators). The flowers are complete, but fruit is produced sporadically and many are aborted during growth. The pods are at first green, then turn brown and harden, eventually dropping to the ground and rotting when the rains come. The developing and mature pods are consumed by a variety of organisms. Larvae of *Anthonomus* weevils (Curculionidae) develop in the bud and the adults emerge from aborted fruit dropped to the ground. At the green fruit stage, *Rhinochenus* weevils lay eggs on the pod, and their larvae kill part or all of the seeds in the attacked pod. Agoutis (*Dasyprocta punctata*) open the mature pods, and these rodents consume the seeds and the dry yellow pulp around the seeds. Peccaries also open and consume the pods. After the seeds have germinated during the rainy season, various rodents bite off the cotyledons and may kill the plant. Young tender leaves are used by leaf-cutter ants and mature leaves are eaten by the fulgorid lantern fly, *Fulgora later-*

naria, and various saturniids (including larvae of *Hylesia lineata*). In addition to reproducing sexually by means of fruit, the trees reproduce asexually from root suckers, and in some areas this seems to be the major source of reproduction (Janzen, 1983; Lee and Langenheim, 1974).

Resin production occurs in all growing stages of *H. courbaril*. Even in the seedlings, resin pockets form in the hypocotyl. In mature trees, resin cavities occur in the cambial zone of both the trunk and roots, and if either of these organs is cut, resin will exude from the cambial zone. Root and trunk resin appear similar (Langenheim, 1967).

Amber from the Dominican Republic contains angiosperm remains in addition to those of *Hymenaea*, including mimosoid and other flowers (Hueber and Langenheim, 1986; Poinar, 1985A) (Figs. 28, 29), leaves (Schlee, 1980, 1984), and tendrils (Schlee, 1980). Flowers representing the families Leguminoseae, Meliaceae, Myristiacaceae, Thymeliaceae, Bombacaceae, and Hippocrateaceae have also been identified in Dominican amber by J. Strother at the University of California Herbarium, Berkeley, and L. Liesner, T. Zarucchi, and R. Gereau of the Missouri Botanical Garden (Figs. 30–33). Spikelets of bamboo grasses belonging to the gen-

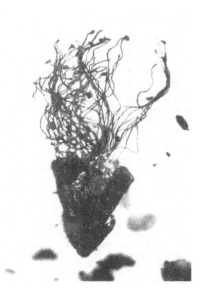

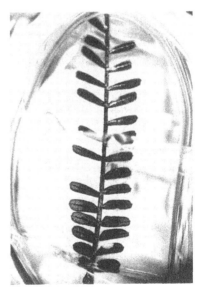

Figure 28. A mimosoid flower with numerous stamens in Dominican amber (Poinar collection).

Figure 29. A mimosoid leaf bearing numerous leaflets in Dominican amber (Poinar collection).

Figure 30. Dominican amber flower tentatively identified as belonging in or near the genus *Prioria* (Leguminoseae) (Poinar collection).

era *Pharus* and ?*Arthrostylidium* have been identified in Dominican amber by J. Travis Columbus.

Kingdom Animalia: Protozoa to Myriapoda

Protozoa

Protozoa have been reported from amber in only one incidence. Legg (1942) found some single-celled organisms associated with a nematoceran fly in Canadian amber. These organisms reminded him of *Paramecium*, and Wichterman (1953) later confirmed his suspicion, label-

Figure 31. Dominican amber flower tentatively identified as belonging in or near the genus *Trichilia* (Meliaceae) (Poinar collection).

Figure 32. Dominican amber flower tentatively identified as belonging in or near the genus *Peritassa* (Hippocrateaceae) (Poinar collection).

Figure 33. Dominican amber flower tentatively identified as belonging in the family Thymeliaceae (Poinar collection).

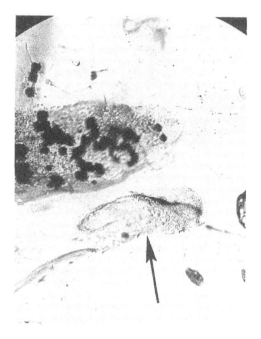

Figure 34. A *Paramecium* (arrow) in Canadian amber (collected and photographed by William M. Legg in 1941; specimen subsequently lost).

ing the organism as a tentative *Paramecium* of the "caudatum" group, based on its size, shape, general morphology, and surface markings (Fig. 34). Because *Alternaria*-like spores and alga-like inclusions were in this same piece, Legg felt that part of an aquatic habitat had been entrapped in the resin.

Nematoda

Nematodes, or roundworms as they are commonly called, constitute a group of numerous yet rarely noticed animals. Their universal dependency on moisture for development and the small size of many of the microbotrophic (microbe-digesting) forms has forced them to enter moist, cryptic habitats such as soil, plants, animals, and lake and ocean bottoms. Those forms visible with the naked eye are mostly parasitic in the bodies of vertebrates and invertebrates. Only nematodes commonly classified as free-living, those that feed on unicellular microorganisms (microbotrophic) or fungi (mycetophagous), are discussed here (Poinar, 1983). Insect-parasitic nematodes are also found in amber but will be discussed in the section on symbiosis.

Microbial feeders in the families Plectidae (*Oligoplectus*) and Diplogasteridae (*Oligodiplogaster*) and the genus *Vetus* (a collective genus of fossil forms unassignable to present-day genera) have been recorded from Mexican and Baltic amber (Menge, 1863; Poinar, 1977). A single mycetophagous species, *Oligaphelenchoides atrebora*, very much resembling present-day *Aphelenchoides* and *Bursaphelenchus*, has been described, together with its probable fungal food source, from Mexican amber (Poinar, 1977) (Fig. 35). Other isolated nematodes in Dominican amber await identification and description (Fig. 36).

Rotifera

The oldest known fossil rotifers are representatives of the class Bdelloidea reported from Dominican amber (Poinar and Ricci, 1992). Bdelloids are ubiquitous microscopic multicellular animals that occur in fresh water, soil, and water films on terrestrial plants. Bdelloid rotifers are the only class of animals that is entirely, obligately parthenogenetic, and their discovery in amber provides evidence of parthenogenetic continuity.

Mollusca

Shells of terrestrial snails (Gastropoda: Pulmonata and Prosobranchia) are rare in amber. However, species of *Electrea* (Cyclophoridae), *Strobilus* (= *Strobilops*; Strobilopsidae), *Vertigo* (Pupillidae), *Baba* (Clausiliidae), *Hyalina* (= *Euconulus*; Euconulidae), *Microcystis* (Euconulidae), and *Parmacella* (Parmacellidae) have been reported from Baltic amber (Klebs, 1886; Sandberger, 1887; Larsson, 1978; Keilbach, 1982; family names updated). Bachofen-Echt (1949) also reported from Baltic amber a *Helix* sp. related to a Recent African species.

Shells (and in some instances partially preserved soft tissue) of terrestrial snails in Dominican amber have been identified as *Strobilops* (*Coelostrobilops* sp.) (Strobilopsidae; Color Plate 8), *Subulina* sp. (Subulinidae), *Spiraxis* sp. (Spiraxidae), a member of the Ferussaciidae, *Varicella* sp. (Oleacinidae) (Fig. 37), a member of the Helicinidae, and a member of the subclass Prosobranchia (Poinar and Roth, 1991).

All of the snail taxa from Dominican amber are indicative of a tropical or subtropical climate and include Hispaniola or other Antillean islands in their present-day ranges. The specimen of *Strobilops* represents

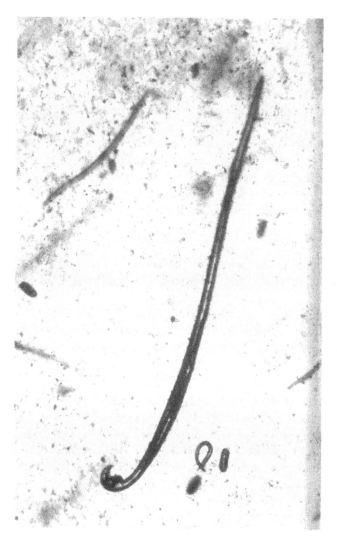

Figure 35. A male, together with eggs and juveniles, of a fungal-feeding nematode, *Oligaphelenchoides atrebora* Poinar, in Mexican amber (Museum of Paleontology collection, University of California, Berkeley).

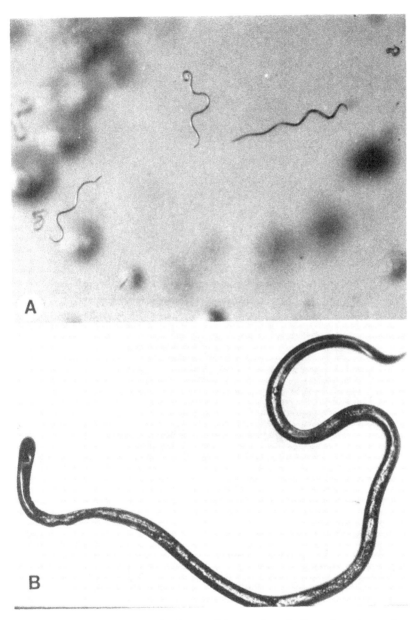

Figure 36. Free-living nematodes in Dominican amber. A. Several specimens that were probably carried to this site by an insect (Poinar collection). B. Detail of specimens in the same amber piece as A.

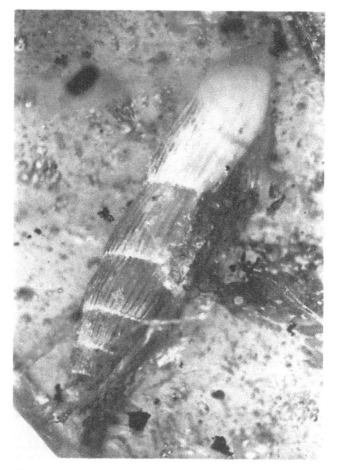

Figure 37. Shell of a land snail of the family Oleacinidae (order Stylommatophora) in Dominican amber (Poinar collection).

the first record of this family in the Western Hemisphere. The species of *Varicella* and the helicinid could be considered arboreal, whereas subulinids are generally predaceous, ground-dwelling forms. Species of *Spiraxis* and *Strobilops* are normally found in plant litter on the ground.

Annelida

Although rare, representatives of the class Oligochaeta have been recovered from amber. In 1863, Menge described a white worm, *Enchytraeus sepultus* (Enchytraeidae), from Baltic amber. Other Baltic amber enchytraeids are cited by Bachofen-Echt (1949) and additional forms occur in the Copenhagen amber collection (Larsson, 1978). Typical earthworms from Baltic amber are figured by Bachofen-Echt (1949), and unidentified forms occur in Dominican amber.

Tardigrada

The only known fossil tardigrade, *Beorn leggi*, was described from Canadian amber by Cooper (1964). It was named in memory of William M. Legg, an early collector and student of Canadian amber, who died in a hunting accident before he could complete his studies (Legg, 1942).

Arthropoda: Crustacea

Representatives of both the amphipods (class Amphipoda) and isopods (class Isopoda) occur in amber. It is especially interesting that amphipods were found in Baltic amber because they are generally associated with an aquatic or semiaquatic habitat. They do occur today in warm, moist terrestrial habitats and that is probably the type of environment utilized by the genus found in Baltic amber, *Palaeogammarus*. The three individuals of *Palaeogammarus* (Crangonycidae) in Baltic amber represent the earliest known amphipod fossils. This genus closely resembles the present-day *Crangonyx* (Just, 1974) and the fossil species is figured in Bachofen-Echt (1949) and Larsson (1978). Extant crangonycids occupy a wide range of habitats including both subterranean and terranean running and standing water.

Isopods are more commonly encountered in amber than are amphipods, with representatives of *Ligidium* (Ligiidae), *Trichoniscoides* (Trichoniscidae), *Oniscus* (Oniscidae), and *Porcellio* (Porcellionidae) having been described from Baltic amber (Keilbach, 1982) (Fig. 38). An oniscoidid from Baltic amber is illustrated in Schlee and Glöckner (1978).

Terrestrial isopods from Dominican amber have been identified and described by Schmalfuss (1980, 1984) (Figs. 39, 40). These include *Protosphaeroniscus tertiarius* Schmalfuss (Sphaeroniscidae), *Trichorhina* sp. (Platyarthridae), *Pseudarmadillo cristatus* Schmalfuss, *P. tuberculatus* Schmalfuss (Pseudarmadillidae), and undescribed representatives of the Philosciidae and Styloniscidae or Truchoniscidae. The occurrence of *Pseudarmadillo* in Dominican amber is the first record of this genus on Hispaniola, although extant forms are known from Cuba (Schmalfuss, 1984). Isopods commonly occur under bark where they are scavengers feeding on decaying plant material. Their habitat would bring them into

Figure 38. A woodlouse (family Trichoniscidae) in Baltic amber (photo by G. Brovad). (Published in *Baltic Amber: A Palaeobiological Study*, by S. Larsson. Scandinavian Science Press Ltd.)

Figure 39. An isopod (family Philosciidae) in Dominican amber (Poinar collection).

Figure 40. Two isopods of the genus *Protosphaeroniscus* (family Sphaeroniscidae) in Dominican amber (Poinar collection).

contact with resin exuding from the trunk or roots of trees. The earliest fossil isopod dates from Pennsylvanian deposits (Schram, 1970).

In their list of arthropods from Canadian amber, McAlpine and Martin (1969) list a possible member of the Copepoda and Ostracoda. Legg's thesis (1942) was cited as a reference to the former report, however, and Legg himself was equivocal, stating that "One specimen in the same piece described under the Protista is possibly a Crustacean, but its crushed condition makes careful determination almost impossible" (p. 55). Thus, until Legg's specimen and the Ostracoda reported above are reexamined, these reports should be considered inconclusive.

Arthropoda: Myriapoda

The myriapods comprise a collective group of many-legged arthropods including the centipedes (Chilopoda), millipedes (Diplopoda), pauropods (Pauropoda), and symphylids (Symphyla).

The centipedes are an ancient group with a fossil record extending back some 375 Ma into the Devonian (Shear et al., 1984). Representatives

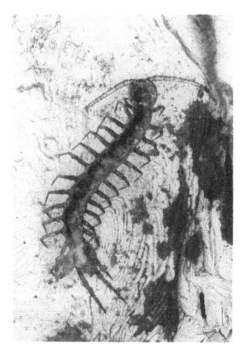

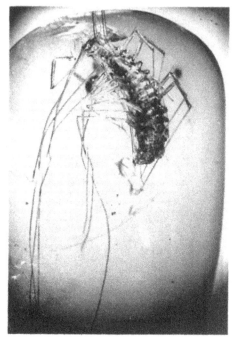

Figure 41. A centipede of the family Litho-biidae in Baltic amber (photo by G. Brovad). (Published in *Baltic Amber: A Palaeobiological Study*, by S. Larsson. Scandinavian Science Press Ltd.)

Figure 42. A "house" centipede of the or-der Scutigeromorpha in Dominican amber (Costa collection, Puerto Plata, Dominican Republic).

of the genera *Geophilus* (Geophilidae), *Cryptops* (Cryptopidae), *Scolopendra* (Scolopendridae), *Lithobius* (Lithobiidae), *Scutigera* (Scutigeridae), and *Euzonus* (Scutigeridae) have been described from Baltic amber (Keilbach, 1982) (Fig. 41). A representative of the genus *Cryptops* (Cryptopidae) was reported from deposits of Dominican amber (Shear, 1987), which also include representatives of the Scutigeromorpha (Fig. 42).

Millipedes, like isopods and centipedes, are also found in amber because of their similar habits of concealment under the bark of trees. Representatives of the genera *Glomeris*, *Polydesmoidea*, and *Chordeumidea* (Glomeridae), *Craspedosoma* (Craspedosomidae), *Julus* (Julidae), *Polyzonium* (Polyzonidae), and *Polyxenus*, *Phryssonotus*, and *Schindalmontus* (Polyxenidae) have been described from Baltic amber (Keilbach, 1982). According to Bachofen-Echt (1949), the nearest relatives of *Schindalmontus* presently occur in South Africa. A color photograph of a polyxenid is shown in Schlee and Glöckner (1978) and a *Polyzonium* is illustrated by Bachofen-Echt (1949).

Millipedes of the genera *Docodesmus* (Pyrgodesmidae) and *Sipho-nocybe* (Siphonophoridae) were described from Dominican amber by Shear (1981) who stated that both closely resemble species now living in the same area. Shear (1987) also briefly mentioned representatives of *Iomus*, *Prostemmiulus*, and a new family of polydesmoids in Dominican amber. *Siphonophora hoffmani* Santiago-Blay and Poinar (Siphonophori-dae) and representatives of the genera *Lophoproctus* (Lophoproctidae), *Glomeridesmus* (Glomeridesmidae), *Epinannolene* (Pseudonannolenidae), *Docodesmus* and *Iomus* (Pyrgodesmidae), *Dasyodesmus*, *Inodesmus* and representatives of the families Stemmiulidae and Chelodesmidae have recently been reported from Dominican amber (Santiago-Blay and Poinar, 1992) (Color Plate 5).

Two specimens of the symphylan *Scolopendrella* (Scolopendrellidae) were reported from Baltic amber by Bachofen-Echt (1942, 1949) and probably represent the only known fossil symphylans. No pauropods have been described from amber.

Kingdom Animalia: Arthropoda: Hexapoda (Insects and Related Arthropod Groups)

The great majority of life forms described from amber are insects. Although it is not possible to treat all of the insect species that have been reported from various amber deposits, an attempt was made to cite all of the insect genera that have been described or reported from amber and to discuss certain details of especially interesting species. The contributions of Keilbach (1982) and Spahr (1981A, 1981B, 1985, 1987, 1988) were most useful, as was Larsson's (1978) discussion of Baltic amber insects. References to descriptions of amber insects that have been omitted from the present work can be found in these publications.

The insect orders covered in this section generally follow the arrangement and terminology found in the fifth edition of *An Introduction to the Study of Insects* (1981) by Borror, DeLong, and Triplehorn and the sixth edition of the same work (1989) by Borror, Triplehorn, and Johnson. This text is useful for identifying amber insects to family, although it covers only North American forms. Unless otherwise noted, remarks on the first appearance of arthropod groups in the fossil record are based on the publications of Crowson et al. (1967) and Carpenter and Burnham (1985). Families and genera in the present work are for the most part arranged alphabetically. Lists of arthropods found in amber from Mexico and the Dominican Republic occur in Appendices A and B, respectively.

Protura (Proturans)

There are no known fossil representatives of this small order, though their habits of living in humus and under bark would seem to make them excellent candidates for fossilization in amber. They may have been missed owing to their small size.

Diplura (Diplurans)

These small, rare arthropods were originally classified as insects in the order Diplura. Today, the Diplura are in a separate class, the Entognatha. Diplurans resemble bristletails but lack a median caudal filament as well as compound eyes and ocelli; the paired cerci (paired appendages on the tip of the abdomen) can be pincherlike, stubby, or relatively long. These small arthropods occur in soil, rotting wood, or other damp areas.

Several fossil representatives of this order occur in amber. Silvestri (1913) described *Campodea darwinii* (Campodeidae) from Baltic amber, a species that closely resembles the extant *C. staphylinus*. Another dipluran occurs in the Gyllenhal collection of Baltic amber in Uppsala, Sweden (Larsson, 1978), a representative of the Procampodeidae occurs in Dominican amber (Fig. 43), and a member of the Japygidae was found in Mexican amber (Fig. 44).

Figure 43. A dipluran (family Procampodeidae) in Dominican amber (Poinar collection).

Figure 44. A dipluran (family Japygidae) in Mexican amber (Museum of Paleontology collection, University of California, Berkeley; photo by P. Hurd).

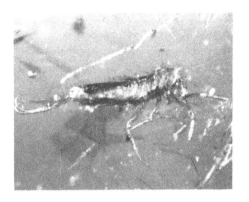

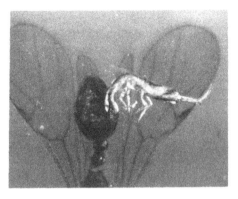

Figure 45. A springtail (order Collembola) of the genus *Seira* (family Entomobryidae) in Dominican amber. Note extended furcula on abdomen (Poinar collection).

Figure 46. A springtail (order Collembola) of the genus *Pseudosinella* (family Entomobryidae) in Dominican amber. Note extended furcula on abdomen (Poinar collection).

Collembola (Springtails)

Although small in size, representatives of this order have been described from Baltic, Mexican, and Canadian amber (Keilbach, 1982) and also occur in Dominican amber (see Appendix B). This group is an ancient one, dating back to *Rhyniella* from the Mid Devonian Rhynie cherts (Scotland) (Lehmann and Hillmer, 1983). Representatives of the families and genera Entomobryidae (*Entomobrya, Lepidocyrtus, Orchesella*), Hypogastruridae (*Hypogastrura*), Isotomidae (*Isotoma, Tetracanthella*), Neonuridae (*Pseudachorutes*), Poduridae (*Podura*), Sminthuridae (*Allacma, Bourletiella, Paidium, Sminthurus*), and Tomoceridae (*Tomocerus*) have been described from Baltic amber. *Isotomurus, Isotoma,* and *Isotomina* of the Isotomidae and *Lepidocyrtus, Lepidocyrtinus, Drepanura, Salina,* and *Paronella* of the Entomobryidae have been recorded from Mexican amber (Christiansen, 1971) (see Appendix A). *Protentomobrya walkeri* Folsom (1938) (Protentomobryidae) was described from Canadian amber.

In a study of the Collembola from Dominican amber, Mari Mutt (1983) described species in the families Isotomidae (*Cryptopygus, Isotoma*), Entomobryidae (*Lepidocyrtus, Pseudosinella, Seira, Salina, Paronella, Cyphoderus*), and Sminthuridae (*Sphyrotheca*). Most common were specimens of *Lepidocyrtus* and *Seira* (Figs. 45, 46). In general, the fauna was modern in character and similar to that reported from Mexican amber.

Springtails are common insects that can be found in moist dark habitats containing decaying plant remains. Most possess a furcula, a stalked forked structure, attached to the ventral side of the abdomen. When trapped in amber, many extend the furcula (Figs. 45, 46). Certain species aggregate and move as a vibrating mass across the soil surface.

In his analysis of springtails from Baltic amber, Handschin (1926) concluded that the climate experienced by these extinct species must have been similar to that of the present-day Mediterranean and North Africa, with many subtropical characteristics.

Thysanura (Bristletails, Silverfish, Firebrats, and Jumping Bristletails)

Bristletails are so named because they have three long filaments or appendages at the tip of the abdomen. They are small to moderate in size, with filiform antennae that often are longer than the body. In most of the forms recovered from amber, at least a portion of the antennae or filaments is missing. Like other members of this primitive group, they are wingless.

Representatives of the families Lepismatidae (*Lampropholis*, *Lepidothrix*, *Pachystylus*), Praemachilidae (*Praemachilis*), and Machilidae (*Machilis*) have been described from Baltic amber. In his analysis of this group, Silvestri (1913) stated that both *Lampropholis* and *Lepidothrix* are extinct genera and that *Lepidothrix pilifera* is more primitive than any present-day species. On the other hand, the species of *Machilis* and *Praemachilis* in Baltic amber are very similar to present-day species.

Lampropholis burmiticus Cockerell (1917F) (Lepismatidae) was described from Burmese amber and a *Neomachilellus* sp. was identified in Mexican amber (Wygodzinsky, 1971). A member of the Nicoletiidae has been identified in Dominican amber (see Appendix B).

The above reports of bristletails in Tertiary amber constitute the earliest fossil record of the Thysanura, a group of primitive, wingless hexapods.

Ephemeroptera (Mayflies)

Because the immature stages (nymphs) of mayflies are aquatic, the presence of this group is an indication of a nearby source of standing or running fresh water. The nymphs, which feed on algae and detritus in the water, are relatively long-lived (1 or 2 years), whereas the adults, which do not feed, usually complete their phase of life within a week. The adults have characteristic swarming flights that can reach cloudlike proportions, and mayflies can become enclosed in amber after being blown into the sticky resin during one of these swarming flights.

The mayflies are a relatively old group, with fossil remains dating back to Upper Carboniferous. The oldest amber fossil, *Cretoneta zherichini* (Leptophlebiidae), was described from Siberian amber (Tshernova, 1971) (Fig. 47). All the remaining descriptions have been from Baltic amber (mostly by Demoulin, 1970; see also Keilbach, 1982) and the specimens closely resemble present-day genera. These comprise representatives of the Ephemeridae, Potamanthidae (*Potamanthus*), Leptophlebiidae (*Paraleptophlebia, Xenophlebia, Blasturophlebia, Choroterpes*), Ametropodidae (*Siphloplecton, Metretopus, Brevitibia*), Siphlonuridae (*Balticophlebia, Baltameletus, Siphlonurus*), Isonychiidae (*Baetis, Cronicus*), Heptageniidae (*Electrogenia, Succinogenia, Cynigma, Heptagenia, Rhitrogena*), and Ephemerellidae (*Ephemerella*).

The description of *Succinogenia larssoni* Demoulin in Baltic amber is based on the only nymph found in amber. In a few amber pieces, the

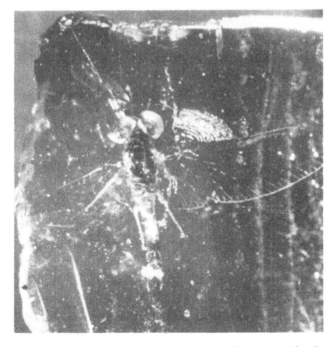

Figure 47. A mayfly, *Cretoneta zherichini* Tshernova (family Leptophlebiidae), in Siberian amber (Yantardak site) (photo by A. P. Rasnitsyn, courtesy of V. V. Zherichin).

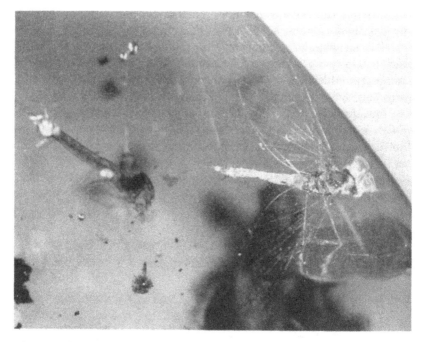

Figure 48. An adult and pre-imago of the mayfly genus *Careospina* (order Ephemeroptera, family Leptophlebiidae) in Dominican amber (Poinar collection).

preadult (preimago) skin is adjacent to the newly emerged adult (mayflies are the only insects that molt once after the wings have become functional). Such an arrangement also occurred with a species of *Careospina* (Leptophlebiidae) in Dominican amber (Fig. 48). Also found in Dominican amber are representatives or near representatives of the present-day genera *Borinquena* (Leptophlebiidae) and *Cloeodes* (Baetidae) (determined by William L. Peters).

Odonata (Dragonflies and Damselflies)

Members of this order are rare in amber. Dragonflies and damselflies are similar to mayflies in having aquatic nymphs, but the adults, which are relatively long-lived (3–8 weeks), do not swarm. This lack of swarming, their larger size, and their stronger flying abilities may account for the rarity of dragonflies and damselflies in amber. Both adults and nymphs are predaceous on a variety of invertebrates, mostly arthropods. The final molt occurs after the nymph has crawled out on land.

The earliest fossils of the Odonata date from the Permian, and all descriptions of this group from amber occur in Baltic deposits. Amber dragonflies (Anisoptera) include an adult and the nymphal skin of *Gomphus* and the species *Gomphoides occultus* Hagen, which was described from a wing fragment. A well-preserved adult damselfly (Zygoptera) of the family Agrionidae is illustrated by Bachofen-Echt (1949). An *Agrion* and a *Platycnemis* (Coenagriidae) constitute the only other known members of this suborder in Baltic amber (Keilbach, 1982). Undescribed Odonata also have been identified in Dominican amber (see Appendix B) and figured in Schlee (1990).

Plecoptera (Stoneflies)

Members of this group are often confused with mayflies. Like the mayflies, stonefly nymphs are aquatic but normally possess only two caudal filaments (mayfly nymphs usually have three). At rest, adult stoneflies fold their wings flat over their abdomen whereas mayflies usually hold their wings vertically.

Like the other groups with aquatic nymphs, stoneflies are rarely found in amber. Most descriptions (based on adults, nymphal skins, and a nymph) were made from Baltic amber, and four families are represented, namely the Nemouridae (*Nemoura*), Perlidae (*Perla*), Taeniopterygidae (*Taeniopteryx*), and Leuctridae (*Leuctra*) (Keilbach, 1982). The nymphs of *Perla* are predaceous whereas those of the other genera feed on algae, diatoms, and decomposing plant matter. A stonefly from Dominican amber, *Dominiperla antiqua* Stark and Lentz (1991), was placed in the Anacroneuriinae (Perlidae). Because extant Antillean Plecoptera are unknown, its presence was a surprise. The stoneflies date back to the Permian and are of interest for their many archaic features.

Orthoptera (Cockroaches, Crickets, Grasshoppers, Mantids, and Others)

This order (here including Phasmida, Mantodea, and Blattaria) contains a variety of insects, including the cockroaches (Blattoidea), the crickets (Gryllidae), the short-horned grasshoppers (Acrididae), the long-horned grasshoppers (Tettigoniidae), the mole crickets (Gryllotalpidae), the walking sticks (Phasmidae), the mantids (Mantidae), the grylloblatids (Grylloblattidae), and a few other families. Many of these groups are now treated taxonomically as separate orders.

The roaches (Blattoidea or Blattaria) evolved morphological and physiological characteristics that have enabled them to survive some 280 million years. First appearing in the Carboniferous Period, where they are the most numerous insect fossils, they were similar in morphology to present-day cockroaches. An organism that was described from the Devonian as a winged relative of the roaches (*Eopterum devonicum* Rohdendorf) was later recognized as being part of a crustacean (Rodendorf, 1961).

Because cockroaches live under the rotting bark of trees today, it is not surprising that they appear in amber. Representatives of the Polyphagidae (*Eutyrrhypha, Holocompsa, Polyphaga*), Blaberidae, Blattidae (*Blatta, Polyzosteria, Periplaneta*), Nyctiboridae (*Nyctibora*), and Blattellidae (*Ectobius, Blattina, Ischnoptera, Ceratinoptera, Blatella, Phyl-*

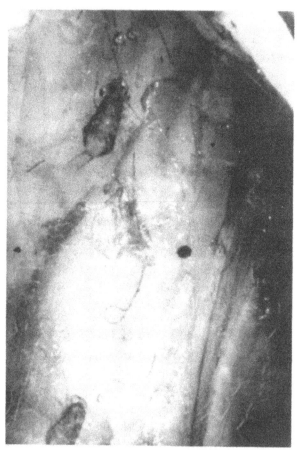

Figure 49. Cockroach nymphs (Blattidae) in Siberian amber (photo by A. P. Rasnitsyn, courtesy of V. V. Zherichin).

Figure 50. Cockroach egg capsule (ootheca) in Dominican amber (Costa collection, Puerto Plata, Dominican Republic).

lodromia, Temnopteryx) have been described from Baltic amber (Keilbach, 1982). Most of the fossil species in the above genera belong to genera that are common today in tropical and subtropical regions. One species from Baltic amber, *Eutyrrhypha pacifica* (Coquebert), survives today in South America, Africa, Madagascar, and Polynesia (Larsson, 1978).

Although cockroaches occur in Mexican, Dominican (Color Plate 6), Lebanese, Siberian (Fig. 49), and French amber, the only other formal description of a roach in amber is *Caenoblattinopsis fushunensis* from Chinese (Fu Shun) amber (Ping, 1931). L. M. Roth determined that one of the Dominican amber roaches in the author's collection belongs to the Corydiinae (Euthyrrhaphidae). Cockroach egg capsules also occur in Dominican amber (Fig. 50).

Representatives of the Gryllidae (crickets) have been formally described only from Baltic amber, although they do occur in Lebanese, Mexican, and Dominican amber (Color Plate 5). Baltic amber representatives of this family include species of *Gryllulus, Nemobius, Heterotrypus, Madasuma, Stenogryllodes, Trichogryllus,* and *Gryllidarum. Stenogryllodes* is now considered extinct and the closest relatives of *Trichogryllus, Heterotrypus,* and *Madasuma* are found in the Old World tropics and subtropics. Two immature crickets in Baltic amber are illustrated in Bachofen-Echt (1949) and an unidentified adult in Dominican amber is presented by Schlee (1980).

Three species of extinct cave crickets (Rhaphidophoridae) of the genus *Prorhaphidophora* have been described from Baltic amber (Chopard, 1936). Bachofen-Echt (1949) provides two photographs of *P. zeuneri* Chopard.

Grasshoppers are rare in amber and only representatives of the long-horned grasshoppers have been described, namely species of *Eomortoniellus* and *Lipotactes* from Baltic amber. The closest living relatives of both occur on the Sunda Islands. A *Locustina* larva and a *Tettigoniidarum* sp. have also been described from Baltic amber, but the material was inadequate for comparison with present-day forms. An unidentified tettigonid in Baltic amber is figured in Bachofen-Echt (1949). One pygmy grasshopper, *Succinotettix chopardi* Piton (Tetrigidae), has been described from Baltic amber. Members of this family have the pronotum (upper part of the prothorax) extended backward over the abdomen (Keilbach, 1982).

In discussing the Lebanese amber insects, Whalley (1981) mentions that "just over 1 percent of the fauna were 'grasshoppers' and crickets but further examination of the very fragmentary 'grasshopper' material is needed."

Mole crickets (Gryllotalpidae) have been identified only in Mexican amber (see Appendix A), but no formal descriptions have been made of this rare material.

Praying mantids (Mantidae) have been recorded only in Tertiary amber from the Baltic, Mexico, and the Dominican Republic (Fig. 51). Because only young juveniles have been recovered, no descriptions have been published. A photograph of a Baltic amber mantid is provided by Bachofen-Echt (1949). These reports are the earliest fossil records of praying mantids.

The walking sticks (Phasmatidae) also occur in Baltic, Dominican (Color Plate 4), and Mexican amber but, again, most of the material consists of juveniles. However, two species from Baltic amber, *Pseudoperla lineata* and *P. gracilipes*, were described by Pictet (in Berendt, 1856). The former species is reported as being relatively common in Baltic amber and its nearest living relatives are found in the South American tropics; photographs of this species occur in Bachofen-Echt (1949). The earliest known representatives of this group have been recorded from the Triassic of Australia.

No representatives of the Grylloblattidae (night crawlers) nor the

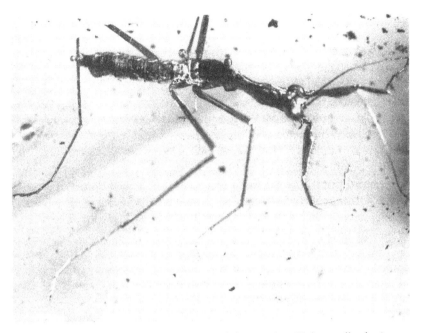

Figure 51. A young praying mantis in Dominican amber (Poinar collection).

Stenopelmatidae (king crickets, wetas) have been described from amber. In fact, their fossil record is poor in general.

Dermaptera (Earwigs)

Members of the Dermaptera similar to present-day forms except for possessing long, multisegmented cerci instead of "pinchers" or forceps were present in the Jurassic. There are, however, no records of this group from Cretaceous amber. Although they occur in Dominican and Mexican amber, most representatives of the Dermaptera are from Baltic amber (Burr, 1911). These include representatives of the Forficulidae (*Forficula, Pygidicrana*) and Labiduridae (*Labidura*). The nearest living relatives of *F. klebsi* occur in Abyssinia; those of *F. pristina* and *F. praecursor* can be found in Africa and India, respectively. Bachofen-Echt (1949) supplies a photograph presumably of *F. baltica* Burr from Baltic amber and Schlee (1980) illustrates an earwig in Dominican amber. Cockerell (1920B) supplies a description of the only other earwig known from amber, *Labidura electrina*, from Burmese amber.

Isoptera (Termites)

The oldest known termite in amber is from the Lebanese deposits that have been judged to be of the Neocomian (Hauterivian Stage) of the Early Cretaceous (Schlee, 1972). This species, however, has not yet been described and, therefore, the earliest described termite is *Valditermes brenanae* Jarzembowski (1981) (Hodotermitidae: Cretatermitinae), which was described on the basis of one forewing found in the Weald clay formation in England. This formation belongs to the same stage as Lebanese amber (Hauterivian Stage of the Neocomian Epoch). These two reports constitute the first fossil records of a termite and of a social insect. Other reports of Cretaceous amber termites were made from the Taimyr deposits in northern Siberia (Zherichin and Sukacheva, 1973) and France (Schlüter, 1978A).

Members of the Isoptera are moderately common in Tertiary Baltic amber where representatives of the families Kalotermitidae, Hodotermitidae, and Rhinotermitidae have been described. Genera include *Electrotermes*, *Proelectrotermes*, *Termopsis*, *Archotermopsis*, *Parastylotermes*, and *Reticulitermes*. Of the above, *Reticulitermes antiquus* (German) was the most commonly encountered species. A revision of *Electrotermes* and *Proelectrotermes* was made by Emerson (1969).

Snyder (1960) studied two termites from Mexican amber. The first, *Kalotermes nigritus* Snyder (Kalotermitidae), still survives in Mexico and Central America today. The second species, *Heterotermes primaevus* Snyder (Rhinotermitidae), is considered extinct, but closely related species occur in the West Indies and in Central and South America. Emerson (1969) reexamined the above material and concluded that the former species was actually an extinct species, not an extant one. He described it as *Incisitermes krishnai* Emerson (Fig. 52) and noted that the closest living termite was *Kalotermes approximatus* from Florida. Emerson (1969) also described *Calcaritermes vetus* (Kalotermitidae) from Mexican amber and mentioned that it constituted the first record of the genus from Tertiary deposits. Later, the same author described *Coptotermes sucineus* from Mexican amber (Emerson, 1971) and Krishna and Emerson (1983) described *Mastotermes electromexicus* (Mastotermitidae) from Mexican deposits. Representatives of the genus *Mastotermes* occur also in Dominican amber, but are found today only in Australia (Krishna and Grimaldi, 1991). Emerson (1969) also redescribed *Kalotermes swinhoei* (Cockerell) and *K. tristis* (Cockerell) from Burmese amber.

Figure 52. A winged adult termite, *Incisitermes krishnai* Emerson (family Kalotermitidae), in Mexican amber (Museum of Paleontology collection, University of California, Berkeley; photo by P. Hurd).

Termites are common in Dominican amber, where they are represented by winged adults (alates) (Color Plate 5), workers, and soldiers. Of the latter, the "nasute" soldiers of colonies of *Nasutitermes* (Nasutitermitinae: Termitidae) are especially interesting (Fig. 53). They are characterized by their bulbous head and long snout that terminate in an opening from which is emitted an irritating and entangling "glue." Containing only vestigial mandibles, these "higher" termites have resorted to a type of chemical warfare to ward off their enemies. The chemical composition of the glue is variable and some products contain an alarm pheromone

Figure 53. A nasute soldier termite of the genus *Nasutitermes* (family Termitidae) in Dominican amber (Poinar collection).

Figure 54. A worker termite with an air bubble extending from its body. The bubble resulted from the activity of gut microbes after the termite was entrapped (C. Werby collection).

that draws other workers to the scene. A mature colony of *Nasutitermes*, which can occur underground, in mounds, or in concealed cavities in trees, is composed of about one million individuals and produces from 5,000 to 25,000 alates each season. Most colonies have a single queen and king, but up to 33 queens and 17 kings can occur in a single nest. The nests are built of feces (with semidigested plant material) and soil, and are enveloped in a thin layer of excreted lignaceous plant material (Edwards and Mill, 1986).

One termite described from Dominican amber, *Coptotermes priscus* Emerson (1971), is closely related to the above-mentioned *C. sucineus* from Mexican amber. Another, *Cryptotermes yamini*, was described from a specimen recovered from a Manhatten jewelry store (Anonymous, 1987). Other Baltic and Dominican amber termites are illustrated in Schlee and Glöckner (1978) and Bachofen-Echt (1949).

Associated with many worker termites in Dominican amber are six-sided fecal pellets and variously shaped air bubbles emerging from the termite's body. Such bubbles represent gases produced by the intestinal microbes of the termites, which continue to metabolize for some time after the termite has been enclosed in the resin (Fig. 54).

Embioptera (Web Spinners)

Members of this unusual order are mainly tropical insects that extend into the warm temperate regions of the world. All individuals, including the wingless females and young, are capable of producing silk, which issues from spinnerets located in the basal segment of the swollen front tarsi. This silk is used to make a network of galleries in debris or soil, on plants, or under the bark of trees. These insects are adapted to confinement and movement within the narrow tunnels that protect them from natural enemies. They feed on living and dead leaves, moss, lichens, and bark and superficially resemble termites in form and habitat. Their communal involvement has prompted some to consider them as primitively social.

Although this group is regarded by some as the only modern descendants of the extinct order Protorthoptera, the earliest known fossils are from Tertiary amber. In 1854, Pictet described *Embia antiqua*, the first fossil embiid from Baltic amber. It was subsequently redescribed and placed in the genus *Electroembia* by Ross (1956), who stated that it was similar to modern-day forms, especially those occurring in the Mediterranean region. This species has apterous males (wings shed after flight) and Ross comments that this condition today is almost universal in regions experiencing prolonged dry seasons.

Burmitembia venosa Cockerell was described from a winged male in Burmese amber, and two winged males of *Oligotoma westwoodi* Hagen were described from Zanzibar copal (Ross, 1956). Web spinners occur in both Mexican (Fig. 55) and Dominican amber. E. S. Ross identified a male of the genus *Mesembia* (Anisembiidae) and males belonging to the Teratembiidae in Dominican amber (Fig. 56). The range of extant *Mesembia* includes Mexico, Central America, and the West Indies.

Zoraptera (Zorapterans)

This is a small order, consisting of a single family containing the single genus *Zorotypus*. Zorapterans are minute insects, 3 millimeters or less in length, that superficially resemble termites. They occur in concealed habitats under bark or wood or in rotting logs, and they have been noted feeding on fungal spores and dead arthropods. Zorapterans consist of three basic morphological types: the first is a pale apterous form with no ocelli or compound eyes; the second is a dark-pigmented winged type with ocelli and compound eyes; and the third is an apterous, dark-pigmented form lacking ocelli and compound eyes. The only reported fossil zorapteran is *Zorotypus palaeus* from Dominican amber (Poinar, 1988B) (Fig. 57).

Figure 55. A male web spinner of the family Teratembiidae (order Embioptera) in Mexican amber (Museum of Paleontology collection, University of California, Berkeley; photo by P. Hurd).

Figure 56. A male web spinner of the genus *Mesembia* (order Embioptera, family Anisembiidae) in Dominican amber (Poinar collection).

Psocoptera (Psocids)

This order was formerly referred to as the Corrodentia and representatives are called "book lice." Members of this group are small (usually under 6 mm in length), soft-bodied insects with four membranous wings (when present). They occur on or under bark in concealed areas where they feed on fungi, pollen, and dead arthropods. Some species are capable of making webs in association with egg deposition or feeding behavior.

The fossil record of the psocids supposedly starts in the Early Permian, although this record may be of a thrips. The first well-documented psocid appears in Jurassic formations (Mockford, personal communication). Representatives of the Amphientomidae (*Amphientomum, Electrentomum*), Liposcelidae (*Liposcelis, Empheria, Trichempheria, Bebiosis, Epipsocus*), Caeciliidae (*Caecilius, Ptenolosia, Palaeopsocus, Kolbea*), Elipsocidae (*Elipsocus*), Philotarsidae (*Philotarsus*), Archipsocidae (*Archipsocus*), Psocidae (*Copostigma, Psocus, Trichadenotecnum*), and Ectopsocidae have been described from Baltic amber (Keilbach, 1982). The majority of

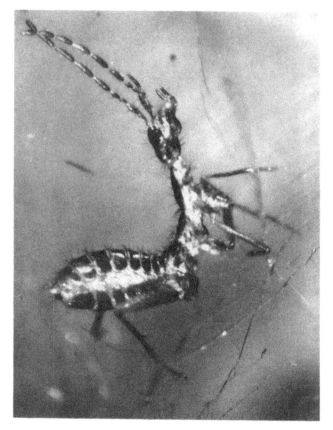

Figure 57. Holotype of the first known fossil zorapteran, *Zorotypus palaeus* Poinar, in Dominican amber (Brodzinsky-Lopez Penha collection, Smithsonian Institution). (Published in *J. New York Entomol. Soc.*)

the abundant and varied psocids from Baltic amber most closely resemble species and genera found today in areas far removed from the Baltic: in the tropics and subtropics of Asia (southern and eastern), America, and Africa. Other species of *Archipsocus* occur in India, South America, and East Africa (Larsson, 1978).

In contrast, most of the genera of psocids from Mexican amber, consisting of representatives of the Epipsocidae, Ectopsocidae, Archipsocidae (Fig. 58), Liposcelidae, Myopsocidae, Psyllipsocidae, and Trichopsocidae (see Mockford, 1969, 1986), still occur in southern Mexico today, and some of the modern species are similar to the fossil ones. Psocids are common in Dominican amber (Fig. 59), and Mockford (1986) has identified representatives of the families Amphientomidae, Archipsocidae, Calciliidae, Cladiopsocidae (*Cladiopsocus, Spurostigma*), Dolabellopsocidae (*Isthmopsocus*), Epipsocidae (*Epipsocus*), Lepidopsocidae (*Echmepteryx*), Liposcelidae (*Liposcelis*), Philotarsidae (*Broadheadia*), Psocidae (*Blaste, Ptycta, Trichadenotecnum*), and Psoquillidae (*Rhyopsocus*) from these deposits. Cockerell described two species from Burmese amber (Keilbach, 1982).

The Psocoptera also occur in Cretaceous amber. About 5 percent of the fauna in Lebanese amber is composed of psocids (Whalley, 1981) and representatives of the Trogiidae (*Eolepinotus*), Psyllipsocidae (*Khatangia*), Amphientomidae (*Proamphientomum*), Elipsocidae (*Cretapsocus*), Lachesillidae (*Archaelachesis*), and Ectopsocidae (*Psocomorpha, Empheriopsis*) have been described from Siberian amber (Vishniakova, 1975). Legg (1942) cites three species of psocids from Canadian amber and provides two photographs of definite representatives of this order. A possible psocid from French Cretaceous amber is described by Schlüter (1978A).

Figure 58. A book louse or psocid (order Psocoptera) of the genus *Archipsocus* (family Archipsocidae) in Mexican amber (Museum of Paleontology collection, University of California, Berkeley; photo by P. Hurd).

Figure 59. A book louse or psocid (order Psocoptera) of the genus *Epipsocus* (family Epipsocidae) in Dominican amber (Poinar collection).

Thysanoptera (Thrips)

Members of this order are small, inconspicuous insects. Most feed on a wide variety of plants or fungal spores, but others are predaceous. The mouthparts are adapted for sucking, but not in the same form as the true bugs. Unique characters include the structure of the four wings (when present), which consist of narrow body extensions fringed with long hairs. Also unique is their metamorphosis. The first two instars (larvae) have no wing pads, and a prepupa and pupa (sometimes in a cocoon) are present. The group is well represented in the fossil record, with the earliest reports coming from the Permian. Although thrips have been found in most amber deposits, actual descriptions are limited to forms in Lebanese and Baltic amber.

Thrips constitute 7 percent of the insect fauna in Lebanese amber (Whalley, 1981). They were described by Zur Strassen (1973) and include representatives of the Rhetinothripidae (*Rhetinothrips, Progonothrips*), Neocomothripidae (*Neocomothrips*), Jezzinothripidae (*Jezzinothrips*), Scudderothripidae (*Scudderothrips, Exithelothrips*), and Scaphothripidae (*Scaphothrips*). All of the above represent new families of genera that are now extinct. Thrips have also been reported from Canadian (Legg, 1942) and Siberian Cretaceous amber (Zherichin and Sukacheva, 1973), and representatives of the families Aeolothripidae, Hemithripidae, Heterothripidae, Thripidae, Merothripidae, and Phlaeothripidae have been described from Baltic amber (Keilbach, 1982). Most of these from Baltic amber are Recent families and about 33 percent of the described genera are extant; most of the species occur in the family Thripidae. Members of the primitive family Aeolothripidae, which includes many predators, are relatively more common in Baltic amber than in Recent fauna (Larsson, 1978). Thrips also occur in Mexican and Dominican amber (Appendices A, B).

Hemiptera (Heteroptera; Bugs)

The common name "true bug" is often used for this group since "bug" is also a nonspecific name for any insect. Characterized by piercing-sucking mouth parts in the form of a beak or rostrum and partially hardened front wings (hemelytra), the true bugs constitute a large, widely distributed group. Their feeding habits are varied; many suck the juices of plants but others are predaceous on arthropods and a few take vertebrate blood.

The earliest appearances of Hemiptera in amber are representatives of the Dipsocoridae, Thaumastellidae, and Anthocoridae in Cretaceous Lebanese deposits (Whalley, 1981).

Anthocoridae (minute pirate bugs). These small, somewhat flattened bugs are mostly predaceous on small insects and mites. They occur on flowers, in decaying vegetation, and under bark. Cretaceous forms are represented in Lebanese (Whalley, 1981) and Canadian amber (McAlpine and Martin, 1969). Tertiary forms occur in Baltic (*Anthocoris* sp.) (Menge, 1856) and Bitterfeld amber (*Xylocoris* sp.) (Barthel and Hetzer, 1982). Unidentified representatives also occur in Dominican amber (Appendix B.)

Aradidae (flat bugs). These small flattened bugs often have shortened wings, probably to accommodate their habit of crawling under tree bark, where they feed on fungi. Four species of *Aradus* and a single species each of *Calisius* and *Mezira* have been described from Baltic amber (Spahr, 1988; Kielbach, 1982; Usinger, 1941). Undescribed representatives occur in Mexican and Dominican amber (Appendices A, B). A representative of the highly modified termite bugs (*Termitaradus* spp.) in the related family Termitaphididae has been found in Mexican amber (Poinar and Doyen, 1992).

Berytidae (stilt bugs). These long-legged slender bugs occur in vegetation, where they feed on a broad variety of plants. Undescribed representatives of *Berytus* from Baltic amber (Spahr, 1988) are the only forms recorded from amber.

Cimicoidea. In 1942, Usinger described two species of the new genus *Electrocoris* from Baltic amber. This annectent genus (having characters intermediate between two groups) could not be placed in any family, and was placed in the superfamily Cimicoidea. The specimens possess several aberrant characters suggesting a relationship with the microphysid-anthocorid complex.

Corixidae (water boatmen). These aquatic bugs are commonly found in ponds and lakes, where most feed on algae and other small, aquatic organisms. Because of their habitat, they are rare in amber and could be trapped only during their brief flying periods between water sources. Thus, it is interesting that Bachofen-Echt (1949) found three wingless nymphs in a piece of Baltic amber.

Cydnidae (burrowing bugs). These small, oval, dark bugs occur beneath stones. Their legs are modified for digging and they feed on plant roots. The first amber representative of this family, *Amnestus guapinolinus* Thomas (1988), was described from Mexican deposits.

Dipsocoridae and Schizopteridae (jumping ground bugs). These minute, oval bugs, some having the ability to jump when disturbed, occur in leaf litter and moist soil. An adult dipsocorid was reported from Lebanese amber (Whalley, 1981), but the only formal description of an amber dipsocorid is that of *Ceratocombus hurdi* Wygodzinsky (1959) from Mexican amber. Members of the schizopterids also occur in Mexican amber (Appendix A) and both families are represented in Dominican amber (Appendix B).

Enicocephalidae (gnat bugs). These small, slender bugs possess entirely membranous front wings and an elongate head. They occur under bark and in concealed places, where they prey on small arthropods. They periodically form large swarms, which may explain their presence in amber. The first descriptions of amber gnat bugs involved three species of *Disphaerocephalus* and one of *Paenicotechys* described by Cocherell from Burmese material (see Stys [1969] for a discussion of these). Gnat bugs also occur in Dominican amber (Fig. 60, Appendix B).

Gerridae (water striders). These long-legged bugs live on the surface of water. Water striders can move rapidly over the water surface, and prey on trapped insects. Baltic amber contains undescribed representatives of *Gerris*, *Halobates*, and *Metrobates* (Spahr, 1988). The presence of *Halobates* in amber is curious because modern representatives of this genus normally live on the surface of the ocean, often far from land. The generic placement of these forms, however, is questionable. *Electrobates spinipes* Andersen and Poinar (1992) in the new subfamily Electrobatinae was described from Dominican amber. This pair, belonging to a now extinct lineage, represents the first record of mate-guarding in a fossil insect (Color Plate 7).

Hebridae (velvet water bugs). These small, squat bugs are covered with velvety hairs that keep them afloat as they search for prey on the surface of pools. A single reference to a hebrid (possibly *Hebrus* sp.) in Mexican amber is the only account of this family in amber (Andersen, 1982) (Fig. 61).

Hydrocorisae. This family includes shore-inhabiting bugs. Schlüter (1978A) reported a hemipteran from this group in French Cretaceous amber.

Figure 60. A gnat bug (family Eni-cocephalidae) in Dominican amber (Poinar collection).

Figure 61. A velvet water bug (order Hemiptera, family Hebridae) in Mexican amber (Museum of Paleontology collection, University of California, Berkeley; photo by P. Hurd).

Hydrometridae (water measurers). These small, slender, normally wingless insects search for prey in marshy areas and are able to walk on the surface of water. Both Larsson (1978) and Spahr (1988) refer to un-described hydrometrids in Baltic amber.

Largidae (cotton strainers). These small, oval, often brightly colored bugs feed on living plants. A single member of this family was reported from Siberian amber (Zherichin and Sukacheva, 1973).

Leptopodidae (spiny shore bugs). These bugs inhabit dry areas along streams in tropical and subtropical areas. The only described representative from amber is *Leptosalda chiapensis* Cobben (1971), which was originally described in the family Saldidae (Spahr, 1988).

Lygaeidae (seed bugs). These common terrestrial bugs occur in vegetation and litter where most feed on seeds. Two species of *Pachymerus* and undescribed representatives of *Rhyparochromus* and *Trapezonotus* have been reported from Baltic amber (Spahr, 1988). Lygaeids also occur in Dominican amber (Appendix B).

Microphysidae. These small bugs occur in moss or lichens where they prey on other invertebrates. Representatives have been reported from Siberian amber (Zherichin and Sukacheva, 1973).

Miridae (plant bugs). These medium-sized bugs are common on a variety of plants, upon which they feed. This is the largest family of the order and a variety of plant bugs occur in amber. Described species of *Ambercylapus, Archeofulvius, Fulvius, Jordanofulvius,* and *Phytocoris* and representatives of *Aetorhinus, Capsus, Dichrooscytus, Hadronema, Harpocera, Homodemus, Hoplomachus, Lopus, Lygus, Miris, Oncotylus, Orthops,* and *Systellonotus* have been found in Baltic amber. Mirids also occur in Dominican amber (Appendix B) and an isometopid (sometimes considered a separate family, the Isometopidae) has been recorded from Mexican amber (Spahr, 1988) (Appendix A).

Nabidae (damsel bugs). These small, slender bugs occur in a wide variety of habitats where they seek out and attack other insects. Several *Nabis* have been described from Baltic amber (Spahr, 1988) and the family is also represented in Dominican amber (Appendix B).

Nepidae (waterscorpions). These predaceous, aquatic bugs have raptorial front legs and a caudal breathing tube formed by the cerci. Their eggs are deposited in the tissues of aquatic plants. Representatives of the genus *Nepa* have been reported from Baltic amber (Bachofen-Echt, 1949; Spahr, 1988), but no other amber waterscorpions are known.

Notonectidae (backswimmers). These small, robust bugs swim on their back, with their long hind legs serving as oars. They feed on insects and small aquatic vertebrates, and can inflict a painful sting when handled. A single undescribed specimen in Baltic amber (Jordan, 1953) is the only known backswimmer in amber.

Pentatomidae (stink bugs). These medium-sized, ovoid bugs feed mostly on plants, but some are predaceous. They are called stink bugs because a characteristic odor is produced when they are handled. Representatives of the genera *Dolycoris, Eurydema,* and *Pentatoma* occur in Baltic amber (Spahr, 1988) and stink bugs also appear in Dominican amber (Appendix B).

Reduviidae (assassin bugs). This large group of predaceous bugs is divided into several subfamilies on the basis of morphology and behavior. One interesting subfamily, the Holoptilinae, contains representatives that are specialized ant predators. Glands opening on the ventral side of the abdomen produce secretions that are attractive to ants, but also have a tranquilizing effect on them. As soon as the ants are quiet, the bug inserts its beak into the victim and sucks out the hemolymph. These bugs tend to be very hairy, which probably protects them from counterattacking ants. One of these bugs, *Proptilocerus dolosus* Wasmann (1933), was found in a piece of Baltic amber, together with the bodies of its probable ant prey, *Dolichoderus tertiarus* Mayr. A new genus of these ant-eating assassin bugs has also been found in Dominican amber (Poinar, 1991) (Color Plate 6).

Thread-legged bugs in the genera *Alumeda, Empicoris*, and *Malacopus* (subfamily Emesinae) have been described in Dominican amber (Popov, 1987A, 1987B, 1989), and *Empicoris* also occurs in Mexican amber (Wygodzinsky, 1966).

Resin bugs of the subfamily Apiomerinae are another group of reduviids. These bugs coat their legs and sometimes portions of their bodies with sticky resin in order to keep their prey from escaping. Representatives occur in Dominican amber, sometimes with resin deposits still on their legs (Fig. 62).

Other reduviids described from Baltic amber have been placed in the genera *Limacis* and *Platymeris* (Spahr, 1988).

Saldidae (shore bugs). These small, robust bugs, which are predaceous in habit, occur along the shores of bodies of fresh or salt water.

Figure 62. An assassin bug (family Reduviidae) in Dominican amber. Note resin deposits on front legs (Poinar collection).

A single species, *Salda exigua* Germar and Berendt, was described from Baltic amber (Spahr, 1988).

Termitaphididae (termite bugs). These highly modified bugs are adapted for survival in termite nests. A single representative has been reported in Mexican amber (Poinar and Doyen, 1992) (Color Plate 8).

Thaumastellidae. A single adult of this family in Lebanese amber (Whalley, 1981) is the only record of thaumastellids in amber.

Thaumastocoridae (palm bugs). These small, flattened bugs have been recorded feeding on palms. Their presence in Dominican amber (Appendix B) constitutes the only report of palm bugs in amber.

Figure 63. A lace bug (family Tingidae) in Dominican amber (Poinar collection).

Tingidae (lace bugs). These small, flattened bugs usually have the dorsal portion of their body adorned with sculptured reticulations resembling lace. They are plant feeders and feed on a variety of trees. Species in the genera *Cantacader, Eotingis, Phatnoma*, and *Tingis* have been described from Baltic amber (Spahr, 1988) and undescribed forms occur in Dominican amber (Fig. 63).

Veliidae (broad-shouldered water striders). These small water striders occur on the surface of streams or ponds where they prey on other insects. A single report of a veliid in Baltic amber (Andersen, 1982) is the only record of this group in amber.

Homoptera (Cicadas, Hoppers, Psyllids, Whiteflies, Aphids, and Scale Insects)

Members of this diverse order resemble the true bugs and sometimes are included as a suborder within the Hemiptera. Homopterans differ basically from the latter in possessing forewings with a uniform texture throughout (leathery or membranous) and in having the beak arise from the posterior part of the head (not the front part as in the Hemiptera). All representatives of this order are plant feeders and their fossil record dates from the Permian where representatives of the Axiidae (planthoppers) were recovered along with some extinct families.

Because of their plant-feeding habits, their small mass, and their weak flight habits, many homopterans occur in amber, probably the result of the wind driving them against the sticky resin. Representatives of the suborder Sternorrhyncha, which includes the aphids, whiteflies, and psyllids, are especially numerous.

In Lebanese amber, some 13 percent of all the insects were homopterans and about 70 percent of these were whiteflies (Aleyrodidae). Schlee (1970) described two new genera (*Bernaea* and *Heidea*) of whiteflies from Lebanese amber on the basis of one male and one female specimen (Color Plate 4). He considered these Lebanese forms as belonging to the stem group of the Aleyrodina and noted that they differed markedly (e.g., had a longer probiscis) from both Recent forms and two Aleyrodina (*Aleurodicus* and *Aleyrodes*) described from Burmese and Baltic amber, respectively (Keilbach, 1982). Representatives of this group also occur in Siberian (Zherichin and Sukacheva, 1973), Mexican, and Dominican ambers (Appendices A, B).

Aphidoidea. Aphids or plant lice are small, soft-bodied insects commonly found on roots, stems, leaves, and tender plant shoots where they suck the plant juices. A review of the fossil aphids, including the amber species, was prepared by Heie (1987). The oldest fossil aphid, *Triassoaphis cubitus* Evans (1956), was recovered from Triassic deposits in Australia. Members of the Canadaphididae (*Canadaphis, Alloambria, Pseudambria*), Palaeoaphididae (*Palaeoaphis, Ambaraphis*), and Drepanosiphidae (*Aniferella*) occur in Canadian amber (Richards, 1966), and representatives of the Aphididae (*Aphidocallis*), Palaeoaphididae (*Palaeoaphis*), Canadaphididae (*Canadaphis*), Tajmyraphididae (*Jantardakhia, Khatangaphis, Retinaphis, Tajmyraphis*), Shaposhnikoviidae (*Shaposhnikovia*), Mindaridae (*Nordaphis*), Drepanosiphidae (*Aniferella*), Elektraphididae (*Antonaphis, Tajmyrella*), and Pemphigidae (*Palaeoforda*) have been described from Siberian amber (Kononova, 1975, 1976, 1977). An undescribed member of the Adelgidae was reported from Alaskan amber (Hurd et al., 1958).

The remaining descriptions of amber aphids, with the exceptions of a *Germaraphis* in Romanian amber (Protescu, 1937) and *Mindazerius dominicanus* Heie and Poinar (1988) in Dominican amber (Fig. 64), are based on material in Baltic amber. Representatives of the Anoeciidae (*Berendtaphis*), Aphididae (*Aphis, Baltichaitophorus, Larssonaphis, Pseudamphorophora*), Drepanosiphidae (*Balticaphis, Balticomaraphis, Conicaudus, Electrocallis, Electromyzus, Megantennaphis, Megapodaphis, Mengeaphis, Oligocallis, Palaeophyllaphis, Palaeosiphon, Sternaphis, Succaphis, Tertiaphis, Zymus*), Elektraphididae (*Antiquaphis, Elektraphis*), Hormaphididae (*Electrocornia*), Mindaridae (*Mindarus*), Pemphigidae (*Germaraphis, Pemphigus, Succinaphis*), and Thelaxidae (*Palaeothelaxes*) occur in Baltic deposits (Heie, 1967, 1972, 1987; Keilbach, 1982; Spahr, 1988).

About 60 percent of the Baltic amber aphids are representatives of the genus *Germaraphis* Heie, and the majority of these belong to *G. dryoides* Germar and Berendt. Curiously, only nymphs and wingless females have been found. Species of *Germaraphis* are characterized by a rostrum that is longer than the entire body, and this feature, together with the well-developed claws, suggests that they lived on the bark of the amber-producing tree, the long rostrum being able to penetrate through the bark to the living cells beneath. The aphids are often found in groups in the amber, sometimes with ants (*Iridomyrmex*), suggesting a symbiotic relationship between these organisms, although Heie (1967) considers these coincidental occurrences. Although the genus *Germaraphis* is considered extinct, it is closely related to the Recent genus *Phloeomyzus* Horvath.

Figure 64. Holotype of an aphid, *Mindazerius dominicanus* Heie and Poinar, in Dominican amber (Poinar collection).

The only representative of the Thelaxidae in Baltic amber, *Palaeothelaxes setosa* Heie, was probably associated exclusively with plants of the oak family (Fagaceae) similar to its nearest extant relative, *Thelaxes*. Representatives of 15 genera in the family Drepanosiphidae occur in Baltic amber and extant members of the family still occur in boreal regions. One of these, *Mengeaphis glandulosa* (Menge), is presumed to have lived on the trunk of the Baltic amber tree, because its rostrum is twice as long as its body. These amber aphids undoubtedly had their enemies, because representatives of the Coccinellidae (*Scymnus, Coelopterus*), Neuroptera (*Hemerobius, Chrysopa*), Syrphidae, and Aphidiidae (*Ephedrus, Aphidius, Propraon*), all of which occur in Baltic amber, mainly attack aphids today.

The aphids as a group are distributed mainly in the northern temperate zone and have adapted to a seasonal climate by producing parthenogenetic summer generations and an autumn sexual generation that results in hibernating eggs. Today, relatively few aphids occur in the tropics, and those that do appear to have lost the sexual generation. In warm climates, treehoppers (Membracidae) and planthoppers (Fulgoroidea) dominate the niches held by aphids in temperate climates. The aphid *Mindazerius dominicanus* Heie and Poinar (1988) is the first fossil aphid from the Neotropics. Placed in the primitive family Drepanosiphidae, it most closely resembles the South American extant genus *Lizerius* (tribe Lizerini), but it is also close to the extant genus *Paoleilla* (Africa, India, and South America). Thus, the tribe Lizerini may have originated in the Southern Hemisphere in the late Mesozoic and populated much of Gondwanaland by the beginning of the Tertiary. Lizerini are associated with the angiosperm families Burseraceae, Combretaceae, Lauraceae, and Nyctaginaceae, and the fossil find suggests that representatives of one of these families might have been present in the amber forest.

Coccoidea (scale insects). In contrast to the aphids, scale insects are more common in tropical and subtropical regions. Active first-instar nymphs crawl around to find a suitable site for settlement. After the second molt, the now legless nymphs begin to feed and secrete a waxy or scalelike covering over their body. The females, wingless and usually legless, remain sessile on plant parts. The males usually possess a pair of forewings, legs, and a spike at the tip of the abdomen but lack mouthparts.

Scale insects have been reported from Lebanese and Siberian amber (Schlee, 1972; Zherichin and Sukacheva, 1973) and *Electrococcus canadiensis* Beardsley (1969) (Coccidae) was described from Canadian amber. Four species of *Monophlebus* (Margarodidae) and the new genera *Arctorthezia*, *Palaeonewsteadia*, *Protorthezia*, and *Newsteadia* were described from Baltic amber (Keilbach, 1982; Spahr, 1988). In the same deposits were representatives of the Ortheziidae (*Orthezia*, *Palaeonewsteadia*, *Protorthezia*), Eriococcidae, Pseudococcidae, Coccidae (*Coccus*), Matsucoccidae (*Matsucoccus*), and Diaspididae (Larsson, 1978; Koteja, 1985, 1987). Cochineal insects (Dactylopiidae) and mealybugs (Pseudococcidae) have been reported from Mexican amber and coccids (Ortheziidae) and soft scales (Coccidae) from Dominican amber (Appendices A, B).

Cicadidae (cicadas). Representatives of this family are relatively large and the males are well known for their songs. The 17-year "locusts" or cicadas belong to this family, but the term "cicada" is sometimes also used for members of the entire suborder Auchenorrhyncha (including also the spittlebugs, leafhoppers, treehoppers, and planthoppers). Representatives of the Cicadidae are rare in amber. Bachofen-Echt (1949) refers to two species of the genus *Cicada* from Baltic amber but descriptions are lacking.

Membracidae (treehoppers). Representatives of this family often possess a large pronotum that covers part of the head and extends back over the abdomen, giving the insect various shapes (Fig. 65). They feed mostly on trees and shrubs and are common representatives of tropical forests. Treehoppers occur in Mexican and Dominican amber, but none has been described (Appendices A, B).

Cercopidae (spittlebugs). As the common name implies, the nymphs of this family produce a frothy spittlelike mass from anal and glandular discharges. This deposit hides the feeding nymphs as they complete their development on various plants. A member of this family has been re-

Figure 65. A treehopper (family Membracidae) in Dominican amber (Poinar collection).

ported in Canadian amber (McAlpine and Martin, 1969) and Whalley (1981) reports one representative of the Stenoviciidae (Cercopoidea) in Lebanese amber. Formal descriptions of cercopids in amber occur only in the genera *Aphrophora* (sometimes placed in the Aphrophoridae) and *Cercopis* from Baltic amber (Keilbach, 1982). Spittlebugs have also been identified in Dominican amber (Appendix B).

Cicadellidae (leafhoppers). Leafhoppers comprise a large group of small homopterans characterized by a row of spines along the tibia of the hind legs. They can be quite colorful and occur on a wide variety of plants. Whalley (1981) reports the presence of nymphal leafhoppers in Lebanese amber, and described leafhoppers from Baltic amber occur in the genera *Tettigonia* (= *Tettigella*), *Jassus* (= *Coelidia*), *Bythoscopus* (= *Iassus*), *Pediopsis* (= *Macropsis*), *Eupteryx*, *Acocephalus* (= *Aphrodes*), *Typhlocyba*, *Thamnotettix*, *Cicadella*, *Cicadula*, and *Deltocephalus* (Keilbach, 1982; Bachofen-Echt, 1949; Spahr, 1988).

Jascopus notabilis Hamilton (Cicadellidae), described from a nymph in Canadian amber, represents an intermediate form between the spittle-bugs and the leafhoppers. Characters typical of the Cercopidae found in *Jascopus* are the length of the hind tibiae, the lack of body armature, the broadly tapered abdomen, and the small tarsal claws. Features typical of the Cicadellidae include the absence of an epistomal suture, the presence of antennal lodges in the nymph, rows of setae on the fore and middle tibiae, narrow tarsal sections, and a double row of ventral setae on the hind first tarsal segment (Hamilton, 1971).

Leafhoppers occur also in Mexican and Dominican amber and J. Sorensen identified forms in the latter deposits as probably belonging to *Xestocephalus* (Xestocephalini) and Iassinae (Appendices A, B).

Fulgoroidea (planthoppers). Representatives of this large group are common in the tropics, where individuals of many species have the head enlarged into a peculiarly shaped snout. Planthoppers have only a few large spines on the hind tibiae and their antennae arise below the com-pound eyes (in front of the eyes of leafhoppers). A cixiid planthopper (Cixiidae) has been reported in Lebanese amber (Whalley, 1981) and Emeljanov (1983) described *Netutela annunciator* from Siberian amber. Baltic amber fulgorids include representatives of the Fulgoridae (*Poeocera* [= *Poiocera*]), Dictyopharidae (*Dictyophara* [= *Pseudophana*]), Flatidae (*Flata*), Ricaniidae (*Tritophania*), Issidae (*Issus*), Cixiidae (*Cixius*, *Oli-arius*), and Achilidae (*Protepiptera*). Fennah (1963, 1987) described two cixiids, *Oeclixius amphion* and *Mnemosyne* sp., and a Flatidae from Mexi-

Figure 66. A planthopper (family Cixiidae) in Dominican amber (Poinar collection).

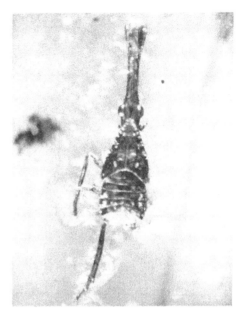

Figure 67. A planthopper (superfamily Fulgoroidea) in Dominican amber. Note elongation of the head (Poinar collection).

Figure 68. A planthopper (superfamily Fulgoroidea) in Dominican amber. Note forward extension of the thorax.

can amber and *Mundopoides aptianus* from Lebanese amber. Cockerell (1917D) described *Liburnia burmitina* from Burmese amber. Representatives of the Cixiidae, Delphacidae, Dictyopharidae, Fulgoridae, and Flatidae (Figs. 66–68) occur in Dominican amber (Appendix B).

Psyllidae (psyllids or jumping plant lice). These are small insects that resemble miniature cicadas but have jumping legs and relatively long antennae. The adults of both sexes possess two pairs of membranous wings.

Figure 69. A psyllid (family Psyllidae) in Dominican amber
(Poinar collection).

Psyllids are rare in amber, but Larsson (1978) mentions three undetermined species in the Copenhagen collection and Bachofen-Echt (1949)
cites a species of *Strophingia*; all were in Baltic amber. Representatives of
this group occur also in Mexican and Dominican amber according to
Schlee (1980) (Fig. 69, Appendices A, B).

Neuroptera (Net-winged Insects)

Included here are a wide assemblage of soft-bodied, relatively large
insects with four membranous wings having many longitudinal and cross
veins. All have complete metamorphosis involving an egg, larva, pupa,
and adult, in contrast to the incomplete (gradual) type of metamorphosis
involving an egg, nymph, and adult. The order Neuroptera is often divided into three suborders, the Megaloptera (alderflies, dobson flies, fish-

flies), the Raphidiodea (snakeflies), and the Planipennia (dusty-wings, lacewings, mantidflies, antlions, owlflies, spongillaflies). Nourishment for all of these representatives is mainly supplied by other arthropods taken as prey. All three suborders date back to the Early Permian (Carpenter and Burnham, 1985).

The Megaloptera are represented in amber by the Sialidae (alderflies) and Corydalidae (dobsonflies). Larvae of this group are aquatic and occur under rocks in streams where they prey on other insects. A single species of *Sialis* was described from Baltic amber by Weidner (1958). This description and an earlier one of *Chauliodes prisca* Pictet (1854) (Corydalidae) are the only records of Megaloptera in amber. The specimen of *Sialis* was a mature larva and probably became entrapped in the resin during its search for a terrestrial pupation site.

Members of the Raphidiodea (snakeflies) are equally rare in amber, being represented by adults of *Rhaphidia baltica* Carpenter (Rhaphidiidae), *Inocellia peculiaris* Carpenter and *Fibla erigena* (Hagen) (Inocelliidae), and some unidentified larvae, all in Baltic amber (Carpenter, 1956).

The Planipennia are represented by *Glaesoconis fadiacra* Whalley (Coniopterygidae), *Banoberotha enigmatica* Whalley (Berothidae), and *Paraberotha acra* Whalley (Berothidae) from Lebanese amber (Whalley, 1980). *Glaesoconis cretica* was described by Meinander (1975) from Siberian amber (Fig. 70) and Schülter (1978A) described *Retinoberotha stuermeri* (Berothidae) from French Cretaceous amber. Evolutionary comparisons of the latter fossil species with Recent species were presented by Schlüter and Stürmer (1984).

Other Planipennia were described from Baltic amber and include representatives of the Coniopterygidae (*Hemisemidalis, Coniortes, Archiconiocompsa, Archiconis, Heminiphetia*), Berothidae (*Proberotha*), Sisyridae (*Sisyra, Rhophalis*), Osmylidae (*Protosmylus*), Psychopsidae (*Propsychopsis*), Hemerobiidae (*Prophlebonema, Hemerobites, Prospadobius, Prolachlanius*), Nymphidae (*Pronymphes*), and Ascalaphidae (*Neadelphus*) (Keilbach, 1982; MacLeod, 1970). A mantidfly (Mantispidae), *Fera venatrix* Whalley (1983), was described from British amber, which probably originated from the western Baltic deposits.

Larvae of present-day Sisyridae (spongillaflies) are aquatic and feed on freshwater sponges. Thus, the presence of *Sisyra* and *Rhophalis* in amber suggests areas of calm water (ponds or swamps) in the original amber forest. The larva of the owlfly *Neadelphus protae* MacLeod (1970) (Color Plate 4) is the only representative of the Ascalaphidae in amber. In

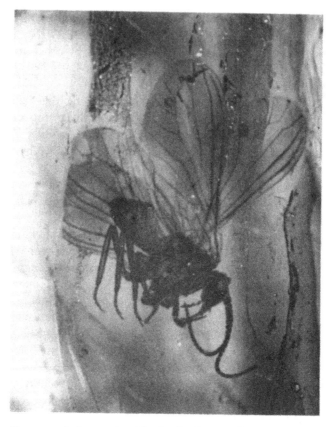

Figure 70. A dusty-wing (family Coniopterygidae), *Glaesoconis cretica* Meinander, in Siberian amber (photo by A. P. Rasnitsyn, courtesy of V. V. Zherichin).

owlflies, as in other Planipennia, the mouthparts are elongated into "pincers" that are inserted into the prey and function as paired hypodermal syringes to draw out the victim's body fluids.

Representatives of *Propsychopsis* are fairly common, both as adults and larvae, in Baltic amber, and yet today the family Psychopsidae is primarily Australian, with a few species in Africa, India, and China (Larsson, 1978).

Representatives of the Coniopterygidae have been reported from Mexican and Dominican amber (Color Plate 8), and chrysopids and hemerobiids also occur in the latter deposits (Fig. 71, Appendices A, B).

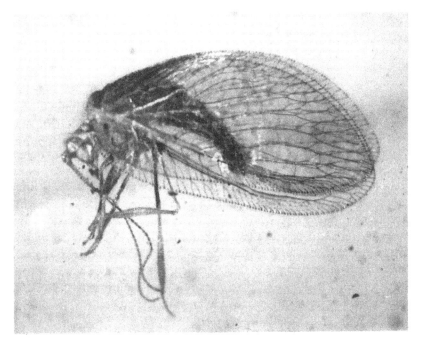

Figure 71. A brown lacewing of the genus *Hemerobius* (family Hemerobiidae) in Dominican amber (Poinar collection).

Mecoptera (Scorpionflies)

The mouthparts of most representatives of this order are prolonged ventrally into a beak, and the males of two families have genitalia that resemble the tail of a scorpion, thus the common name. The larvae tend to be grub- or caterpillarlike and their feeding habits are mostly saprophytic. The four membranous wings are similar to those of the preceding order.

The scorpionflies date back to the Early Permian, where they were represented by nearly twice as many families as exist now. Mecopterans have only been recovered and described from Baltic amber, where they are represented by the genera *Bittacus* and *Electrobittacus* in the Bittacidae, the genus *Panorpa* in the Panorpidae, and *Panorpodes* in the Panorpodidae (Carpenter, 1954; Spahr, 1989). One species, *Bittacus validus* Hagen, was recently assigned to the Trichoptera (Carpenter, 1976). Existing species of *Panorpodes* are restricted today to eastern Asia. Photographs of a *Panorpa* sp. in Baltic amber occur in the works of Katinas

(1983) and Bachofen-Echt (1949) and the latter author provides two excellent photographs of a *Bittacus* sp. in the same deposits.

Coleoptera (Beetles)

Because this is the largest order of insects, with over 250,000 described species, it is not surprising that they are commonly found in amber. Beetles normally have two pairs of wings, the front hard and chitinized (elytra) and the back membranous. Only the rear wings are used for flight, the front serving as protective shields. Beetles undergo complete metamorphoses and the larvae may or may not possess three pairs of true legs. The mouthparts are always of the chewing type even though the snout may form a rostrum as in the weevils (Curculionidae). Beetles have varied food habits: most are scavengers or plant feeders, but others are predaceous and a few are parasitic. The most common beetles found in amber are small, cryptic forms associated with the bark of trees. Many of these are fungus feeders. A list of rare beetles in amber is presented in Table 9. In the following account, the families will be grouped into suborders and then described in alphabetical order.

Suborder Archostemata. The fossil record of the earliest known beetles is from the Permian, where representatives of the suborder Archostemata were discovered. This group is considered by some as the most primitive branch of the Coleoptera and is represented today by only two families, both of which occur in amber. Species of *Priacma* and *Cupes* of the Cupedidae (reticulated beetles), whose most closely related extant species occur in North America, have been described from Baltic amber (Bachofen-Echt, 1949). *Micromalthus debilis* LeConte, reported from Mexican amber, is the only identified fossil of a Micromalthidae, but it has been assigned to an extant species, the only one known in the family, *M. debilis* (Rozen, 1971). This species has an unusual life cycle involving parthenogenesis and different types of larvae, and Rozen (1971) based his diagnosis of the fossil on the three first instars, which resembled present-day forms except for an apparent difference in the length of the tibia and tarsus. A possibly first-instar larva of *Micromalthus* has been reported from Lebanese amber (Whalley, 1981). Representatives of this family also occur in Dominican amber (Appendix B).

Suborder Adelphaga. Most beetles in the suborder Adelphaga are predaceous and possess filiform antennae and a 5–5–5 (on the first, second, and third pairs of legs, respectively) tarsal arrangement.

Table 9. Coleoptera of rare occurrence in amber

Family	Taxon or taxa	Amber source	Reference
Acanthocnemidae	*Acanthocnemoides*	Siberian	Spahr, 1981A
Aspidiophoridae	undetermined	Baltic	Larsson, 1978
?Boganiidae	undetermined	Lebanese	Whalley, 1981
Brachypsectridae	undetermined	Dominican	Appendix B herein
Cerophytidae	*Aphytocerus*	Siberian	Arnoldi et al., 1977
Circaeidae	*Circaeus*	Baltic	Yablokov-Khnzorian, 1961
Corylophidae	*Corylophus*	Baltic[a]	Spahr, 1981A
Dryopidae	*Palaeoriohelmis samlandica* Bollow	Baltic	Bollow, 1940
Georyssidae	undetermined	Baltic	Williamson, 1932
Heteroceridae	undetermined	Baltic	Larsson, 1978
Lyctidae	*Lyctus*	Baltic[a]	Spahr, 1981A
Lymexylonidae	?*Atractocerus, Hylecoetus,* ?*Lymexylon*	Baltic	Spahr, 1981A
Malachiidae	*Anthocomus, Apalochrus, Attalus, Colotes, Ebaeus, Malachius*	Baltic	Spahr, 1981A
Mycteridae	*Neopolypria nigra* Abdullah	Baltic[a]	Spahr, 1981A
Oedemeridae	*Oedemera*	Baltic	Spahr, 1981A
Rhizophagidae	*Rhizophagus*	Baltic[a]	Spahr, 1981A
Rhysodidae	undetermined	Baltic[a]	Williamson, 1932
Scaphidiidae	*Scaphidium*	Baltic[a]	Spahr, 1981A
Thorictidae	undetermined	Baltic	Williamson, 1932

[a] Family also represented in Dominican amber; see Appendix B herein.

Carabidae (ground beetles). The most commonly encountered members of the Adelphaga in amber are the ground beetles. Representatives of this large family tend to be secretive, hiding under stones or debris during the day and attacking other invertebrates during the night. Because many live under tree bark, they could easily come into contact with resin.

Most descriptions of amber ground beetles are from Baltic amber, where representatives of the genera *Dromius* and *Lebina* predominate. Other genera containing described species include *Chlaenius, Polyderis, Dyschiriomimus, Bembidion, Trechus, Trechoides, Protoscalidion, Cymindoides, Philorhizus,* and *Agatoides*. A color photo of a *Chlaenius* sp. can be found in Bachofen-Echt (1949). Erwin (1971) described the new genus *Tarsitachys* from Baltic amber and concluded, on the basis of its bilobed tarsal structure and habit, that it represented an extinct lineage of ar-

Figure 72. Holotype of a carabid beetle, *Polyderis antiqua* Erwin (family Carabidae), in Mexican amber (Museum of Paleontology collection, University of California, Berkeley; photo by P. Hurd).

boreal beetles. Although carabids have been found in Dominican amber, the only other described amber specimen is *Polyderis antiqua* Erwin (1971) from Mexican amber (Fig. 72). The latter species resembles present-day species of the worldwide genus *Polyderis*.

Cicindelidae (tiger beetles). Adult tiger beetles (often placed in the Carabidae) have been recovered only from Baltic amber. The species *Tetracha carolina* Linnaeus occurs today in the southern United States, West Indies, and Central America. The only other described tiger beetle from amber is *Pogonostoma chalybaeum* Handlirsch. Also reported as occurring in amber are representatives of the genus *Collyris*, which today can be found from India to Indonesia, and *Odontochila* sp. (Spahr, 1981A). The adults of these beetles probably preyed on insects that lived under the bark of the amber tree. The larvae, like those of other tiger beetles, probably lived in burrows in the soil or plant stems and preyed on passing invertebrates.

Dytiscidae (predaceous diving beetles). Both the larvae and adults of the predaceous diving beetles feed on aquatic invertebrates in ponds and quiet streams. The appearance of adults in amber might have resulted from flights, possibly when the water table dropped. Therefore, it is of interest that larvae of *Rhantus* and *Hyphydrus* have been found in Baltic amber (Weidner, 1958), as have adults resembling *Agabus*, *Laccophilus* and *Hydaticus* (Larsson, 1978; Spahr, 1981A). Unidentified dytiscids have also been found in Dominican amber (Appendix B).

Gyrinidae (whirligig beetles). A single whirligig beetle, *Gyrinoides limbatus* Motschulsky, has been described from Baltic amber, although species of *Gyrinus* and *Orechtochilus* from other Baltic amber collections have been mentioned (Bachofen-Echt, 1949).

Haliplidae (crawling water beetles). These beetles are small, convex forms that live in or near water. The adults feed on algae and other vegetation and the larvae are predaceous. Representatives of this family occur in Dominican amber (Appendix B), but none of the amber forms has been described.

Paussidae. This family comprises an interesting group of beetles adapted to life in ant nests. They are similar to and were included in the ground beetle family (Carabidae), but because of their modified morphology and specialized behavior, they are now considered a separate family. The Baltic amber representatives of this group were studied by Wasmann (1929) who described 12 species of *Arthropterus* and additional species in the genera *Paussoides, Arthropterillus, Cerapterites, Protocerapterus, Arthropterites*, and *Eopaussus*. Representatives of the genera *Megalopaussus, Paussus*, and *Pleuropterus* also have been reported from Baltic amber (Spahr, 1981A). All of the above genera are considered extinct except for representatives of *Arthropterus*, which now occur in Australia (Larsson, 1978). A photograph of an unidentified paussid in Baltic amber is provided by Bachofen-Echt (1949). This family is also represented in Dominican amber (Fig. 73).

Suborder Polyphaga. The habits of beetles in this suborder are varied but the great majority feed on plants. A structural character separates these beetles from the Adephaga. The margin of the first visible abdominal sternum extends completely across the abdomen and is not divided by the hind coxae.

Aderidae or Euglenidae (antlike leaf beetles). These beetles lack an anterior extension of the pronotum and have emarginate (not smoothly oval) eyes. They are small, rarely over 3 millimeters in length, and occur on flowers and foliage. Representatives of *Aderus* have been reported from Baltic amber (Spahr, 1981A) and examples of this family occur also in Mexican and Dominican amber (Appendices A, B).

Alleculidae (comb-clawed beetles). These medium-sized (4–12 millimeter), elongate-oval beetles are found on foliage, under bark, and in rotting vegetation. Species of *Allecula, Cteniopus, Gonodera, Hymenalia*,

Figure 73. A paussid beetle (family Paussidae) in Dominican amber (Poinar collection).

Isomira, Mycetochara, and *Mycetocharoides* are represented in Baltic amber (Spahr, 1981A). *Hymenorus chiapasensis* Campbell (1963) from Mexican amber is similar to the extant *H. tibialis* Champion from Guatemala. This family is also represented in Dominican amber (Appendix B).

Anobiidae (death-watch beetles). These are small oval forms in which the head is deflexed and the three terminal antennal segments are often enlarged. The larvae and adults commonly occur in the wood of a variety of trees and are not uncommon in amber. Species or representatives of *Anobium, Coelostethus, Coenocara, Crichtonia, Dorcatoma, Dryophilus, Ernobius, Eucrada, Gastrallus, Hedobia, Lasioderma, Mesocoelopus, Mesothes, Nicobium, Oligomerus, Petalium, Ptilinus, Theca, Xestobium,* and *Xyletinus* occur in Baltic amber (Spahr, 1981A). Spilman (1971) identified an undescribed species of *Cryptorama* and described *Stichtoptychus mexambrus* from Mexican amber. The only known extant species of *Stichtoptychus* was collected in Texas.

Anthicidae (antlike flower beetles; including Pedilidae). In these small forms the head is deflexed and the pronotum often extends over the head.

The adults can be found on flowers and foliage or in concealed habitats, especially in association with fungi. Species of *Protomacratria* and *Macratria* have been described, and representatives of *Amblyderes*, *Anthicus*, *Endomia*, *Notoxus*, *Pedilus*, *Steropes*, and *Tomoderus* occur in Baltic amber (Spahr, 1981A). Although there are representatives in Mexican and Dominican amber, the only other report of this family in amber is *Eurygenius wickhami* Cockerell (1917D) from Burmese amber.

Artematopidae. Artematopids are small, elongate, pubescent beetles with a deflexed head and filiform antennae. Two extinct genera have been described from Baltic amber: *Electribius oligocenicus* Crowson is moderately common in these deposits and *Protartematopus electricus* Crowson is described from a single specimen (Crowson, 1973). This family of beetles has not been reported from other amber deposits.

Bostrichidae (branch and twig borers). The head of most branch and twig borers is deflexed, and the antennae bear a terminal club of three or four segments. These small beetles tend to be elongate and the tips of the elytra (hardened forewings) often bear spines or teeth. They are rare in amber but do occur in Mexican and Dominican collections. Representatives of the genera *Apate*, *Bostrichus*, and possibly *Rhizopertha* occur in Baltic amber (Spahr, 1981A).

Brachypsectridae. Discovering a larva of one of these beetles in Dominican amber is interesting, because the present distribution of this rare family is California, Australia, southern India, and Singapore (Fig. 74). All members of the family have been placed in the single genus *Brachypsectra*. The larvae of the California species, *B. fulva*, occur under *Eucalyptus* bark, where they apparently feed on spiders. Observations showed that spiders were attracted to quiescent larvae, and when contact was made, the larva would turn up, grasp the spider with its mandibles, and impale the prey on the pointed tip of the terminal abdominal segment. After the spider had been killed, the larva would begin feeding (Crowson, 1973).

Bruchidae (seed beetles). These small, stout beetles possess a short snout and abbreviated elytra. Most larvae feed and pupate inside seeds, especially of the family Leguminoseae. They are represented, but undescribed, in Baltic and Dominican amber (Appendix B).

Buprestidae (metallic wood-boring beetles). Adults of the metallic wood-boring beetles tend to be compactly built and to possess shiny, me-

Figure 74. Flattened larva of a rare beetle of the family Brachypsectridae in Dominican amber (Cardoen collection, Miami,
Florida).

tallic blue, green, or black coloration. The larvae can become large as they
develop and pupate in the wood of various trees and shrubs and occasionally in herbaceous plants. Only *Mastogenius primaevus* Obenberger
and *Electrapate martynoir* Yablokov-Khnzorian have been described from
Baltic amber. The latter species was originally placed in a separate family,
the Electrapatidae, but is now considered a buprestid (Larsson, 1978).
Also mentioned as occurring in Baltic amber are representatives of the
genera *Agrilus, Anthaxia, Buprestis, Phaenops,* and *Poecilonota* (Spahr,
1981A). Cockerell (1917D) described *Acmaeodera burmitina* as an elaterid
(click beetle) from Burmese amber, but this species is now placed in the
Buprestidae (Spahr, 1981A). Metallic wood borers also occur in Mexican
and Dominican amber (Fig. 75, Appendices A, B).

Figure 75. A metallic wood-boring beetle (family Buprestidae) in Dominican amber (Poinar collection).

Figure 76. Holotype (male) of the soldier beetle *Silis chiapasensis* Whittmer (family Cantharidae) in Mexican amber (Museum of Paleontology collection, University of California, Berkeley; photo by P. Hurd).

Byrrhidae (pill beetles). These medium-sized, oval forms with a deflexed head are found under debris in a sandy shore habitat. Representatives of the genera *Byrrhus, Limnichus*, and *Syncalypta* occur in Baltic amber (Spahr, 1981A), but there are no records of this group in other amber deposits.

Cantharidae (soldier or leather-winged beetles). The adults usually are found on flowers and the predaceous larvae occur in the soil or other concealed habitats (under bark, moss, etc.). Species in the genera *Cacomorphocerus, Cantharis*, and *Malchinus* have been described, and representatives of the genera *Rhagonycha, Malthodes, Silis, Absidia*, and *Malthinus* have been reported from Baltic amber (Spahr, 1981A). Although soldier beetles have been reported from Dominican and Mexican amber, only the species *Silis chiapasensis* Wittmer (1963) has been briefly described from the latter deposits (Fig. 76).

Cerambycidae (long-horned or longicorn beetles). This family contains over 1,200 species of large, mostly elongate, often brightly colored beetles with long antennae. Although the adults of some species occur on flowers, the larvae develop in or on the roots, seeds, or stems of plants. Reports of this group in Cretaceous amber only include mention of a "Cerambycoidea" in Canadian amber (McAlpine and Martin, 1969). Descriptions of species or representatives of the genera *Acanthocinus*, *Aenictosoma*, *Anaglyptus*, *Callidium*, *Cerambyx*, *Dorcadionoides*, *Dorcaschema*, *Gracilia*, *Grammoptera*, *Lamia*, *Leptura*, *Molorchus*, *Necydalis*, *Nothorrhina*, *Obrium*, *Pachyta*, *Palaeoasemum*, *Parmenops*, *Pogonocherus*, *Saperda*, *Spondylis*, *Stenocorus*, *Strangalia*, and *Tetropium* occur in Baltic amber (Spahr, 1981A). Larsson (1978) notes that, at present, a single species of *Nothorrhina* occurs mainly in the Mediterranean and adjacent central European countries where it feeds on pine. With two described species and possibly others (including some larvae), this genus dominates in Baltic amber, and it is supposed that these representatives developed in the amber tree. A possible *Nothorrhina* larva is illustrated in Bachofen-Echt (1949). Representatives of the genera *Eburia*, *Merostenus*, *Methia*, *Plectromerus*, *Pentomacrus*, *?Eugamandus*, and *Elaphidion* have been identified in Dominican amber by J. Chemsak (Fig. 77).

Figure 77. A long-horned beetle of the genus *Merostenus* (family Cerambycidae) in Dominican amber (Poinar collection).

Chrysomelidae (leaf beetles). This is a large, widespread group of medium-sized beetles in which all developing stages feed on higher plants, mostly angiosperms. Although moderately abundant in Tertiary amber, they have not been reported from Cretaceous amber, although the extinct subfamily Protoscelinae was described from Mesozoic deposits (Crowson, 1981). Species or representatives in the genera *Anisodera, Cassida, Chalepus, Chrysomela, Colasposoma, Criocerina, Crioceris, Cryptocephalus, Donacia, Electrolema, Eumolpus, Galeruca, Galerucella, Hadroscelus, Haemonia, Haltica, Hispa, Inclusus, Lema, Luperus, Melasoma, Monolepta, Nodostoma, Ochrosis, Oposispa, Pachnephorus, Protanisodera, Pseudocolaspis*, and *Sucinagonia* occur in Baltic amber (Spahr, 1981A). Because it is doubtful that any of the above-mentioned genera had representatives that fed on the Baltic amber tree, these chrysomelids were probably trapped in resin during flight.

Gresset (1963) described a new genus and species of leaf beetle, *Profidia nitida*, from Mexican amber, and remarked that because of our incomplete knowledge of neotropical chrysomelids, the genus may not actually be extinct. Incomplete knowledge is certainly one of the major problems in attempting to compare fossil species with representatives in the present-day fauna. Considering that there are one million species of beetles (a conservative estimate) and only about a quarter of these have been described, dependable comparisons of extinct with extant groups can be made only when complete collections of the group in question have been made and studied worldwide. In 1971, Gresset described a second chrysomelid, *Crepidodera antiqua*, from Mexican amber. Although no closely related extant species are known, the genus occurs today in Mexico and Guatemala. Unidentified chrysomelids also occur in Dominican amber (Appendix B).

Ciidae (minute tree-fungus beetles; also known as Cisidae). These small forms are mainly associated with wood-rotting fungi in the family Polyporaceae (shelf or brachen fungi). Because such fungi are so common on a variety of trees, it is surprising that the Ciidae are not more widely represented in amber. Representatives of the genus *Cis* have been reported from Baltic amber (Spahr, 1981A) and the family is represented in Dominican amber (Appendix B).

Clambidae (fringe-winged beetles). These minute, oval, dark, pubescent beetles occur in decaying plant material, where they feed on primitive funguslike organisms, the Myxomycetes (Blackwell, 1984). Records of *Clambus* in Baltic amber (Spahr, 1981A) and unidentified forms in Mexican amber (Appendix A) constitute the presence of this family in amber deposits.

Cleridae (checkered beetles). These elongate, pubescent, often brightly colored predaceous beetles are commonly associated with standing or decaying tree trunks. Adults of described species in the genera *Clerus* and *Prospinoza* and undescribed representatives of *Corynetes*, *Necrobia*, *Necrobinus*, *Opilo*, *Tillus*, *Tarsostenus*, *Trichodes*, and *Trogodendron* have been reported from Baltic amber, as have unidentified larvae (Spahr, 1981A; Larsson, 1978). The larva illustrated by Bachofen-Echt (1949) as a member of the Cantharidae actually belongs to the subfamily Phyllobaeninae of the Cleridae (Larsson, 1978). This family is also represented in Dominican amber (Appendix B).

Coccinellidae (ladybird beetles). Most ladybird beetles are predaceous both as adults and as larvae, and the adults of many hibernate under bark. Because they must have been feeding on the aphids associated with the Baltic amber forest, it is strange that none has been described from amber. However, representatives of the genera *Coccinella*, *Coelopterus*, *Pharoscymnus*, *Platynaspis*, and *Scymnus* have been reported from Baltic amber (Spahr, 1981A). This family is also represented in Mexican and Dominican amber (Appendices A, B).

Colydiidae (cylindrical bark beetles). These hard, shiny beetles commonly occur under bark in association with fungi. Although the majority of species probably feed on ascomycetous fungi, some feed on lichens, and others are found together with bark beetles, probably feeding on fungi associated with them (Crowson, 1984). Representatives or species of *Bothrideres*, *Cicones*, *Colydium*, *Coxelus*, *Diodesma*, *Endophloeus*, *Murmidius*, *Rhopalocerus*, *Synchita*, and *Xylolaemus* occur in Baltic amber (Spahr, 1981A). The family is also represented in Mexican and Dominican amber (Fig. 78, Appendices A, B) and Whalley (1981) reports a possible member of this group from Lebanese amber.

Cryptophagidae (silken fungus beetles). These small, elongate-oval beetles commonly occur in decaying vegetable matter. Many are found under bark where they apparently feed on a variety of fungi, especially ascomycetes (Crowson, 1984). Representatives of *Antherophagus*, *Atomaria*, *Cryptophagus*, *Emphylus*, *Micrambe*, and *Telmatophilus* occur in Baltic amber (Spahr, 1981A), and *Nganasania khetica* Zerichin has been

Figure 78. A cylindrical bark beetle (family Colydiidae) caught in a spider's web in Dominican amber (Poinar collection).

described from Siberian amber (Arnoldi et al., 1977). The family is also represented in Dominican amber (Appendix B).

Cucujidae (flat bark beetles). These small to medium-sized, dorso-ventrally flattened forms are found under bark. They were considered to be predaceous on other arthropods in the same bark community, but Crowson (1984) provided evidence that they are mainly fungivorous. *Airaphilus denticollis* Ermisch and representatives of *Cucujus, Europs, Nausibius, Passandra, Platisus,* and *Silvanus* occur in Baltic amber (Spahr, 1981A). Keilbach (1982) also includes ?*Prostomis mandibularis* Liedtke from Baltic amber in this family. Representatives also occur in Mexican and Dominican amber (Appendices A, B).

Curculionoidea (weevils). Representatives of the large group Curculionoidea include the fungus weevils (Anthribidae), the straight-snouted weevils (Brentidae), and the snout beetles (Curculionidae). The great majority are plant feeders with larvae concealed in the soil or plant tissue. The adults have the anterior portion of their head prolonged into a snout with the mandibles at the tip.

The fungus weevils, which have the least modified snout, have a blunt beak with non-elbowed antennae. The larvae develop in or on fungi, seeds, and deadwood. The anthribids are uncommon in amber and have been recorded only from Baltic deposits; species of *Inclusus* and *Pseudomecorhis* have been described and representatives of *Tropideres* reported from Baltic amber (Keilbach, 1982). Spahr (1981A) believes that the genus *Inclusus* belongs to the Chrysomelidae, so further clarification is required regarding this taxon.

The brentids are narrow, elongate, mainly tropical forms that have a straight projecting snout. The larvae occur in wood. Undescribed forms occur in Dominican amber (Appendix B).

The curculionids (snout beetles or weevils) constitute a large group of over 40 subfamilies. They have varied habits and habitats and are well represented in Tertiary amber, but, curiously, there are no reports of snout beetles in Cretaceous amber. Most reports and descriptions of amber weevils are from Baltic deposits, where the following genera are represented: *Acalles, Ampharthropelma, Anchorthorrhinus, Anthonomus, Apion, Archimetrioxena, Bagous, Calandra, Car, Ceuthorrhynchus, Choerorrhinus, Cleonus, Curculio, Dorytomus, Dryophthorus, Electrotribus, Erirhinoides, Erirrhinus, Hylobius, Hypera, Involvulus, Isalcidodes, Lixus, Magdalis, Mecinus, Mesites, Necrodryophthorus, Notaris, Otiorrhynchus, Paleopissodes, Paonaupactus, Phyllobius, Phytonomus, Pissodes, Polydrosus, Protonaupactus, Pseudostyphlus, Ptochus, Rhinoncus, Rhynchites, Rhyncolus, Sitona, Succinacalles, Synommatus, Thryogenosoma, Thylacites,* and *Trachyphloeus* (Spahr, 1981A). A comparison of the above extant genera with their present-day host plants indicates that many of the weevils (*Hypera, Phytonomus, Sitona*) were trapped in resin during flight. Only species of *Magdalis* and *Pissodes* presently breed in conifers and could have used the amber tree as a source of food. There is the possibility, of course, that host plants changed over the years and more of the above-mentioned weevils may have been associated with conifers in the Tertiary. The most abundant genus in Baltic amber was *Phyllobius* and Larsson (1978) assumed that it probably fed on oak leaves.

Zimmerman (1971) described, from Mexican amber, a species of

Zygops and two species of *Cryptorhynchus*, and cited another undetermined species similar to the latter genus. He noted that similar species occur in Mexico today, although the fossil specimens may represent extinct species. A wide assemblage of weevils occurs in Dominican amber (Appendix B) (Color Plate 7).

Dascillidae (soft-bodied plant beetles). These beetles are associated with aquatic vegetation, where the larvae feed on plant roots. Representatives of *Dascillus* and *Pseudodactylus* occur in Baltic amber (Larsson, 1978) and the family is also represented in Dominican amber (Appendix B).

Dermestidae. Dermestid beetles are small, compact, usually hairy, oval forms with short, clubbed antennae. They are well-known scavengers found in a variety of habitats including under bark. The oldest occurrence of this family in amber is from Lebanese deposits (Whalley, 1981) and this record, incidentally, is the earliest for any member of the superfamily Dermestoidea (Crowson et al., 1967). Representatives of the genera *Anthrenus*, *Attagenus*, *Dermestes*, *Globicornis*, *Phranodoma*, and *Trinodes* occur in Baltic amber (Spahr, 1981A; Keilbach, 1982). *Dermestes larvalis* Cockerell (1917A) was described from Burmese amber and *Cryptorhopalum electron* Beal (1972) from Mexican amber. Dermestids are also represented in Dominican amber (Appendix B).

Elateridae (click beetles). These beetles have the protective ability to spring violently into the air when knocked onto their back. This motion is often accompanied by a clicking sound, and eventually results in the insect landing right side up. Click beetles are brown or black, medium-sized, elongate, robust forms. The larvae (wireworms) live a concealed life in soil, rotting wood, or fungi, but the adults are moderately common on vegetation.

Whalley (1981) mentions the presence of this family in Lebanese amber, which is the only record of click beetles in Cretaceous amber. Species or representatives of the following genera occur in Baltic amber: *Adelocera*, *Adrastus*, *Aeolus*, *Ampedus*, *Anchastus*, *Athous*, *Betarmon*, *Cardiophorus*, *Colaulon*, *Corymbites*, *Crioraphes*, *Cryptohypnus*, *Ctenicerus*, *Denticollis*, *Diaraphes*, *Dipropus*, *Drasterius*, *Ectamenogonus*, *Elater*, *Elatron*, *Holopleurus*, *Hypnoidus*, *Idolus*, *Limonius*, *Ludius*, *Mecynocanthus*, *Megapenthes*, *Melanoxanthus*, *Neotrichophorus*, *Orthoraphes*, *Pheletes*, *Plagioraphes*, *Porthmidius*, *Procraerus*, *Sericus*, *Silesis*, *Sternopes*, *Synaptus*, and *Tetraraphes* (Spahr, 1981A). The Elateridae is one of the most commonly occurring beetle groups in Baltic amber. Because many of the above have contemporary relatives that develop in meadow habitats, their occurrence in amber is probably related to some characteristic (and perhaps frequency) of their flight behavior. Present-day *Elater* and *Adelocera* include species that live in conifers and the fossil species may have fed on the amber tree.

Elater burmitinus Cockerell (1917D) was described from Burmese amber, and the genus *Mionelater* and species of *Agriotes* and *Glyphonyx* were described by Becker (1963) from Mexican amber (Fig. 79). Click beetles also occur in Dominican amber (Appendix B).

Endomychidae (handsome fungus beetles). These small, shiny, oval beetles are found under bark and in rotting wood, where they apparently feed on fungi. A single species, *Phymaphoroides antennatus* Motschulsky, and representatives of the genera *Hylaia, Liesthes, Lycoperdina, Mycetaea, Mycetina, Symbiotes*, and *Trochoideus* occur in Baltic amber (Spahr, 1981A). This family is also represented in Dominican amber (Appendix B).

Erotylidae (pleasing fungus beetles). These small to medium-sized, shiny, oval beetles are found in fungi, around sap, and in rotting wood. None has been formally described from amber, but representatives of the genera *Cryptophilus, Dacne, Diplocoelus*, and *Tritoma* have been reported from Baltic amber (Spahr, 1981A). The family is also represented in Dominican amber (Appendix B).

Eucnemidae (false click beetles). These are related to the click beetles, but are much rarer and occur in decaying wood. A few can jump like click beetles and most have the habit of vibrating their antennae. Representatives of the genera *Dirhagus, Eucnemis, Hypocoelus, Microrhagus, Phlegon*, and *Xylobius* occur in Baltic amber (Spahr, 1981A). The species *Throscogenius takhtajani*, which Yablokov-Khnzorian (1962) described as a Throscidae, is considered by Cobos (1963) to belong in the Eucnemidae. This family is also represented in Dominican amber (Appendix B).

Helodidae (marsh beetles). The small, oval adults of these beetles are found in areas of wet vegetation, and the larvae are aquatic. Representatives of the genera *Brachelodes, Cyphon, Cyphonogenius, Helodes, Helodopsis, Hydrocyphon, Microcara, Plagiocyphon, Prionocyphon*, and *Scirtes* occur in or have been described from Baltic amber (Keilbach, 1982; Spahr, 1981A). The family is also represented in Dominican amber. The large number of these amphibious beetles in Baltic amber (10–20 percent of all Coleoptera according to Larsson, 1978) indicates that the amber trees were associated with a wet biotype at least part of the year.

Figure 79. A. Holotype (female) of the click beetle *Agriotes succiniferus* Becker (family Elateridae) in Mexican amber. B. Holotype (female) of the click beetle *Glyphonyx punctatus* Becker (family Elateridae) in Mexican amber. (Both in the Museum of Paleontology collection, University of California, Berkeley; photos by P. Hurd.)

Histeridae (hister beetles). These beetles are associated with decaying organic matter but apparently prey on other arthropods in this habitat. They are recorded from Mexican and Dominican amber, and the genera *Abraeus, Acritus, Carcinops, Hister*, and *Platysoma* have been identified in Baltic amber (Larsson, 1978, Bachofen-Echt, 1949).

Hydrophilidae (water scavenger beetles). This is one of the most common families of beetles associated with water. The adults, which are similar to the predaceous diving beetles in shape, are scavengers and the larvae are predaceous. The larvae have only a single tarsal claw in contrast to dytiscid larvae, which have a pair on each tarsus. These beetles occur in Baltic and Dominican amber. Baltic amber forms include *Sphaeridium melanarium* Gistl and representatives of the genera *Hydrophilus* and *Cercyon*. The latter genus also includes species that occur in rotting wood and fungi (Bachofen-Echt, 1949; Spahr, 1981A).

Lagriidae (long-jointed beetles). These medium-sized, dark metallic, slender beetles have an elongate apical antennal segment. The larvae breed in rotting plant debris and the adults occur on vegetation. Representatives of the genera *Lagria* and *Statira* occur in Baltic amber (Larsson, 1978) and the family is also represented in Dominican amber (Appendix B).

Lathridiidae (brown scavenger beetles). These minute, elongate-oval forms are found in decaying vegetation, where they apparently feed on fungi (Crowson, 1984). Although they are relatively frequent in Baltic amber, none has been described and most appear to belong to the genera *Corticaria, Enicmus, Holoparamecus, Lathridius*, and *Melanophthalma* (Larsson, 1978). *Succinimontia infleta* Zherichin (in Arnoldi et al., 1977) has been described from Siberian amber. The family is also represented in Dominican amber (Appendix B).

Leiodidae (round fungus beetles). These oval-shaped beetles have the ability to roll up into a ball. Most feed on fungi (Crowson, 1984), but others occur in decaying vegetation and some seek out carrion and ant nests. Although they are rare in amber, records exist of representatives of *Anisotoma* and *Leiodes* in Baltic amber (Spahr, 1981A) and undetermined forms in Dominican amber (Appendix B).

Limulodidae (horseshoe crab beetles). These small, oval beetles are shaped like horseshoe crabs. They lack hind wings and compound eyes, and are modified for living in the nests of ants and termites. They have been reported from Dominican and Mexican amber (Appendices A, B).

Lucanidae (stag beetles). These moderately large, robust beetles are usually associated with trees, and the larvae develop in decaying wood. Species of *Dorcasoides, Paleognathus,* and *Platycerus* have been described from Baltic amber, and the family is also represented in Siberian and Dominican amber (Spahr, 1981A) (Appendix B).

Lycidae (net-winged beetles). In this group, which is poorly represented in amber, the adults occur on vegetation and the larvae are predaceous in the soil and decaying plant remains. A single species, *Pseudaplatopterus scheelei* Kleine, and representatives of the genera *Calopteron* (= *Dictyoptera*), *Lygistopterus* and *Lycus* occur in Baltic amber (Spahr, 1981A).

Melandryidae (false darkling beetles). These elongate, flattened beetles occur on vegetation and under bark where, apparently, at least some feed on fungi (Crowson, 1984). The oldest records of this family in amber are of the species *Archizylita zherichini* Nikitsky (1977) and *Pseudohallomenus cretaceous* Nikitsky (1977) from Siberian deposits. A single species, *Abderina helmii* Seidlitz, and representatives of the genera *Abdera, Anisoxya, ?Carida, Dircaea, Eustrophus, Hallomenus, Hypulus, Melandrya, Orchesia, Osphya, Phloeotrya,* and *Serropalpus* occur in Baltic amber (Spahr, 1981A). The occurrence of *Osphya* larvae in Baltic amber may indicate a preference for the amber tree, and periods of larval activity that correlate with seasonal resin flows (Larsson, 1978). This family is also represented in Mexican and Dominican amber (Appendices A, B).

Meloidae (blister beetles). These moderately large, elongate beetles are often seen on flowers. The larvae are predaceous or parasitic, and infection is initiated by an active first-instar "triungulin" (active, highly mobile stage) that searches for a host. Some species are known to develop in bee nests and feed on bee larvae. In such cases, the triungulin waits on a flower, climbs on a visiting bee, and is carried back to the nest. A possible meloidid triungulin was reported in Baltic amber (Larsson, 1978) and another occurs in Dominican amber (Fig. 136). In the latter case, the triungulin was still attached to the "neck" region of a worker bee (*Proplebeia dominicana*) and was apparently being transported back to the nest when both were trapped in the resin. Both the wedge-shaped beetles (Rhipiphoridae) and the Strepsiptera also have triungulin larvae that are known to associate with bees in a similar fashion. No species of blister beetles in amber have been described, although representatives of the genera *Lytta* and *Meloe* occur in Baltic amber (Spahr, 1981A).

Melyridae (soft-winged flower beetles) (including Malachiidae). These small to medium-sized, oval-elongate, often brightly colored beetles commonly occur on flowers. Both adults and larvae are mostly predaceous on insects that live under bark and in wood. Examples of the genera *Apalochrus, Attalus, Colotes, Cerallus, Dasytes, Dasytina, Ebaeus, Haplocnemus, Malachius, Melyris, Microjulistus, ?Psilothrix,* and *Zygia* have been reported from Baltic amber, and the family is also represented in Siberian and Dominican amber (Spahr, 1981A) (Appendix B).

Mordellidae (tumbling flower beetles). These small to medium-sized beetles have a characteristically curved body and an extended abdominal tip. Although the adults are found on flowers, the larvae live mostly in decaying wood. Species or representatives of the genera *Anaspis, Glipostena, Mordella, Mordellina,* and *Mordellistena* occur in Baltic amber (Spahr, 1981A). *Anaspis antica* Guérin was described from Sicilian amber, and the family is also represented in Canadian, Siberian, Mexican, and Dominican amber (Spahr, 1981A) (Appendices A, B).

Mycetophagidae (hairy fungus beetles). These small, flattened, oval forms occur in shelf fungi and decaying vegetation. Species or representatives of the genera *Berginus, Crowsonium, Litargus, Mycetophagus, Pseudotriphyllus, Triphyllus,* and *Typhaea* occur in Baltic amber (Spahr, 1981A). The presence of mycetophagid larvae in the Danish Baltic amber collections is mentioned by Larsson (1978) as evidence that shelf fungi (Polyporaceae) attacked the amber tree. This family is also represented in Dominican amber (Appendix B).

Nitidulidae (sap beetles). These small, robust beetles are found in a variety of habitats associated with decaying vegetable matter. The larvae breed in fungal fruiting bodies, in fermenting sap flows from damaged trees, and under the bark of dead trees. All of these habitats produce the yeast or other fungi that seem to be the major food source (Crowson, 1984). A single species, *Omositoidea gigantea* Schaufuss, and representatives of the genera *Carpophilus, Cryptarcha, Cyllodes, Epuraea, Glischrochilus, Nitidula, Omosiphora,* and *Pria* occur in Baltic amber (Spahr, 1981A). The family is also represented in Dominican amber (Appendix B).

Figure 80. A shining flower beetle (family Phalacridae) in Dominican amber (Poinar collection).

Phalacridae (shining flower beetles). The larvae and adults of these small, dark, oval beetles can be found in flowers. The developmental stages of many phalacrids apparently feed on fungi (especially the spores of rust and smut) that attack the inflorescences of a variety of plants, especially monocots; some species feed on the pollen and sap of plants (Steiner, 1984). Only two genera, *Olibrus* and *Phalacrus*, are present in Baltic amber, but representatives also occur in Mexican and Dominican amber (Spahr, 1981A) (Fig. 80, Appendices A, B).

Platypodidae (platypodid or flat-footed beetles). Platypodids resemble narrow, cylindrical bark beetles. The adults tunnel into the wood of both hardwood and softwood trees. There, the larvae feed on ambrosia fungi (yeasts) that line the tunnel galleries and eventually pupate in the tree. Because of their tunneling activities, platypodids are considered damaging to lumber, but apparently they rarely attack healthy trees. They are more common in tropical and subtropical climates.

Although *Platypus* has been recorded several times from Baltic amber, none of these has apparently been formally described. Platypodidae are much more common in Mexican and Dominican amber (Color Plate 5) because of the warmer climate and the predominance of angiosperms in these sites. Schedl (1962) described three new species from Mexican amber—*Cenocephalus succinicaptus*, *C. hurdi*, and *C. quadrilobus* (now considered by some to be in the genus *Mitosoma*)—and Schawaller (1981B) described *Mitosoma rhinoceroide* from Dominican amber. Specimens of the latter species had pseudoscorpions attached to them (Fig. 113). Guérin (1838) provided a brief description of the new species *Platypus maravignae* from Sicilian amber.

Pselaphidae (short-winged mold beetles). These small beetles have reduced elytra similar to those of rove beetles, but mold beetles possess only three tarsal segments (most rove beetles have five) and usually have clubbed antennae. They live in concealed habitats (under bark, in logs, in animal nests) where they feed on mites and other small arthropods (Newton, 1984). Species of the following genera, most of which are extinct, have been described from Baltic amber: *Barybryaxis, Batrisus, Bryaxis, Bythinus, Ctenistodes, Cymbalizon, Dantiscanus, Deuterotyrus, Eupines, Euplectus, Euspinoides, Faronus, Greys, Hagnometopias, Hetereuplectus, Monyx, Nugaculus, Nugator, Pammiges, Pantobatrisus, Parabryaxis, Pselaphus, Tmesiphoroides, Trimium, Tyrus,* and *Tychus* (Keilbach, 1982; Spahr, 1981A). Representatives of this family also occur in Dominican amber, and Oke (1957) described *Eupines setifera* from Late Cenozoic (probably Pleistocene) resin from Australia.

Ptiliidae (feather-winged beetles). In these minute forms, seldom over 1 millimeter in length, the hind wings bear a long fringe of hairs that sometimes stick out beneath the elytra. The adults and larvae occur in rotting plant and animal remains where they feed mainly on fungi. Although feather-winged beetles have been reported from Mexican and Dominican amber, only two species have been described, *Ptinella oligocaenica* Parsons and *Microptilium gustantsi* Dybas, both from Baltic amber (Larsson, 1978). Representatives of the genus *Ptenidium* are mentioned by Bachofen-Echt (1949) as also occurring in Baltic amber.

Ptilodactylidae. These beetles occur on aquatic vegetation and in rotting wood. A single species, *Ptilodactyloides stipulicornis* Motschulsky, was described from Baltic amber, and representatives also occur in Mexican and Dominican amber (Larsson, 1978) (Appendices A, B).

Ptinidae (spider beetles). These beetles possess long legs and are spiderlike in appearance. They occur in a variety of habitats and feed on plant and animal products. Representatives of the genera *Niptus* and *Ptinus* occur in Baltic amber, but representatives also occur in Dominican amber (Spahr, 1981A) (Appendix B).

Pyrochroidae (fire-colored beetles). These medium-sized, elongate forms often have a brightly colored pronotum, serrate or pectinate antennae, and large eyes. The larvae and adults occur under the bark of trees, where at least some are known to be associated with ascomycetous fungi. *Palaeopyrochroa crowsoni* Abdullah and *Pyrochroa* sp. occur in Baltic amber, and the family is also represented in Dominican deposits (Spahr, 1981A) (Appendix B).

Rhipiphoridae (wedge-shaped beetles). These forms have characteristic antennae that are pectinate in males and serrate in females. The adults can be found on flowers and the larvae are parasitic, many in the nests of wasps and bees. The first-stage larva is an active triungulin that is phoretically carried by worker bees or wasps back to their nests. Some species utilize cockroaches as hosts. *Myodites burmiticus* Cockerell (1917B) has been described from Burmese amber and *Rhipidius primordialis* Stein from Baltic amber. Representatives of *Pelecotoma* and *Rhipiphorus* also occur in Baltic deposits, and the family is represented in Canadian and Dominican amber (Spahr, 1981A) (Appendix B; Color Plate 6).

Salpingidae (narrow-waisted bark beetles; including Pythidae). Representatives of this family have the basal portion of the pronotum constricted, similar to the case in ground beetles. These medium-sized, elongate beetles occur under bark and in decaying vegetation where all stages are considered predaceous, although some are associated with fungi (Crowson, 1984). Representatives of *Lissodema* and *Salpingus* occur in Baltic amber (Spahr, 1981A) and the family is represented in Lebanese amber (Whalley, 1981).

Scarabaeidae (scarab beetles). These comprise a large, morphologically diverse group of robust beetles that occur in a variety of habitats. Many are associated with dung, while others feed on living or decomposing plants, and some occur in the nests of ants and termites. The family dates back to the Jurassic (Crowson et al., 1967) and is represented in Lebanese Cretaceous amber (Whalley, 1981). Species and representatives in the genera *Aphodius*, *Ataenius*, ?*Rhyssemus*, *Saprosites*, *Scarabaeus*, and *Trox* occur in Baltic amber, and scarabs also occur in Dominican material (Spahr, 1981A) (Appendix B). The Dominican material contains representatives of the subfamily Acanthocerinae (= Acanthoceratidae or Ceratocanthidae), which are known for their ability to roll up into a ball. They occur on flowers, under bark, in rotting wood, and in termite nests.

Scolytidae (bark and ambrosia beetles; also known as Ipidae). These dark, small (6–8 millimeters long) cylindrical beetles possess short, geniculate antennae with a large club. The bark beetles live and feed within the bark of trees, whereas the ambrosia beetles feed on "ambrosial" fungi that they carry and cultivate in galleries within a tree. Because of their habitats, this group is well represented in amber. Described species or representatives of the genera *Charphoborites, Hylastes, Hylastites, Hylates, Hylescierites, Hylesinus, Hylurgops, Hylurgus, Myelophilites, Myelophilus, Phloeophthorus, Phloeosinites, Phloeosinus, Polygraphus, Taphramites, Taphrorychus, Tomicus, Trypodendron, Xyleborus, Xylechinites,* and *Xylechinus* occur in Baltic amber (Spahr, 1981A; Keilbach, 1982); some of these genera are thought to be extinct. The most abundant genus is *Hylurgops,* whose present-day species, together with those of *Hylastes,* are associated with pines. Several pines have been described from Baltic amber, and these insects were probably feeding on at least some of these. Bark beetles occur also in Mexican and Dominican amber, and the species *Cryphalites rugosissimus* Cockerell (1917A) has been reported from Burmese amber and *Dryocoetes* aff. *autographus* from Romanian amber (Protescu, 1937).

Scraptiidae. Representatives of this family, often included in the Melandryidae, are found under bark or in logs. Species of *Archescraptia, Palaeoscraptia, Scraptia,* and *Trotomma* have been described from Baltic amber, and the family is represented in Siberian amber (Spahr, 1981A).

Scydmaenidae (antlike stone beetles). Representatives of this family are antlike in shape and have clavate femurs and moderately long antennae. The larvae and adults are predaceous on mites and other small arthropods in litter and other concealed habitats (Newton, 1984). Although antlike stone beetles are present in Lebanese (Whalley, 1981), Canadian (McAlpine and Martin, 1969), Mexican, and Dominican amber, only Baltic amber forms have been described. These include species of the genera *Clidicus, Cryptodiodon, Cyrtoscydmus, Electroscydmaenus, Euconnus, Hetereuthia, Heuretus, Palaeomastigus, Palaeothia, Semnodioceras,* and *Scydmaenoides.* Other genera reported in Baltic amber include *Cephennium, Euthia, Mastigus, Neuraphes,* and *Scydmaenus* (Spahr, 1981A). Oke (1957) mentions a *Megalederus* sp. in a piece of Upper Cenozoic resin (probably Pleistocene) from Australia.

Silphidae (carrion beetles). These beetles are associated with decaying animal matter, although some occur in fungi or in the nests of various animals. The two described species in amber are *Ptomaphagus germari* Schlechtendal and *Nemadus colonoides* Jeannel from Baltic amber (Spahr, 1981A). The latter species is thought to have been associated with the nests of birds and mammals (Larsson, 1978).

Staphylinidae (rove beetles). Staphylinids are moderately frequent in amber, probably because individuals are small and the group is one of the largest of the beetle families. Their frequent occurrence in amber may also be due to their active habits and widely various habitats, especially in relation to decaying animal and plant material. Rove beetles have been reported in Lebanese and Siberian amber (Whalley, 1981; Zherichin and Sukacheva, 1973), and Schlüter (1978A) described *Stenus inerpectatus* from French Cretaceous amber.

Descriptions of rove beetles from Baltic amber include species of the genera *Bembicidiodes, Lathrobium, Osorius, Oxypoda, Oxyporus, Pseudo-lesteua, Pseudolesteva,* and *Stenus,* whereas undescribed representatives occur in the genera *Aleochara, Anthobium, Anthophagus, Atheta, Bledius, Bolitobius, Bryocharis, Carpalimus, Conosoma, Cryptobium, Eusphalerum, Gyrophaena, Homalota, Hypocyptus, Lathrobium, Leptacinus, Leptocherus, Leptusa, Medon, Micropeplus, Mycetoporus, Oligota, Olophrum, Omalium, Oxyporus, Oxytelus, Paederus, Philonthus, Phloeocharis, Planeustomus, Proteinus, Quedius, Rugilus, Scopaeus, Sepedophilus, Staphylinus, Stichoglossa, Stilicus, Tachinus, Tachyporus, Trogophloeus, Xantholinus,* and *Zyras* (Keilbach, 1982; Larsson, 1978; Bachofen-Echt, 1949; Hieke and Pietrzeniuk, 1984; Spahr, 1981A).

From Mexican amber, Seevers (1971) described the genera *Palaeopsenius* and *Paracyptus* as belonging to specialized groups. *Palaeopsenius mexicanus* presumably lived in termite nests, and *Paracyptus minutissima,* considered one of the smallest known species of rove beetles, probably lived in dead wood associated with termites. *Palaeopsenius mexicanus* is limuloid in form (i.e., shaped somewhat like a horseshoe crab), which is characteristic of some beetle species inhabiting termite and ant nests. Members of the Trichopseniinae (termitophilous forms) occur in Dominican amber (Fig. 81) and have been confused with the true horseshoe crab beetles (Limulodidae) that occur in ant nests. Staphylinoid beetle larvae have been found associated with mammalian hair in Dominican amber (Poinar, 1988A) (Fig. 82).

Tenebrionidae (darkling beetles). This is a large group of beetles with diverse shapes, habits, and sizes, although most forms are dark and smooth. The tenebrionids feed on a wide variety of plants and plant products (also fungi) and some are predaceous on other insects. The earliest known fossil representatives of this family occur in brown coal deposits of the Mid Eocene. These are roughly contemporary with the Baltic amber deposits that include representatives of the genera *Bolitophagus,*

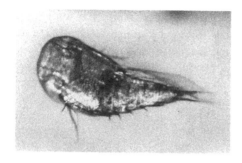

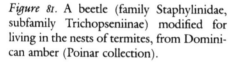

Figure 81. A beetle (family Staphylinidae, subfamily Trichopseniinae) modified for living in the nests of termites, from Dominican amber (Poinar collection).

Figure 82. A beetle (superfamily Staphylinoidea) modified for living in the hair of mammals, from Dominican amber (Poinar collection).

Helops, Hopatrum, Hypophloeus, Laena, Leichenum, Palorus, Sepidium, Tribolium, and *Uloma* (Spahr, 1981A). The above genera are mostly forms that today occur in shelf fungi and rotting wood. The extant tenebrionid *Platycilibe brevis* Cart. was identified in Victorian copal (Hills, 1957) and Kaszab and Schawaller (1984) described *Hesiodobates antiquus,* the first fossil tenebrionid from the New World, from Dominican amber.

Throscidae. The throscids comprise a small group of minute, dark, oblong beetles that occur in living and decaying vegetation. Species or representatives of the following genera occur in Baltic amber— *Aulonothroscus, Palaeothroscus, Throscites, Throscogenius,* and *Throscus* (Spahr, 1981A)—although Cobos (1963) stated that the extinct genera *Palaeothroscus* and *Throscites* do not differ from the Recent *Throscus* and *Aulonothroscus,* respectively. This family is also represented in Dominican amber (Appendix B).

Trogossitidae (bark-gnawing beetles; also known as the Ostomatidae and the Ostomidae). This family comprises elongate forms that prey on other arthropods under bark (Trogossitinae) and oval-elliptical forms that feed mainly on fungi (Peltinae). None from amber has been formally described, although representatives of the genera *Calitys, Grynocharis, Lophocateres, Ostoma,* and *Trogosita* have been reported from Baltic amber and the family is represented in both Mexican and Dominican amber (Spahr, 1981A) (Fig. 83, Appendices A, B).

Figure 83. A bark-gnawing beetle (family Trogossitidae) in Dominican amber (Poinar collection).

Strepsiptera (Twisted-winged Parasites)

The insects in this small order are unusual, and their relationship with other Hexapoda has baffled entomologists for years. They are often grouped with the beetles because male strepsipterans possess functional hind wings. The fore wings are reduced to appendages that resemble the halteres (reduced hind wings) of flies. All females are wingless; most are also legless and never leave their insect host. The winged males search out the females and mate. The females produce active triungulin larvae, which search out new hosts. All species are parasites of insects, mostly Hymenoptera, Hemiptera, Homoptera, Orthoptera, and Thysanura.

A single strepsipteran, *Mengea tertiaria* (Menge), represented now by only two of the eight original specimens, has been described from Baltic amber (Menge, 1863; Kulicka, 1978). The family Mengeidae, in which it is placed, is considered primitive because the wingless females come

Figure 84. A male strepsipteran (order Strepsiptera) in Dominican amber. Note raspberrylike eyes and greatly reduced front wings (Brodzinsky-Lopez-Penha collection, Smithsonian Institution).

out of their host and exist free in the soil. Recorded hosts for extant members of this family include members of the Thysanura and Hemiptera.

In 1979 Kinzelbach described *Protelencholax schleei* from Dominican amber. This interesting new genus and species is the first fossil record of the family Elenchidae, and Kinzelbach considers it a stem group of that family, thus indicating that other subgroups of the Strepsiptera probably existed simultaneously. A strepsipteran from Dominican amber is illustrated in Fig. 84.

Trichoptera (Caddis Flies)

Caddis flies are mothlike in appearance, but when at rest, their four hairy membranous wings form a "roof" over the abdomen. The antennae are long and the mouthparts are adapted for chewing, although the adults consume mainly liquids. The larvae resemble caterpillars, but they are aquatic or semiaquatic and possess filamentous gills on the sides of the abdomen. Many make protective cases from plant debris or stones, and pass the developmental period in streams or lakes. Because the earliest known fossils of this group date back to the Early Permian (Carpenter and Burnham, 1985), it is not surprising to find caddis flies in Cretaceous deposits of amber.

The two specimens of caddis flies reported by Legg (1942) from Canadian amber were misidentified, but true Trichoptera were discovered from the Medicine Hat site in Canada. These were described in a new genus, *Electralberta*, in the new family Electralbertidae (Botosaneanu and Wichard, 1983.) The same authors also studied a collection of Trichoptera in Siberian amber, and erected the new family Taymyrelectronidae, with the type genus *Taymyrelectron*, and the new genera *Palaeohydrobiosis* (Hydrobiosidae), *Archaeopolycentra* (Polycentropodidae), *Praeathripsodes* (Leptoceridae), and *Calamodontus* (Calamoceratidae or Odontoceridae). Also from the same source were new species of *Rhyacophila* (Rhyacophilidae) and *Holocentropus* (Polycentropodidae) and representatives of the Philopotamidae and Hydropsychoidea. The only other known Cretaceous caddis fly was described by Cockerell (1917C) as ?*Dolophilus praemissus*. This species is known only from the distal part of a fore wing in Late Cretaceous amber from the Emscherien of Tennessee (part of the Atlantic Coastal Plain amber), and Botosaneanu and Wichard (1983) place the specimen in the family Philopotamidae.

In analyzing the Trichoptera found in Late Cretaceous amber, Botosaneanu and Wichard (1983) noted that the fauna tended to have a modern character but, unlike in other groups, there was no sharp line of extinction or sudden appearances at the Cretaceous-Tertiary border. They noted that most of the Late Cretaceous caddis flies were small forms, and notwithstanding the fact that these may have been more easily trapped in the resin, they concluded that small caddis flies were the norm at that time. Indeed, Botosaneanu and Wichard point out that *Rhyacophila antiquissima*, *Taymyrelectron suratshevae*, and *Calamodontus grandaevus* (all described by Botosaneanu and Wichard) are the smallest known representatives of their respective families. They also comment on the fact that the forms they described from the Upper Cretaceous, which are all from the Northern Hemisphere, have descendants that today are entirely or mainly confined to the Southern Hemisphere, especially to Australia, Asia, and South America.

Our knowledge of the Trichoptera from Baltic amber stems from the monograph by Ulmer (1912) who cited 152 species distributed among 56 genera and 12 families. Species have been described in the following families and genera: Beraeidae (*Bereodes*); Brachycentridae (*Brachycentrus*); Calamoceratidae (*Ganonema*, *Georgium*); Ecnomidae (*Archaeoti-*

nodes); Glossosomatidae (*Electragapetus, Glossosoma*); Goeridae (*Goera, Lithax, Silo*); Helicopsychidae (*Electrohelicopsyche, Helicopsyche, Palaeohelicopsyche*); Hydropsychidae (*Diplectrona, Electrodiplectrona, Hydropsyche, Potamyia*); Hydroptilidae (*Agraylea, Allotrichia, Electrotrichia, Hydroptila, Palaeagapetus*); Lepidostomatidae (*Archaeocrunoecia, Electraulax, Electrocrunoecia, Maniconeurodes, Palaeocrunoecia, Palaeolepidostoma*); Leptoceridae (*Erotesis, Setodes, Triplectides*); Molannidae (*Molanna, Molannodes*); Odontoceridae (*Electrocerum, Electropsilotes, Marilia, Plecophlebus*); Philopotamidae (*Dolophilus, Electracanthinus, Philopotamus, Ulmerodina, Wormaldia*); Phryganeidae (*Phryganea*); Polycentropodidae (*Archaeoneureclipsis, Cyrnus, Holocentropus, Neureclipsus, Nyctiophylax, Nyctiophylacodes, Phylocentropus, Plectrocnemia, Polycentropus*); Psychomyidae (*Lype*); Rhyacophilidae (*Rhyacophila*); Sericostomatidae (*Aulacomyia, Pseudoberaeodes, Sphaleropalpus, Stenoptilomyia*); Stenopsychidae (*Stenopsyche*) (Ulmer, 1912; Wichard, 1986A; Spahr, 1989).

The relatively large number of caddis flies in Baltic amber (more than twice as many as the Lepidoptera and approximately 5.6 percent of the total Königsberg collection) supports the concept of humid, subtropical conditions in the Baltic amber forest, with an adequate supply of both still and running waters. Furthermore, Ulmer (1912) noted a difference between the distribution of Baltic amber and present-day caddis flies: the Limnephilidae, to which a quarter of the present-day caddis flies belong, is not known at all from Baltic amber. He added that then, as today, representatives of the Limnephilidae may have avoided a subtropical climate. He also noted that 44 percent of all the species (67 percent of all the individuals) found in Baltic amber belong to the Polycentropidae, a family that at present constitutes only about 6 percent of the species in the order and is probably in decline. Larsson (1978), however, commented that Ulmer was comparing a predominantly subtropical fauna (Baltic forms) with the worldwide multiclimatic forms of today.

The Trichoptera in Dominican amber have been investigated by Wichard (1981, 1983A, 1983B, 1985, 1986B, 1987, 1989) and Wells and Wichard (1989). They include species or representatives in the following families and genera: Atopsychidae (*Atopsyche*); Glossosomatidae (*Campsiophora, Cubanoptila*) (Fig. 85); Helicopsychidae (*Helicopsyche*); Hydropsychidae (*Leptonema, Palaehydropsyche*); Hydroptilidae (*Alisotrichia, Leucotrichia, Ochrotrichia, Oxyethira*); Leptoceridae (*Nectopsyche*); Philopotamidae (*Chimarra*); Polycentropodidae (*Antillopsyche, Cernotina*)

Figure 85. Holotype of the caddis fly *Cubanoptila poinari* Wichard (family Glossomatidae) in Dominican amber (Poinar collection).

(Fig. 86). Although the caddis flies in these deposits make up less than o.i percent of the total fossils (in contrast to 5.6 percent in the Baltic amber collections), the biogeographical data are interesting (Wichard, 1986B). Both the present-day and the Dominican amber caddis fly fauna contain Nearctic, Neotropical, and Ethiopian elements. At present, almost half of the amber forms belong to *Antillopsyche oliveri* Wichard, a genus presently restricted to the Antilles and the Ethiopian and Oriental regions. Wichard (1987) considers *Antillopsyche* to be a relict from the Mesozoic when Africa was in contact with South America. Another half of the specimens, constituting the vast majority of the genera occurring

Figure 86. A caddis fly, *Antillopsyche oliveri* Wichard (family Polycentropodidae), in Dominican amber (Poinar collection).

in Dominican amber, are Neotropical forms that immigrated to the Greater Antilles from South America. The single species *Palaehydropsyche fossilis* Wichard, which belongs to the hydropsyche branch of the Hydropsychidae, is considered to be of Nearctic origin and a late (Oligocene-Miocene) colonizer of the Greater Antilles.

Botosaneanu (1981) investigated Trichoptera in Burmese amber and described *Burminoptila bemeneha*, considered to be the most primitive representative of the subfamily Hydroptilinae of the family Hydroptilidae. He also reported that the specimen described by Cockerell (1917D) as a new genus and species of caddis fly from Burmese amber, *Plecophlebus nebulosus*, is a member of the Homoptera. Unidentified caddis flies also occur in Mexican amber (Appendix A).

Lepidoptera (Butterflies and Moths)

This large order is characterized by adults that possess scales on all four wings and have mouthparts, a proboscis, normally adapted for sucking up nectar. The larvae, many of which are commonly called caterpillars, feed mostly on plants although some feed on animal and plant products. Microlepidoptera (small lepidopterans) mainly have been re-ported from amber. Early studies by Rebel (1934, 1935) and Kuznetsov (1941) on moths in Baltic amber have resulted in most of the described amber species. More recently, Skalski (1976, 1977, 1985) has made contri-butions to this area.

Micropterigidae (mandibulate moths). Some early fossil records of unquestionable Lepidoptera (Paratrichoptera from the Triassic, de-scribed by Riek [1976], are disputed Lepidoptera) consist of five moth specimens from Lebanese amber (Whalley, 1977, 1978). Three of these specimens, described as *Parasabatinca aftimacrai* Whalley, closely re-semble the present-day genus *Sabatinca* of the family Micropterigidae, which is generally regarded as the most primitive family of Lepidoptera. Adults in this family possess mandibles instead of a proboscis, and present-day representatives feed on pollen and spores. Whalley (1981) supposed that the fossil forms fed on spores because pollen has not been recovered from Lebanese amber. The larvae of extant species feed on mosses and liverworts. Whalley (1978) reexamined the specimens of *Mi-cropteryx pervetus* Cockerell (1919) from Burmese amber and *Micropteryx proavitella* Rebel (1935) from Baltic amber, and concluded that both spe-cies belong to the genus *Sabatinca*. This genus is presently restricted to Australia, New Zealand, and New Caledonia, and some species possess a primitive venation very similar to that of the Trichopteran *Rhyacophila*.

Previous to Whalley's report (1977), early fossil records of Lepidop-tera were a larval head capsule in Canadian amber (MacKay, 1970), mi-cropterigid scales from French amber (Kühne et al, 1973), and the mi-cropterigid *Undopterix sukatshevae* Skalski (1985) from Siberian amber.

Eriocraniidae. These small moths resemble micropterigids but the adults lack mandibles. Cockerell (1919) described *Dyseriocrania perveta* from Burmese amber and Kuznetsov (1941) described *Electrocrania im-mensipalpa* from Baltic amber.

Incurvariidae. Females in this family of small moths contain a pierc-ing ovipositor. Some larvae in this family are leaf miners when young, then build protective cases when older and exposed to predators. The case is carried around as the larva feeds. The earliest fossil record of this family is represented by two specimens, in Lebanese amber, that were placed by Whalley (1978) in the genus *Incurvarites*. The only other fossil representatives of this family in amber are two species of *Prophalonia* and *Incurvarites alienella* Rebel from Baltic amber (Keilbach, 1982).

Tineidae (clothes moths and others). The tineids are mostly small moths with larvae that often build protective cases during all or a portion of their development. Representatives of the genera *Adelites, Architinea, Dysmasiites, Electromeessia, Glessocardia, Martynea, Monopibaltia, Palaeoinfurcitinea, Palaeoscardiites, Palaeotinea, Proscardiites, Pseudocephitinea, Scardiites, Simulotinea, Tillyardinea, Tinea, Tineites, Tineolamima,* and *Tineosemopsis* have been described from Baltic amber (Keilbach, 1982; Kozlov, 1987). Bachofen-Echt (1949) provides a photograph of *Tineolamina aurella* Rebel. Representatives of this family also occur in Dominican amber (Appendix B).

Psychidae (bagworm moths). The bagworm moths are so named because of the characteristic cases made and carried about by the larvae of this family. The females are wingless and legless, and pupate, mate, and oviposit in the case. Although no adults or larvae have been recovered, the presence of the characteristic cases in Baltic amber is well documented; three are illustrated in Bachofen-Echt (1949). On the basis of one of these cases, Rebel (1934) described ?*Sterrhopteryx pristinella.*

Lyonetiidae. The larvae of these small, narrow-winged moths are either leaf miners or web spinners. A single species, *Prolyonetia cockerelli* Kusnezov, has been described from Baltic amber (Skalski, 1976).

Oecophoridae. The oecophorids are small to medium-sized moths whose larvae often spin webs on plant leaves or occur in decaying vegetation or under bark, where they may feed on fungi (Rawlins, 1984). Although recorded from Mexican amber (Appendix A), only forms from Baltic amber have been described. These include species of the genera *Borkhausenites, Depressarites, Epiborkhausenites, Glesseumeyickia, Hofmannophila, Microsymmocites, Neoborkhausenites, Palaeodepressaria, Paraborkhausenites, Schiffermuelleria, Symmocites,* and *Tubuliferola* (Keilbach, 1982; Larsson, 1978; Skalski, 1985; Spahr, 1989).

Elachistidae (grass miner moths). These small, lanceolate-winged moths are mostly found in meadows where the leaf-mining larvae make blotchy holes in grasses. Baltic amber species were described in the genera *Anybia, Elachistites, Microperittia, Palaeoelachista,* and *Praemendesia* (Rebel, 1934; Kozlov, 1987).

Scythrididae and Symmocidae. A scythriid, *Scythropites balticella* Rebel (1936), and a representative of the closely related Symmocidae, *Oego-*

coniites borisjaki Kuznetsov (1941), were described from Baltic amber. Representatives of the related Blastobasidae occur in Dominican amber (Appendix B).

Cosmopterigidae, Gelechiidae, and Plutellidae. Representatives of the Cosmopterigidae and Gelechiidae are known from Dominican and Mexican amber, and a representative of the latter family has been reported from Baltic amber (Kuznetsov, 1941; Spahr, 1989). The Plutellidae is possibly represented by two larvae in Baltic amber (MacKay, 1969).

Yponomeutidae (ermine moths). These small, often brightly colored moths include leaf miners and web spinners usually found on trees. Working with Baltic amber, Rebel (1934, 1935) described four species of *Epinomeuta* and, in the closely related family Argyresthiidae, two species of *Argyresthites*.

Tortricidae. In this large family of small moths, the larvae are commonly known as leaf rollers, although they also occur in fruit, attack buds, and so on. Although common now, only species of *Electresia, Spatalistiforma, Tortricibaltia, Tortricidrosis,* and *Tortrix* have been described from Baltic amber (Keilbach, 1982; Spahr, 1989). One reason so few tortricids have been found in amber is that they avoided becoming entangled in the resin; they were not poorly represented in the amber forests. Indeed, Skalski (1975) commented that while examining a present-day biotype where the tortricids were one of the most common families of the Microlepidoptera, only a single example could be found in tree resin. Thus, representation in amber depended on the behavior of the insects.

Other Lepidoptera. A single representative of the Pyralidae, *Glendotricha olgae,* was described from Baltic amber by Kuznetsov (1941). Skalski (1975) figures a representative of the Chrysoesthiidae (= Heliodinidae) and, in 1976, a *Homoneura* sp. from Siberian amber. The latter genus is also represented in French and Middle East ambers (Spahr, 1989).

Bachofen-Echt (1949) cites some Macrolepidoptera from Baltic amber. These include representatives of the Sphingidae (*Sesia, Sphinx*), Arctiidae (*Arctia*), Noctuidae (*Triphaena*), Papilionidae (*Papilio*), and Lycaenidae (*Lycaenites*). Other reports of Lepidoptera in amber include species of *Ectoedemia* and *Johanssonia* (Nepticulidae), *Gracillariites* (Gracillariidae), and *Adela* and *Adelites* (Adelidae) in Baltic amber; also reported are representatives of the Ethmiidae and Walshiidae in Mexican amber and a Mnesarchaeidae in Siberian amber (Spahr, 1989). An adult gracillarid and an adult tineid, both with attached parasitic erythraeid mites, were reported from Dominican amber (Poinar et al., 1991), and adult Lycaenidae and/or Riodinidae have been recovered from Dominican amber (Color Plate 7).

Diptera (True Flies)

The Diptera represent the second largest order of insects (after the Coleoptera) with some 120,000 species worldwide. All are characterized by having only a single anterior pair of functional wings (when wings are present). The hind wings are reduced to small knoblike balancing organs called halteres. Of all the other insects, only some male scales (Homoptera) also have their hind wings reduced to halterlike structures. The mouthparts of adult flies are variable, ranging from piercing-sucking to sponging-lapping. Those of the larvae range from chewing (with mandibles) to rasping (with mouth hooks). The larvae are usually legless and maggotlike and may or may not have a distinct head capsule. Larvae in the suborder Nematocera normally have a well-developed head with laterally chewing mandibles, whereas those of the suborder Brachycera have a sclerotized head in which vertically moving mouth hooks have replaced the mandibles. In the order Cyclorrhapha, the head lacks sclerotization and the mouth hooks are often the only distinct feature of the head.

Developing stages of Diptera occur in a variety of terrestrial and aquatic habitats. Most are detrivores and scavengers, but many are herbivores, predators and parasites. Adult flies feed on fluids obtained from plant and animal sources. Many of the blood-sucking forms are medically important pests of man, both as parasites and vectors of pathogens.

The oldest known Diptera occur in the Early Triassic of France. They include representatives of the extinct suborder Archidiptera and the extant suborder Nematocera (Crowson et al., 1967). Rohdendorf (1964) feels that because of their specialized wing venation, all of these Triassic flies may belong to a separate extinct suborder. The oldest known representatives of the suborder Brachycera were found in Jurassic deposits in Germany and the oldest known Cyclorrhapha (a member of the Chloropidae) was reported from Lebanese amber.

In the following treatment of Diptera from amber, the families are arranged alphabetically under each of the three suborders, the Nematocera, Brachycera, and Cyclorrhapha. For detailed nomenclatural information, the works of Keilbach (1982) and Spahr (1985) can be consulted. Dipterous families of rare occurrence in amber are listed in Table 10.

Suborder Nematocera (long-horned flies). So named because of the usually long, many-segmented antennae, these slender-legged flies breed mostly in water or moist habitats. They include the midges and fungus gnats, which are some of the most common insects found in amber. A list of Baltic amber Nematocera in the Warsaw Earth Museum has been prepared by Kulicka et al. (1985).

Anisopodidae (wood gnats). The adults of this family occur in moist habitats, and often appear in large swarms. The larvae occur in decaying organic matter and in fermenting sap. Amber forms include species of *Sylvicola* (= *Rhyphus*) and *Mycetobia* in Baltic amber (Spahr, 1985) and *Caloneura plectiles* Hong (1981) from Chinese amber. The family is also represented in Canadian (McAlpine and Martin, 1969), Mexican, and Dominican amber (Appendices A, B) (Grimaldi, 1991).

Bibionidae (March flies). Named because of their appearance in early spring, adult March flies are of small to medium size, and have short antennae. They can be abundant during certain periods and may occur in large swarms. The larvae feed on plant roots and decaying vegetation.

The oldest known species, *Plecia myersi*, was described by Peterson (1975) from Canadian amber. Some species of *Chrysothemis, Electra, Penthetria,* and *Plecia* have been described from Baltic amber (Keilbach, 1982;Spahr, 1985); the first two genera are placed by Spahr in the family Rachiceridae. Upon describing *Plecia pristina* from Mexican amber, Hardy (1971) commented that this species closely resembled modern-day forms and that the genus has exhibited little or no apparent evolutionary change over the past millions of years. Bibionids also occur in Siberian amber (Zherichin and Sukacheva, 1973) and in Dominican amber (Appendix B).

Cecidomyiidae (gall midges; also Lestremiidae). These small, fragile-appearing flies are characterized by relatively long antennae and legs and by reduced wing venation. The larvae occur in plants, where they can cause gall formation on the leaves and stems; others are fungivores or detrivores.

What may be the earliest fossil record of this family was recorded from Lebanese amber (Schlee and Dietrich, 1970). Other Cretaceous amber records of gall midges include three species described by Gagné (1977) from Canadian amber—*Cretowinnertzia angustala, Cretocatocha*

Table 10. Diptera of rare occurrence in amber

Family	Taxon or taxa	Amber source	Reference
Anthomyiidae	Anthomyia sp.	Baltic	Spahr, 1985
Asteiidae	Succinasteia carpenteri Hennig	Baltic	Keilbach, 1982
Aulacigastridae	Protaulaciagaster electrica Hennig	Baltic[a]	Hennig, 1965
Camillidae	Protocamilla succini Hennig	Baltic	Hennig, 1965
Carnidae	Meoneurites enigmaticus Hennig	Baltic[a]	Hennig, 1965
Chamaemyiidae	Procremifania electrica Hennig	Baltic	Hennig, 1965
Chironomapteridae	none specified	Siberian	Zherichin, 1978
Chyromyidae	Gephyromyiella electrica Hennig	Baltic	Hennig, 1965
Clusiidae	Electroclusiodes meunieri (Hendel), E. radiospinosa Hennig	Baltic[a]	Hennig, 1965
Conopidae	Palaeomyopa tertiaria Meunier	Baltic	Keilbach, 1982
Cryptochetidae	Phanerochaetum tuxeni Hennig	Baltic	Hennig, 1965
Cylindrotomidae	Cylindrotoma succini Loew	Baltic	Spahr, 1985
Cypselosomatidae	Cypselosomatites succini Hennig	Baltic	Hennig, 1965
Diadocidiidae	Diadocidia sp.	Baltic	Spahr, 1985
Diastatidae	Pareuthychaeta electra Hennig, P. minuta (Meunier)	Baltic	Hennig, 1965
Ditomyiidae	Symmerus balticus Edwards	Baltic	Spahr, 1985
Dryomyzidae	Palaeotimia lhoesti Meunier, Prodryomyza electrica Hennig	Baltic	Hennig, 1965
Ironomyiidae	Cretonomyia pristina McAlpine	Canadian	McAlpine, 1973
Megamerinidae	Palaeotanypeza spinosa Meunier	Baltic	Hennig, 1965
Neurochaetidae	Anthoclusia gephyrea Hennig, A. remotinervia Hennig	Baltic	Spahr, 1989

Family	Species	Amber	Reference
Odiniidae	*Protodinia electrica* Hennig	Baltic[a]	Hennig, 1965
Oestridae	*Novoberendtia baltica* (Townsend)	Baltic	Spahr, 1985
Otitidae	none specified	Baltic	Spahr, 1985
Pachyneuridae	none specified	Dominican	Schlee and Glockner, 1978; Appendix B herein
Pallopteridae	*Morgea freidbergi* McAlpine, *Morgea mcalpinei* Hennig, *Glaesolonchaea electrica* Hennig, *Pallopterites electrica* Hennig	Baltic	Hennig, 1967; McAlpine, 1981
Periscelidae	*Periscelis annectans* Sturtevant	Mexican[a]	Sturtevant, 1963
Phyllomyzidae	none specified	Mexican	Appendix A herein
Platystomatidae	*Scholastes foordi* Cockerell	Baltic (British)	Spahr, 1985
Pleciomimidae	none specified	Siberian	Spahr, 1985
Proneottiophilidae	*Proneottiophilum extinctum* Hennig	Baltic	Spahr, 1985
Pseudopomyzidae	*Eopseudopomyza kuehnei* Hennig	Baltic	Keilbach, 1982
Psilidae	*Electrochyliza succini* Hennig	Baltic	Hennig, 1965
Rhagionempididae	none specified	Soviet	Spahr, 1985
Sarcophagidae	*Sarcophaga* sp.	Baltic	Spahr, 1985
Scatophagidae	*Scatophaga* sp.	Baltic	Spahr, 1985
Sepsidae	*Protorygma electricum* Hennig	Baltic	Hennig, 1965
Sphaeroceridae	*Copromyza* sp.	Baltic[a]	Spahr, 1985
Tanyderidae	*Macrochile spectrum* Loew (extinct genus)	Baltic	Spahr, 1985
Tephritidae	none specified	Dominican	Appendix B herein
Tethinidae	none specified	Mexican	Appendix A herein

[a]Family also represented in Dominican amber; see Appendix B herein.

mcalpinei, and *Cretocordylomyia quadriseries*—and undescribed forms in Siberian and French amber (Zherichin & Sukacheva, 1973; Schlüter, 1978A).

Most of the Cecidomyiidae recorded from amber are from Baltic deposits. These include representatives or species of the genera *Asynapta*, *Bryocrypta*, *Camptomyia*, *Campylomyza*, *Cecidomyia*, *Colomyia*, *Colpodia*, *Dasineura*, *Dirhiza*, *Epidosis*, *Joannisia*, *Ledomyiella*, *Lestrema*, *Meunieria*, *Miastor*, *Monardia*, *Monodicrana*, *Oligotrophus*, *Palaeocolpodia*, *Palaeospaniocera*, *Ruebsaamenia*, and *Winnertzia* (Spahr, 1985; Keilbach, 1982).

Cockerell (1917A) described *Winnertziola burmitica* from Burmese amber and Gagné (1973) described representatives of *Lestodiplosis*, *Henria*, *Contarinia*, *Clinodiplosis*, *Bremia*, *Phaenolautia*, and *Monardia* from Mexican amber. The family is also represented in Dominican amber (Appendix B) and in Sicilian and Austrian amber (Guérin, 1838; Schlee, 1984).

Ceratopogonidae (biting midges). The adults of this family tend to be minute, stoutish flies that hold the wings flat over the abdomen at rest, in contrast to the chironomids, which hold them rooflike at rest. The adults suck blood from insects and vertebrates, and can inflict a painful bite. The larvae are detrivores in aquatic and semiaquatic habitats. Biting midges are commonly found in amber and, for unknown reasons, females are found at a higher rate than males.

The earliest Ceratopogonidae in amber is reported from Lebanese amber (Schlee and Dietrich, 1970), and this is also the oldest undoubted member of the Ceratopogonidae from any source (Szadziewski, 1988). Boesel (in Carpenter et al., 1937) described species of *Atrichopogon*, *Ceratopogon*, *Dasyhelea*, *Lasiohelea*, and *Protoculicoides* from Canadian amber, and Remm (1976) described species of *Atriculicoides*, *Baeohelea*, *Ceratopogon*, *Culicoides*, and *Leptohelea* from Siberian amber. Also reported from Cretaceous amber are representatives of *Forcipomyia* in Canadian deposits (Downes and Wirth, 1981), *Culicoides casei* in New Jersey amber (Grogan and Szadziewski, 1988), and undescribed forms in Alaskan (Langenheim et al., 1960) and French (Schlüter, 1978A) material. A study of the Ceratopogonidae from Baltic amber revealed 101 species distributed among the following 24 genera: *Alluaudomyia*, *Atrichopogon*, *Bezzia*, *Brachypogon*, *Ceratoculicoides*, *Ceratopalpomyia*, *Ceratopogon*, *Culicoides*, *Dasyhelea*, *Eohelea*, *Forcipomyia*, *Fossihelea*, *Gedanohelea*, *Leptoconops*, *Mantohelea*, *Meunierohelea*, *Monohelea*, *Nannohelea*, *Neurohelea*, *Palpomyia*, *Physohelea*, *Serromyia*, *Stilobezzia* and *Wirthohelea* (Szadziewski, 1988). Females of *Eohelea sinuosa* (Meunier) show an interesting wing modification not known on extant forms. This consists of a series of transverse bars on the wing tips that has been interpreted as a stridulatory organ. A honeycomb type of stridulating organ also occurs on *E. pe-trunkevitchi* Szadziewski (1988) from Baltic amber.

Other Tertiary biting midges in amber include *Ceratopogon* sp. from Sicilian amber, *Johannsenomyia swinhoei* Cockerell from Burmese amber, *Palpomyia unca* Hong (1981) from Chinese amber, *Heteromyia* sp. from Dominican amber (G. Gradhaus, personal communication), and unidentified forms in Mexican amber (Spahr, 1985) (Appendix A).

Chaoboridae (phantom midges). These flies resemble mosquitoes, but possess a short proboscis and do not take blood. The aquatic larvae are transparent and phantomlike, and are well known for their predaceous habits. Although phantom midges are not common in amber, representatives have been reported from Lebanese (Whalley, 1981) and Siberian Cretaceous amber (Zherichin and Sukacheva, 1973). Species of *Mochlonyx* and *Corethra* have been described from Baltic amber (Keilbach, 1982), *Trichia gracilis* Hong (1981) from Chinese amber, and a *Chaoborus* sp. from Burmese amber (Spahr, 1985). The family is also represented in Dominican deposits (Appendix B).

Chironomidae (true midges). These abundant small, fragile flies are often confused with mosquitoes. They do not bite, however, and the males usually differ from females in having more plumose antennae. The adults commonly swarm during mating, often emitting an audible sound. Although most midge larvae are aquatic scavengers, some are predaceous and others feed on living plants or decaying vegetable matter.

Midges are represented in all major fossiliferous amber deposits, including Lebanese, New Jersey, Canadian, Siberian, Burmese, Baltic, Mexican, Dominican, and Chinese. The great majority of described amber midges occur in Baltic deposits and have been placed by Meunier (1904, 1916) in the genera *Camptocladius, Chironomus, Cricotopus, Cricotopiella, Euhycnemus, Palaeotanypus, Tanypus,* and *Tanytarsus.* The earlier descriptions of Loew (1850) are considered incomplete. Other Baltic amber midges occur in the genera *Sendelia* and *Jentzschiella* (Spahr, 1985).

Other Tertiary amber forms include species of *Aspinus, Fushunitendipes,* and *Microtendipes* from Chinese amber (Hong, 1981), representatives of Orthocladiinae, Chironominae, and Tanypodinae from Dominican amber, and undescribed forms from Burmese and Mexican amber (Spahr, 1985). Two new genera and species, *Cretodiamesa taimyrica* (Fig. 87) and *Electrotenia brundini,* were described by Kalugina (1976,

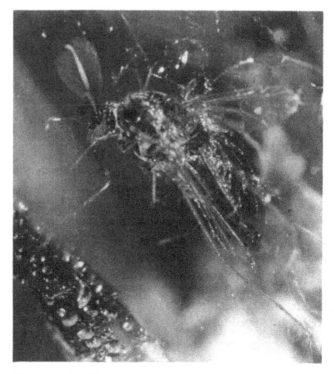

Figure 87. A midge, *Cretodiamesa taimyrica* Kalugina (family Chironomidae), representing one of the most abundant insect families in Siberian amber (photo by A. P. Rasnitsyn, courtesy of V. V. Zherichin).

1980) from Siberian amber, and Boesel (see Carpenter et al., 1937) described *Spaniotoma conservata*, *Smittia vetus*, and *Metriocnemus cretatus* from Canadian amber. The oldest known fossil amber midge is *Libanochlites neocomicus* Brundin (1976) from Lebanese amber.

Midges have provided evidence of paleosymbiosis in the form of phoretic and parasitic associations with mites and nematodes (see Chapter 5).

Culicidae (mosquitoes). The females of this widespread family are well known for their blood-sucking character and ability to vector some of man's most deadly pathogens. The adults are small, fragile insects, and like midges, male mosquitoes tend to have more plumose antennae than do females. Both sexes have a proboscis, which the males (and sometimes females) use for obtaining plant juices, but the females are the blood suckers. Mosquito larvae are able to develop in a wide range of aquatic habitats, such as tree holes, ponds, and marshes, where they feed on organic debris.

Mosquitoes are exceedingly rare in amber and some of the early reports of mosquitoes in Baltic amber were actually chaoborid midges (Larsson, 1978). A culicid has been found by T. Pike in Canadian amber and *Culex* species have been noted in Baltic amber (Spahr, 1985). The Baltic amber record establishes the Culicidae in the Late Eocene, which parallels the earliest fossil record of this group (in Eocene Green River Wyoming deposits) (Edwards, 1923). Keilbach (1982) also cites a record of a male and female of *Culex pipiens* Linnaeus from Baltic amber. The only other amber mosquitoes have been reported from Mexican and Dominican amber (Fig. 88, Appendices A, B). The fossil mermithid nematode

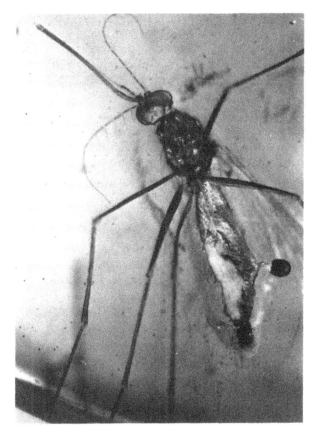

Figure 88. A mosquito (family Culicidae) resembling the genus *Culex* in Dominican amber (Poinar collection).

Heydenius dominicus Poinar (1984B) was suspected of parasitizing a mosquito in Dominican amber.

Dixidae (dixid midges). These midges resemble mosquitoes but do not bite and lack wing scales. The aquatic larvae feed at the surface of the water and move by bending and straightening their body. Baltic amber species in the genera *Dixa* and *Paradixa* are the only descriptions of this group in amber (Keilbach, 1982). The family is also represented in Dominican amber (Appendix B).

Mycetophilidae (fungus gnats; sometimes known as Fungivoridae). Fungus gnats are small, delicate insects with long legs, a result partly of their elongated coxae. The larvae occur in damp habitats where they feed on decaying vegetation, fungi, and occasionally plant parts; a few are predaceous on other insects. The adults, which in nature occur near the soil or in the undergrowth, can become numerous in greenhouses and are often considered as pests. Their large numbers in amber indicate their abundance, preference to breeding under decaying bark, and weak flight habits (Color Plate 5). Larsson (1978) suggests that the fresh resin served as an attractant for the adults.

Cretaceous mycetophilids have been reported from Canadian amber (McAlpine and Martin, 1969), Siberian amber (Zherichin and Sukacheva, 1973), and New Jersey amber (Grimaldi, personal communication). In 1981, Matile, examining French amber, described *Schlueterimyia cenomanica*, a member of the subfamily Keroplatinae, which is considered by Spahr (1985) to be a separate family.

Some 150 species have been described from Baltic amber, the great majority by Meunier (1904) and Loew (1850). These fall into the genera *Acnemia, Allodia, Anaclileia, Anaclinia, Archaeboletina, Asindulum, Azana, Boletina, Coelosia, Cordyla, Dianepsia, Docosia, Dziedzickia, Ectrepesthoneura, Empalia, Hesperodes, Kelneria, Keroplatus, Leia, Loewiella, Macrocera, Manota, Mycetophila, Mycomya, Mycothera, Palaeoanaclinia, Palaeoboletina, Palaeodocosia, Palaeognoriste, Palaeoplatyura, Palaeosynapha, Phronia, Platyura, Polylepta, Proanaclinia, Proboletina, Proceroplatus, Proneoglaphyroptera, Rondaniella, Rymosia, Sciobia, Sciomorpha, Sciophila, Scudderiella, Synplasta, Syntemna, Tetragoneura,* and *Trichonta* (Spahr, 1985; Keilbach, 1982). Hong et al. (1974) and Hong (1981) described species of *Boletina* and *Eosciophila* from Chinese amber; Gagné (1980) described *Manota* sp. and identified representatives of Mycetophilinae from Mexican amber; Cockerell (1917D) described *Burmacrocera petiolata* from Burmese amber; and Schmalfuss (1979), in what is the first fossil record of the extant genus *Proceroplatus*, described *Proceroplatus hennigi* from Dominican amber.

Psychodidae (moth and sand flies). Representatives of this family are small, slow-flying forms that are covered with noticeable hairs; some (the Psychodinae) hold their wings rooflike over their back. The larvae feed on decaying vegetable matter in moist habitats and the adults are found in similar habitats. Adults in the subfamily Phlebotominae suck the blood of vertebrates and are known to vector several disease-causing pathogens of humans (Color Plate 8).

The earliest known moth flies are *Phlebotomites brevifilis* and *P. longifilis* described by Hennig (1972) from Lebanese amber (Fig. 89). The author remarked that these forms do not differ significantly from present-day representatives of the family, which is also represented in Siberian amber (Zherichin and Sukacheva, 1973). Baltic amber species were described in the genera *Eatonisca, Nemopalpus, Pericoma, Phalaenomyia, Phlebotomus, Posthon, Psychoda, Sycorax,* and *Trichomyia* (Spahr, 1985). Present-day species of *Trichomyia*, the most numerous species in amber, breed in rotted wood, and this may account for their relative frequency in

Figure 89. A sand fly (family Psychodidae) in Lebanese amber. (Courtesy of Paul Whalley and the British Museum of Natural History.)

amber. Larvae of *Pericoma* and *Psychoda* are amphibian, suggesting the presence of standing water in or near the amber forest. The blood-sucking species of *Phlebotomus* probably fed on mammals or reptiles in the amber forest.

Quate (1961, 1963) described species of *Trichomyia, Phlebotomus* (= *Lutzomyia*), *Telmatoscopus* (Fig. 90), *Philosepedon, Psychoda,* and *Brunettia* from Mexican amber. He commented that, in general, all of the fossils had a modern appearance and there had been no major evolutionary changes during the past 30 million years in the examined forms. Species of *Trichomyia* predominate in Mexican amber as they do generally in amber. Quate considered *Trichomyia* as an ancient genus that flourished during the Tertiary and has apparently declined, although today it is found in all regions of the world except the Orient. The genus is currently represented by 22 described species, and Quate believed that the biased representation of *Trichomyia* in amber is due to their habits of association with tree bark and sap. At present, *Brunettia* is primarily an Indo-Malayan genus, although two species are known from North America. It has not been collected in the Neotropical region indicating that *Brunettia* was more widespread during the Tertiary than at present (Quate, 1961). Cockerell (1917B, 1920B) described *Trichomyia swinhoei* and *Eophlebotomus connectens* from Burmese amber, and Schlüter (1978B) described *Nemopalpus hennigianus* from Dominican amber.

Scatopsidae (scavenger flies). These minute, dark flies have a stoutish body and short antennae. The larvae are scavengers and develop in rotting organic matter and excrement. Originally placed with the bibionids, scatopsids differ in size and preference for developmental sites. The adults of some species share with the bibionids the behavior of remaining in copula for relatively long periods (several hours) and mating pairs can be found in amber (Fig. 91). Representatives of this family occurring in Canadian (McAlpine and Martin, 1969) and Siberian (Zherichin and Sukacheva, 1973) Cretaceous amber are the earliest known fossil scatopsids. Described forms are limited to species of *Scatopse* from Baltic amber (Keilbach, 1982) and species of *Scatopse, Swammerdamella,* and *Procolobostema* from Mexican amber (Cook, 1971). The Mexican species represent the first described fossil scatopsids from the Western Hemisphere. Cook later identified some copulating pairs in Dominican amber as belonging to the Neotropical subgenus *Neorhegmoclemina* of the genus *Rhegmoclemina* (Boucot, 1990).

Sciaridae (dark-winged fungus gnats). This family is biologically and morphologically similar to the Mycetophilidae. Adult sciarids, however,

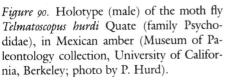

Figure 90. Holotype (male) of the moth fly *Telmatoscopus hurdi* Quate (family Psychodidae), in Mexican amber (Museum of Paleontology collection, University of California, Berkeley; photo by P. Hurd).

Figure 91. A mating pair of scavenger flies (family Scatopsidae) in Dominican amber.

are distinguished by two features: the dorsal portion of their eyes extend toward and almost touch each other (except in *Pnyxia*) and the wing cross vein designated the r-m is not recognizable as a separate connecting vein as it is in the Mycetophilidae.

The sciarids are a common group, and because of their habits, they are abundantly represented in amber (Fig. 92). The family is represented in Cretaceous Lebanese amber (Whalley, 1981), Canadian amber (McAlpine and Martin, 1969), and Siberian amber (Zherichin and Sukacheva, 1973). Species of *Bradysia, Corynoptera, Heterotricha, Lycoriella, Neosciara, Pseudosciara, Ruebsaameniella, Sciara, Trichosia,* and *Zygoneura* have been described from Baltic amber (Spahr, 1985). Hong (1981) described species of *Sciara* (= *Lycoria*) from Chinese amber, Cockerell (1917B) described *Sciara burmitina* from Burmese amber, and Gagné (1980) identified a *Bradysia* sp. in Mexican amber. The group is also represented in Dominican amber (Appendix B).

Simuliidae (blackflies). These small, black flies are stoutish and have a bent-over appearance. They possess broad wings and short legs and antennae. Most female blackflies are bloodsuckers and some species are

Figure 92. A fungus gnat (family Sciaridae) in Baltic amber (photo by G. Brovad). (Published in *Baltic Amber: A Palaeobiological Study,* by S. Larsson. Scandinavian Science Press Ltd.)

known to transmit vertebrate parasites. All blackfly larvae are aquatic, live in flowing water, and are able to tolerate a wide temperature range, as shown by their occurrence in the arctic as well as the tropics. The oldest fossil blackflies occur in the middle Jurassic of Central Asia (Crosskey, 1990).

Blackflies are rare in amber, and no mention has been made of this family in Cretaceous deposits, although a *Simulium*-like fly found in Lebanese amber is now being studied (Whalley, personal communication). Five species of *Simulium* have been described from Baltic amber (Spahr, 1985), and Guérin (1838) mentions a *Simulium* sp. from Sicilian amber. According to Crosskey (1990), *S. cerberus* (End.) from Baltic amber really belongs to the North American genus *Ectemnia*. Representatives also occur in Dominican amber (Appendix B).

Tipulidae (crane flies; including Limoniidae and Cylindrotomidae). The crane flies, which constitute the largest family of Diptera, look something like giant mosquitoes or daddy longlegs with wings. Their long legs and elongate body are characteristic, as is their slow, heavy flight pattern. The adults gather in damp, shaded habitats and do not bite. They feed on nectar and other plant products. The larvae occur in aquatic, semiaquatic, or terrestrial situations where they scavenge decaying plant material or attack the roots of living plants.

Crane flies are represented in Cretaceous Lebanese amber (Whalley, 1981), Siberian amber (Limoniinae) (Zherichin and Sukacheva, 1973), Canadian amber (Tipuloidea) (McAlpine and Martin, 1969), and French amber (Schlüter, 1978A). Species in the genera *Limonia, Macalpina,* and *Trichoneura* have been described from Canadian amber (Krzeminski and Teskey, 1987). Described species or representatives of the genera *Adelphomyia, Allarithmia, Austrolimnophila, Brachypremna, Calobamon, Ceratocheilus, Critoneura, Cylindrotoma, Dasymolophilus, Dicranomyia, Dicranoptycha, Electrolabis, Elephantomyia, Eriocera, Erioptera, Gnophomyia, Gonomyia, Helius, Hexatoma, Limnophila, Limonia, Macromastix, Ormosia, Palaeoerioptera, Palaeopoecilostola, Pilaria, Polymera, Pseudolimnophila, Rhabdomastix, Styringomyia, Tanymera, Tanysphyra, Thaumastoptera, Tipula, Trentepohlia, Trichoneura,* and *Tricyphona* occur in Baltic amber (Spahr, 1985; Keilbach, 1982). Most of the surviving genera in the above list are Holarctic in distribution. However, *Limonia, Dicranoptycha, Pseudolimnophila,* and *Adelphomyia* are also Ethopian, *Eriocera* is pantropic, and *Styringomyia, Trentepohlia,* and *Ceratocheilus* are palaeotropic (Larsson, 1978). The Baltic amber limoniids in the collection of the Warsaw Earth Museum were studied by Krzeminski (1985A), and *Pseudolimnophila siciliana* Krzeminski and Skalski (1983) was described from Sicilian amber. Descriptions of Limoniinae in Bitterfeld amber include species in the genera *Cheilotrichia, Dicranomyia, Palaeopoecilostola, Rhabdomastix, Tanysphyra,* and *Trichoneura* (Schumann and Wendt, 1989). Crane flies have been reported also from Mexican and Dominican amber (Appendices A, B). Schlee and Glöckner (1978) illustrated a pseudoscorpion attached to the legs of a crane fly in Baltic amber.

Trichoceridae (winter crane flies). These slender, medium-sized flies often can be seen flying on mild winter days. They inhabit dark, humid areas such as caves and cellars, and can occur in large swarms. The larvae feed on decaying plant matter. This small family, which consists of some 100 extant species, is considered by some to be the most primitive family in the Diptera.

Only two species have been described from amber, both from Baltic deposits. The two, *Trichocera antiqua* (Dahl) and *Trichocera primaeva* (Dahl), were originally described in the genus *Oligotrichocera,* now considered a subgenus (Dahl, 1971). A third specimen resembling *T. antiqua* has recently been described by Krzeminski (1985B).

Suborder Brachycera. Adult Brachycera tend to be large, stoutish flies with usually three (fewer than five) antennal segments. If an elongate process protrudes from the terminal segment, it is usually spinelike (a style) and not bristlelike (an arista). Most Brachycera, both as adults and larvae, are predators or parasites on other insects. The larvae have a reduced, partially sclerotized, retractile head containing vertically moving mouthparts. The larva forms a rigid pupal case from which the adult emerges through a T-shaped opening.

Acroceridae (small-headed flies). The adults of this family resemble blackflies in general shape and in having the head bent down under the enlarged pronotum, but the small head and reduced mouthparts are distinctive. The larva develop as internal parasites of spiders and usually pupate nearby, often in the spider's web. Only four described species of amber acrocerids exist, all from Baltic amber. These include *Glaesoncodes completinervis* Hennig, *Eulonchiella eocenica* Meunier, *Prophilopota succinea* Hennig, and *Villalites electrica* Hennig (Spahr, 1985; Keilbach, 1982). Present-day relatives of *P. succinea* can be found in the Neotropics, Mediterranean region, South Africa, North India, and China (Larsson, 1978). The family is also represented in Dominican amber (Appendix B).

Asilidae (robber flies). These are generally large flies with an elongate abdomen. The adults, which are moderately hairy, bear a depression on the top of the head. The predaceous larvae occur in soil and rotting vegetation. *Holopogon pilipes* (Loew) and species of *Asilus* have been described from Baltic amber (Keilbach, 1982), and Schumann (1984) described *Protoloewinella keilbachi* from Bitterfeld amber. Guérin (1838) cited a *Dasypogon* sp. from Sicilian amber, and Eric Fischer (personal communication) identified representatives of the genera *Ommatius* and *Wilcoxia* from Dominican amber (Fig. 93) and *Pilica* from Mexican amber.

Bombyliidae (bee flies). Because of their robust shape, hairy body, and habit of visiting flowers, these flies are often confused with bees. The adults often possess a long proboscis for obtaining nectar and can dart rapidly from plant to plant. The larvae are parasitic on various stages of other insects. The family is represented in Cretaceous Siberian amber by *Proplatypygus rohdendorfi* Zaitzev (1981) and *Procyrtosia sukatshevae* Zaitzev (1987), but most descriptions are from Baltic amber material. These include species of the genera *Amictites*, *Bombylius*, *Glaesamictus*,

Figure 93. A robber fly of the genus *Wilcoxia* (family Asilidae) in Dominican amber (American Museum of Natural History collection).

Palaeoamictus, Paracorsomyza, Proglabellula, Proplatypygus, and *Zarzia* (Spahr, 1985). To date, a single species, *Glabellula kuehnei* Schlüter, (1976), has been described from Dominican amber.

 Dolichopodidae (long-legged flies). The small, bristly adults of this family are frequently colored metallic green or blue-green. They possess an arista and a short, fleshy proboscis, and are relatively long-legged. The adults are predaceous and occur in damp habitats in association with heavy plant growth. The larvae, which also seem to be predaceous in large part, occur in decaying vegetation, under bark, and in water. Their relative abundance in amber may be due to their occurrence under bark and their flight habits.

 Dolichopodids have been reported from Canadian amber (McAlpine and Martin, 1969), and Negrobov (1978) has described *Archichrysotus hennigi, A. minor,* and *Retinitus nervosus* from Siberian amber. Most dolichopodids in amber occur in Baltic material, and have been identified or described in the following genera: *Anepsius, Argyra, Camp-*

sicnemus, Chrysotus, Diaphorus, Dolichopus, Gheynius, Gymnopternus, Hygroceleuthus, Lyroneurus, Medetera, Nematoproctus, Neurigona, Palaeochrysotus, Poecilobothrus, Porphyrops, Prochrysotus, Psilopus, Rhaphium, Thinophilus, Thrypticus, Wheelerenomyia, and *Xiphandrium* (Spahr, 1985).

Negrobov (1976) described *Prosystenus zherichini* in Sakhalin amber from the lower Tertiary, and Hong (1981) described species of *Dolichopus* and *Septocellula* from Chinese amber. This family is also represented in Mexican and Dominican ambers (Appendices A, B).

Empididae (dance flies). The adults of these small to medium-sized flies occur in damp, shady situations and characteristically form swarms in which the individuals collectively move up and down. The larvae occur under bark, in water, or in rotting vegetation. Both adults and larvae are mainly predaceous on other insects. Empidids are common in amber, perhaps because many breed under bark and the adults fall against the resin during their swarming flights.

Two species of dance flies, *Trichinites cretaceus* Hennig (1970) and *Microphorites extinctus* Hennig (1971), have been described from Lebanese amber. *Cretomicrophorus rohdendorfi* Negrobov (1978), *Archiplatypalpus cretaceous* Kovalev (1974), and *Cretoplatypalpus archaeus* Kovalev (1978) were described from Siberian amber. *Ecommocydromia difficilis* Schlüter (1978A) was described from French amber, and records of the family also occur in Canadian (McAlpine and Martin, 1969) and Alaskan amber (Langenheim et al, 1960; Hurd et al., 1958).

Most empidids in Tertiary amber have been described from Baltic material (Fig. 94). These include species or representatives of the following genera: *Brachystoma, Chelifera, Chelipoda, Drapetiella, Drapetis, Dysaletria, Empis, Euhybos, Euthyneuriella, Gloma, Hemerodromia, Hilara, Hybos, Leptopeza, Meghyperiella, Micrempis, Microphorus, Oedalea, Oustaletmyia, Palaeoedalea, Palaeoleptopeza, Palaeoparamesia, Parathalassiella, Phyllodromia, Platypalpus, Ragas, Rhamphomyia, Tachydromia, Tachypeza, Thirza, Timalphes,* and *Trichopeza* (Spahr, 1985). Cockerell (1917B, 1917F) described *Burmitempis halteralis* and *Electrocyrtoma burmanica* from Burmese amber, Hong (1981) described *Lochmocola osterosa* and *Symballophthalmus clavilabrosus* from Chinese amber, and Hennig described a representative of the family from Miocene Austrian amber (Bachmayer and Schulz, 1978). The family is also represented in Dominican and Mexican amber (Appendices A, B).

Rhagionidae (snipe flies). These medium- to large-sized flies generally have an elongated abdomen, longish legs, and spotted wings. The

Figure 94. A dance fly (family Empididae) in Baltic amber (photo by G. Brovad). (Published in *Baltic Amber: A Palaeobiological Study*, by S. Larsson. Scandinavian Science Press Ltd.)

adults occur in damp undergrowth and on foliage where they prey on other insects. The larvae, also predaceous, occur in decaying vegetation or water. The presence of adults in amber has been attributed to their habit of resting on tree trunks while waiting for prey (Larsson, 1978).

The only Cretaceous amber records of this family are from Siberian amber (Zherichin and Sukacheva, 1973). Species descriptions of snipe flies have been made from Baltic material and include representatives of the genera *Atherix, Bolbomyia, Chrysopilus, Protovermileo, Rhagio, Succinatherix*, and *Symphoromyia* (Spahr, 1985; Keilbach, 1982). When describing *Succinatherix*, Stuckenberg (1974) placed it in a new family, Athericidae, which he had erected earlier; his paper should be consulted for a discussion of this group. This family is also represented in Dominican amber (Appendix B).

Stratiomyidae (soldier flies). The medium- to large-sized adults of this family are often brightly colored and occur on flowers. Some resemble bees and wasps, and as such are similar to hover flies. The larvae occur in water, in decaying matter, or under bark, where they feed on plant material or are predaceous on other invertebrates.

The oldest known representative of this family is *Cretaeogaster pyg-*

maeus Teskey (1971) from Canadian amber. Although considered extinct, its closest relative, *Parhadrestia*, occurs in Chile, where it is represented by two extant species. The similarity between the two genera is discussed by Woodley (1986) who placed both in the new subfamily Parhadrestinae.

Species of *Beris, Cacosis, Chrysothemis, Hermetiella, Lophyrophorus,* and *Rhachicerus* have been described from Baltic amber (Keilbach, 1982; Spahr, 1985). James (1971) described *Pachygaster antiqua* from Mexican amber and noted that it closely resembled the American extant species *P. pulchra* Loew. Representatives of *Pachygaster* occur in the temperate and tropical regions of the world. The larvae develop under bark, where they may be predaceous on other insects. Soldier flies also occur in Dominican amber, from which Norman Woodley (personal communication) has identified two specimens belonging to or near the genus *Nothomyia* in the tribe Prosopochrysini (Color Plate 6).

Tabanidae (horseflies, deerflies). The females of these medium- to large-sized, robust flies suck blood from a variety of vertebrates, whereas the males feed on nectar and pollen. The larvae are aquatic or semiaquatic and prey on other invertebrates. The oldest tabanids are from the Eocene, which corresponds with the age of Baltic amber material. The latter include *Silvius laticornis* (Loew) and *Haematopota pinicola* Stuckenberg (1975). Stuckenberg (1975) remarked that *H. pinicola* is similar to Recent species, which supports his theory that the Tabanidae originated in the early Mesozoic and diverged in the Tertiary. On the basis of the wing structure of *H. pinicola*, Stuckenberg suggested that this fly lived in a nonforested, warm savanna habitat, not unlike that found in the Mediterranean today. The family is also represented in Dominican amber by two species of the genus *Stenotabanus* (Lane et al., 1989; Fairchild and Lane, 1989) (Fig. 95).

Therevidae (stiletto flies). These medium-sized, densely pubescent flies possess a slightly elongated abdomen and slender legs. The adults occur in open grassy areas, whereas the larvae are predaceous in the soil or decaying vegetation. Stiletto flies are rare in amber, but species of *Glaesorthactia, Psilocephala,* and *Thereva* have been described from Baltic amber, and Cockerell (1920A) described *Psilocephala electrella* from Burmese amber (Spahr, 1985).

Xylomyidae. The adults of this small family are slender and wasplike. The larvae occur under bark and in decaying vegetation. A single species, *Solva nana* (Loew), has been described from Baltic amber, and represen-

Figure 95. The horsefly *Stenotabanus brodzinskyi* Lane, Poinar, and Fairchild (family Tabanidae) in Dominican amber (Brodzinsky Lopez-Penha collection, Santo Domingo, Dominican Republic). (Published in *The Florida Entomologist*, Vol. 71, No. 4.)

tatives of the family have also been reported from Siberian amber (Spahr, 1985).

Xylophagidae. These somewhat variably shaped flies of medium to large size inhabit forests. The larvae are predaceous and occur in decaying vegetation or soil. They are a small family, 15–20 extant species, distributed at present mainly in the northern Holarctic. The family has been reported to occur in Siberian amber, and species in the genera *Arthropiella*, *Habrosoma*, and *Xylophagus* have been described from Baltic amber (Spahr, 1985). Species of *Chrysothemis*, *Electra*, and *Lophyrophorus*, often considered in the separate family Rachiceridae, have been described from Baltic amber (Spahr, 1985).

Suborder Cyclorrhapha. Flies in this suborder have three-segmented antennae and an arista or bristle on the terminal segment. The larvae are legless and maggotlike, lack a sclerotized head, and possess vertically moving mouth hooks.

Agromyzidae (leaf miner flies). The larvae of these small, dark, wide-spread flies are leaf miners on a wide variety of plants. All of the flies originally described from Baltic amber as agromyzids have been transferred to other families: species of *Agromyza* and *Dizygomyza* are now in the Chloropidae and Clusiidae, and *Napomyza* is now in the Sciadoceridae (Spahr, 1985). The family is, however, represented in Mexican and Dominican amber (Appendices A, B).

Anthomyzidae. These small flies are found in grass and wet meadows where the larvae feed on grasses and sedges. Species of *Anthoclusia*, *Protanthomyza*, and *Xenanthomyza* from Baltic amber have been placed in this family and in other families within the Anthomyzoidea (Spahr, 1985; Keilbach, 1982).

Calobatidae. These are slender, elongate flies with long legs and wings that usually have dark markings. The adults occur in moist meadows and the larvae live in decaying material. A *Calobata* sp. and three species of *Electrobata* occur in Baltic amber (Spahr, 1985), and the family is also represented in Mexican amber (Appendix A).

Chloropidae. These small, naked, often brightly colored flies are abundant in meadows and fields. The larvae mostly develop on plants, but some are scavengers and a few are predators or parasites. A member of this family has been found in Lebanese amber and constitutes the oldest known representative of the schizophoran (muscoid) Diptera (Whalley, 1981). Another member of this family has been reported from Canadian amber (McAlpine and Martin, 1969), but this record has since been disclaimed (D. Grimaldi, personal correspondence). *Protoscinella electra* Hennig, representatives of *Chlorops* and *Oscinis*, and "*Agromyza*" *aberrans* Meunier occur in Baltic amber (Hennig, 1965; Keilbach, 1982). The family is also represented in Mexican and Dominican amber (Appendices A, B).

Diopsidae (stalk-eyed flies). This mainly tropical family includes flies that have the compound eyes at the end of lateral stalklike head extensions. Little is known concerning their biology. Species of *Diopsis*, *Prosphyracephala*, and *Sphyracephala* occur in Baltic amber (Spahr, 1985) but, curiously, no representatives have been reported from Neotropical amber.

Drosophilidae (small fruit flies). The small adults of this family are commonly seen around spoiled fruit and vegetation, in which the larvae of many species develop. Some larvae prefer mushrooms, and others have

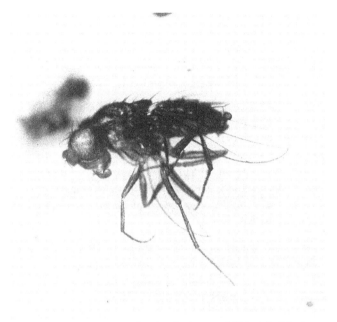

Figure 96. A small fruit fly (family Drosophilidae) in Mexican amber (Museum of Paleontology collection, University of California, Berkeley; photo by P. Hurd).

become predaceous or parasitic on insects. All known fossil drosophilids occur in amber or copal and a review of these, along with descriptions of new taxa, was presented by Grimaldi (1987).

The first described fossil drosophilid was *Drosophila berryi* Cockerell from copal deposits in Colombia, South America. The first amber drosophilids were two specimens from Mexican amber that were placed by Wheeler (1963) in the genus *Neotanygastrella* and subsequently described as *N. wheeleri* by Grimaldi (1987) (Fig. 96). A single drosophilid, *Electrophortica succini*, was described from Baltic amber by Hennig (1965).

Grimaldi (1987) described two new genera—*Miomyia* and *Protochymomyza*—and seven new species of drosophilids from Dominican amber. These include, in addition to single species in the above two genera, species in the extant genera *Chymomyza*, *Drosophila*, and *Scaptomyza*, and two species whose generic affinities could not be ascertained but that have been assigned to the Drosophilinae.

The Dominican amber drosophilids also offer two examples of sym-

biotic relationships involving phoresis and parasitism. The former examples consist of three relatively large macrochilid mites attached to the
abdomen of *Protochymomyza miocena* Grimaldi (Fig. 134) and the latter of
allantonematid nematode parasites emerging from the body of *Chymomyza primaeva* Grimaldi (Poinar, 1984A) (Fig. 139).

Ephydridae (shore flies). The small adults of these flies occur along
the shores of ponds, streams, and oceans, sometimes collecting in large
masses. The aquatic larvae feed on decaying vegetation and are able
to tolerate waters with a high saline or alkaline content. An *Ephydra* sp.
had been cited by Loew (1850) from Baltic amber but Hennig (1965) considered that it was probably *Protocamilla succini* Hennig of the Camillidae
(see Table 10). The family is represented in Mexican and Dominican amber (Appendices A, B).

Heleomyzidae. These small to medium-sized flies occur in damp,
shady areas and the larvae develop in decaying organic matter. Although
they are the most common family of acalyptrate Diptera, they are represented only in Baltic amber. Described species occur in the genera
Chaetohelomyza, Electroleria, Heleomyza, Heteromyza, Leria, Protosuillia,
and *Suillia* (Keilbach, 1982).

Lauxaniidae. These small, stoutish flies occur in damp, shady places
and the larvae develop in decaying vegetation. Species of *Chamaelau-
xania, Hemilauxania*, and *Sapromyza* have been reported from Baltic amber (Hennig, 1965; Spahr, 1985) and *Tryaneoides ellipticus* was described
from Chinese amber (Hong, 1981). The family is also represented in Dominican amber (Appendix B).

Milichiidae. These small flies normally occur in fields and meadows
where the larvae are scavengers and feed on decaying organic matter. The
first description of a milichiid in amber was *Phyllomyza hurdi* from Mexican amber (Sabrosky, 1963) (Fig. 97). *Pseudodesmometopa succineum* Hennig (1971) and *Phyllomyza jaegeri* Hennig (1967) were later described from
Baltic amber (Keilbach, 1982). The family is also represented in Dominican amber (Appendix B) from which one specimen was identified as
Phyllomyza sp. by D. Grimaldi.

Muscidae. This large, cosmopolitan group of flies includes the noto-
rious housefly. Muscids, as with other calyptrate Diptera, are rare in amber. All of the early reported "*Musca*" from Baltic amber (see Spahr, 1985)
apparently belong to other families. A single representative, *Fannia scalaris* Fabricius, is the only true muscid described in Baltic or any other

Figure 97. Holotype (male) of the milichid fly *Phyllomyza hurdi* Sabrosky (family Milichiidae), in Mexican amber (Museum of Paleontology collection, University of California, Berkeley; photo by P. Hurd).

Figure 98. A scuttle fly, *Dohrniphora poinari* Disney (family Phoridae), in Dominican amber (Poinar collection).

amber deposit (Hennig, 1966). Hennig remarked that this species did not differ enough from extant *F. scalaris* to be considered a new species, and the fossil was possibly in copal, which would make it fairly recent in age. Muscids are reported from Dominican amber (Appendix B) and what appear to be muscoid larvae have been found associated with decaying frog bones in Dominican material.

Phoridae (scuttle flies). This is a group of small flies that demonstrate a short, sporadic flight pattern and run quickly over foliage. The adults are characterized by having the anterior wing veins more strongly developed than the posterior ones and by having laterally flattened hind femora. The larvae have varied developmental cycles and feeding habits, ranging from saprophagic in fungi and dead organic matter, to parasitic in invertebrates and vertebrates, to commensal in the nests of ants and termites.

The family is represented in Canadian amber by *Prioriphora canadambra, P. intermedia, P. longicostalis,* and *P. setifemoralis* (McAlpine and Martin, 1966; Brown and Pike, 1990) and in New Jersey amber by *Metopina goeleti* (Grimaldi, 1989). Species from Baltic amber occur in the genera *Anevrina, Aphiochaeta, Chaetopleurophora, Conicera, Diplonevra,*

Dohrniphora, Electrophora, Hypoceridites, Megaselia, Phora, Protophorites, Protoplatyphora, Spiniphora, Stephanostoriscus, and *Triphleba* (Spahr, 1985; Keilbach, 1982). All species of *Dohrniphora* have been subsequently transferred to the genus *Diplonevra* (Borgmeier, 1968). From Chinese amber, *Chaetopleurophora gastracanthoidis* and *Rhoptrocera eocenica* were described by Hong (1981), and species of *Dohrniphora* and *Megaselia* were described by Disney (1987) from Dominican amber (Fig. 98). Grimaldi (1989) also reported the presence of *Metopina* in amber from the Dominican Republic and Mexico.

Pipunculidae (big-headed flies). These small flies have an oversized or swollen head containing large compound eyes. The adults frequently hover around flowers and the larvae are endoparasites of homopterans. Although the family has been reported in Canadian amber (McAlpine and Martin, 1969), all described amber species occur in Baltic deposits. These include representatives of the genera *Cephalosphaera, Metanephrocerus, Nephrocerus, Pipunculus, Protonephrocerus,* and *Verrallia* (Spahr, 1985). The family also occurs in Dominican amber (Appendix B).

Platypezidae (flat-footed flies). So named because the hind tarsi are flattened or broadened; the males of this family swarm during mating. The larvae develop in mushrooms and other fungi. Representatives of this family have been reported from Lebanese (Whalley, 1981) and Siberian amber (Zherichin and Sukacheva, 1973). A single species, *Oppenheimiella baltica* Meunier, has been described from Baltic amber (Keilbach, 1982). The family has also been reported from Dominican amber (Appendix B).

Sciadoceridae. Representatives of this little-known family have been reported from Siberian amber (Zherichin and Sukacheva, 1973), and *Sciadophora bostoni* was described by McAlpine and Martin (1966) from Canadian amber. A single species, *Archiphora robusta* (Meunier), has been described from Baltic amber (Keilbach, 1982) and *Eosciadocera helodis* Hong (1981) was described from Chinese amber.

Sciomyzidae (marsh flies). These are yellow-tawny colored, stoutish, medium-sized flies that occur in wet or damp areas. The larvae are predaceous on snails and slugs. Adults are represented in amber only by species of *Palaeoheteromyza, Prophaeomyia, Prosalticella,* and *Sepedonites* from Baltic deposits (Keilbach, 1982; Hennig, 1965).

Syrphidae (syrphid flies). The adults of these medium- to large-bodied flies can be found hovering around flowers. The larvae have invaded many different habitats, including plant surfaces (as insect predators, notably on aphids), pond bottoms and decaying vegetation (as scavengers), and nests of social insects (as symbionts). Species have been described from Baltic amber, including representatives of the genera *Cheilosia, Cheilosialepta, Criorhina, Doliomyia, Eoxylota, Megaxylota, Myolepta, Palaeoascia, Palaeopipiza, Palaeosphegina, Pipiza, Protorhingia, Pseudosphegina, Ptilocephala, Spheginascia, Syrphus, Tropidia,* and *Volucella* (Spahr, 1985). Larsson (1978) attributes the relative abundance of these flies in amber to the presence of aphid prey on the trunk of the amber-bearing trees.

Tachinidae (tachinid flies). Tachinids are stout-bodied, bristly, dark flies that occur in many different habitats. The larvae are parasitic or predaceous on other insects. Few tachinids have been reported from amber. *Electrotachina smithii* Townsend and *Paleotachina smithii* Townsend have been described from Baltic amber (Spahr, 1985) and the family is represented in Dominican amber (Appendix B).

Siphonaptera (Fleas)

Fleas are small, laterally flattened, highly modified insects adapted for living in the hair of mammals or the plumage of birds. The antennae are short and located in grooves of the head, the coxae are enlarged, and the reduced head bears piercing and sucking mouthparts. The adults take blood from mammal and bird hosts, whereas the larvae feed on organic remains, often in the nest of the host.

The earliest known, definite fossil fleas and the only described amber fleas are *Palaeopsylla klebsiana* Dampf and *Palaeopsylla dissimilis* Peus from Baltic amber (Keilbach, 1982). Extant species of this genus, which is in the family Hystrichopsyllidae, occur on shrews and moles. R. Traub (personal communication) placed three adult fleas from Dominican amber in the family Rhopalopsyllidae (Fig. 99). Although extant fleas of this family occur on rodents and marsupials, Traub believes that the Dominican fossils were bird fleas. They possess certain characters (comparatively larger eyes, thinner bristles, more spines on the pronotal comb, and fewer grooves on the tarsal claws) that are found on extant bird fleas.

Hymenoptera (Sawflies, Ants, Wasps, and Bees)

Winged members of this order contain two pairs of membranous wings with reduced venation. The hind wings can be interlocked with the fore wings by means of small chitinous hooks. The mouthparts are usually mandibulate and adapted for biting, but the labium and maxillae

Figure 99. A rare flea (order Siphonaptera) in Dominican amber (Traub collection).

may be modified for lapping and sucking. An ovipositor is always present and may be used for sawing, piercing, or stinging. The order is divided into two suborders, the Symphyta, or sawflies, and the Apocrita, or bees, ants, and wasps. In the Symphyta, the adult abdomen is broadly joined to the thorax and the larvae are caterpillarlike. In the Apocrita, the adult abdomen is basally constricted and the first segment is fused with the metathorax, and the larvae are grub- or maggotlike. Both of the suborders are represented in Triassic deposits, which contain the earliest known fossil Hymenoptera. Additional references to species of the genera listed below can be obtained from Keilbach (1982) and Spahr (1987).

Suborder Symphyta (sawflies and horntails). The members of this suborder, commonly called sawflies and horntails, feed on plants, except for one family, the Orussidae, which is a rare group whose larvae are parasitic on beetles of the family Buprestidae. The majority of the larvae, which resemble the caterpillars of moths and butterflies, feed exposed on vegetation, although a few are concealed within stems, fruits, or wood. Representatives of this suborder have been reported mainly from Baltic amber, but the fossil record dates back to the Early Triassic.

Argidae. Members of this family can be recognized by their three-segmented antennae. On males of some genera the third antennal segment is bifurcate along its entire length, as shown in a representative sawfly from Dominican amber (Fig. 100). This specimen and others in Dominican amber are the only known fossil argids.

Cephidae (stem sawflies). The larvae of these slender, laterally compressed sawflies bore into the stems of grasses and other plants. *Electrocephus stralendorffi* Konow and an undescribed species of *Cephus* occur in Baltic amber (Menge, 1856; Spahr, 1987).

Cimbicidae. These large, robust sawflies, characterized by clubbed antennae, are rare in amber. A single larva was mentioned by Menge (1856) as occurring in Baltic amber.

Diprionidae (conifer sawflies). The larvae of these medium-sized sawflies feed on conifers. The males have pectinate or bipectinate antennae. A larva of *Diprion* (= *Lophyrus*) was identified by Menge (1856), and Bachofen-Echt (1949) also noted two species of this genus in Baltic amber.

Figure 100. A sawfly (family Argidae) in Dominican amber. The males have branched antennae (Poinar collection).

Electrotomidae. This family was erected to contain *Electrotoma succini* Rasnitsyn (1977A) described from Baltic amber.

Pamphiliidae (web-spinning and leaf-rolling sawflies). The medium-sized, stoutish adults of this family occur on vegetation whereas the larvae may live within coiled-up leaves. A single larva in Baltic amber identified by Menge (1856) as belonging to the genus *Pamphilius* is the only record of this family in amber.

Siricidae (horntails). The larvae of these large wood wasps burrow in the wood of trees. *Urocerus klebsi* Brues (1926) was described from Baltic amber and Klebs earlier reported a *Sirex* from the same source (Spahr, 1987).

Tenthredinidae (sawflies). In this most common family of the suborder, the adults are wasplike and frequently brightly colored. They occur on foliage or flowers whereas the larvae are mostly external feeders on the leaves of trees and shrubs. No species of this family have been described from amber, although adults and/or larvae of the genera *Tenthredo* and *Allantus* have been reported from Baltic amber (Larsson, 1978; Spahr, 1987).

Suborder Apocrita (bees, wasps, and ants). The larvae of this suborder are grublike or maggotlike in appearance and have a wide variety of feeding habits. Some are internal parasites whereas others are external parasites of arthropods. Still others are predators and some are plant feeders. The forms developing on plants are chiefly concealed in stems, seeds, or galls. Food habits of the social Hymenoptera are varied and the larvae are dependent on material collected by the workers. The suborder Apocrita is sometimes divided into nonstinging forms (the Parasitica) and stinging forms like bees and ants (the Aculeata). Except for ants and bees, most of the Apocrita are commonly referred to as wasps, although this term is applied by specialists only to certain members of the Aculeata. Below, the families in this suborder are alphabetically arranged under the superfamilies Ichneumonoidea, Chalcidoidea, Cynipoidea, Evanioidea, Pelecinoidea, Proctotrupoidea, Bethyloidea, Formicoidea, Scolioidea, Vespoidea, Sphecoidea, and Apoidea.

Parasitic Hymenoptera (Parasitica). Representatives of this group are mainly parasites in or on the bodies of insects and other arthropods. The females generally possess a long ovipositor, which allows them to deposit eggs in or on various growth stages of invertebrate hosts in concealed habitats (stems, seeds, soil). The ovipositor is rarely used as a sting for defense. The Parasitica are important regulators of insect populations, and many are used as biological control agents of insect pests.

Superfamily Ichneumonoidea. All members of this superfamily are parasitic (endoparasitic or ectoparasitic) on other invertebrates. Although sometimes wasplike in appearance, they rarely sting humans. The adults have long filiform antennae (usually of more than 16 segments) and two-segmented hind trochanters (leg segments); a costal cell in the wing is absent (except in the Stephanidae) and the ovipositor arises anterior to the tip of the abdomen.

Braconidae (braconids). This is a large family, composed of relatively small, stout-bodied insects that play an important role as parasites of plant pests, especially Lepidoptera. Thanks to the detailed studies by Brues (1923, 1933A, 1939) the Baltic amber Braconidae have been carefully described and analyzed. Species of the genera *Agathis* (= *Microdus*), *Anacanthobracon, Ascogaster, Aspicolpus, Austrohelcon, Blacus, Brachistes, Cantharoctonus, Chelonohelcon, Chelonus, Clinocentrus, Coeloreuteus, Diachasma, Digastrotheca, Diodontogaster, Diospilites, Doryctes* (= *Ischiogonus*), *Doryctomorpha, Elasmosomites, Electrohelcon, Eocardiochiles, Eubazus* (= *Eubadizon*), *Eumacrocentrus, Helcon* (= *Gymnoscelus*), *Hormiellus, Ichneutes, Macrocentrus, Meteorites, Meteorus, Microctonus, Microtypus, Miracoides, Neoblacus, Onychoura, Orgilus, Palaeorhyssalus, Parasyrrhizus, Phanerotoma, Polystenus, Prochremylus, Promonolexis, Pygostolus, Rogas* (= *Rhogas*), *Rhyssalus, Semirhytus, Snellenius, Taphaeus,* and *Triaspis* have been described from Baltic amber (Brues, 1923, 1933A, 1939; Spahr, 1987). In his analysis of the above, Brues (1933A) cited some 15 genera as now being extinct; of the remaining 28 genera, 18 occur in the Palearctic, 17 are Nearctic, 15 are Holarctic, 12 are Neotropical, 9 are Ethopian, 11 are Oriental, and 11 occur in Australia.

In other Tertiary amber deposits, Guérin (1838) recorded a *Bracon* sp. from Sicilian amber (Keilbach, 1982), Hong et al. (1974) described *Sinobracon speciosus* from Chinese amber, and Muesebeck (1960) described *Ecphylus oculatus* from Mexican amber. The family is also represented in Dominican amber (Appendix B) by a variety of forms, including *Phanerotoma* sp. (Cheloninae) (Fig. 101), identified by L. Caltagirone.

Braconids also exist in Cretaceous amber, and Brues (1937) described *Diospilus allani, Neoblacus facialis,* and *Pygostolus patriarchicus* from Canadian amber. Species of *Dirrhope, Heterospilus,* and *Parahormius* have been reported from Siberian amber by Zherichin (1978).

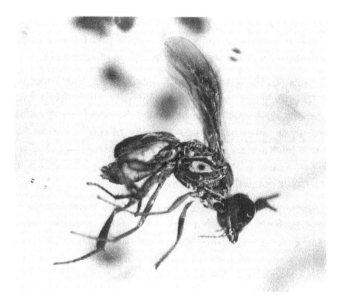

Figure 101. A braconid of the genus *Phanerotoma* (family Braconidae) in Dominican amber (Poinar collection).

Ichneumonidae (ichneumonids). The adults of this large family often have a slender, wasplike appearance but, in contrast to wasps, they possess an ovipositor that cannot be withdrawn into the body. Ichneumonids attack a wide variety of insects and other arthropods, and the parasitic larvae usually develop inside the immature stages of the host.

Although probably as abundant in amber as the Braconidae, the fossil forms of this family have been little studied. Only two species in the genera *Pimpla* and *Astiphromma* have been formally described from Baltic amber, but nonspecific representatives of *Bassus, Cryptus, Hemiteles, Ichneumon, Lampronota, Mesochorus, Mesoleptus, Mesostenus, Pezomachus, Phygadeuon, Porizon,* and *Tryphon* have been reported from the same deposits (Spahr, 1987). The family is also represented in Mexican and Dominican amber (Appendices A, B; Color Plate 8). An ichneumonid ovipositing into a caterpillar in Baltic amber was illustrated by Wunderlich (1986). From Siberian Cretaceous amber, Townes (1973) described the species *Catachora minor, Eubaeus leiponeura,* and *Urotryphon pussilus.*

Megalyridae. Brues (1923) described from Baltic amber two species in the genus *Prodinapsis* belonging to this small family, and Rasnitzyn (1977A) described *Cretodinapsis caucasica* from Cretaceous amber from the southwestern portion of the Soviet Union (Spahr, 1987). Extant species of *Prodinapsis* occur in South Africa.

Stephanidae. This small family comprises parasites that resemble ichneumonids in general appearance and are known to attack larvae of wood-boring beetles. Three species of *Electrostephanus* in Baltic amber (Brues, 1933A) are the only reported stephanids in amber.

Superfamily Chalcidoidea. This is a large group of very small, sometimes minute, wasps characterized by the pattern of wing venation, the usually elbowed antennae with less than 13 segments, and the structure of the thorax. Most chalcidoids parasitize the eggs or immature stages of Lepidoptera, Coleoptera, Diptera, and Homoptera. A few feed on plants (Color Plate 5).

Chalcididae. Members of this family normally range from 2 to 7 millimeters in length and are characterized by a swollen hind femur that is often toothed. Amber chalcidids have, unfortunately, been little studied. No species descriptions exist although representatives probably occur in most of the fossiliferous deposits. The family is represented in Mexican and Dominican amber (Appendices A, B), and a *Brachymeria* sp. from the latter material has been identified by L. Caltagirone. A member of this family has also been reported from Lebanese amber (Whalley, 1981).

Encyrtidae, Eulophidae, Eurytomidae, and Eupelmidae. Representatives of these families are small wasps, usually 1–3 millimeters in length, that resemble each other in general appearance. The larvae of most parasitize insects, but a few consume plants. All four families occur in one or more deposits of Tertiary amber, but only the single species *Propelma rohdendorfi* Trjapicyn has been described from Baltic amber; Keilbach (1982) places this species in the Eupelmidae whereas Spahr (1987) puts it in the Encyrtidae. A new genus in the Eulophidae has been discovered in Alaskan amber, but a description has not yet appeared (Langenheim et al., 1960). L. Caltagirone has identified representatives of the Encyrtidae (Copidosomatini), Eulophidae, and Eupelmidae from Dominican amber.

Mymaridae (fairyflies). These tiny wasps, which are often less than 1 millimeter in length, possess elongate hind wings with nearly parallel sides. Often, both wings are lined with long hairs giving the insect a mysterious appearance. All representatives parasitize insect eggs.

Mymarids occur in Cretaceous deposits of Siberian (Zherichin and Sukacheva, 1973) and Lebanese amber (Schlee and Glöckner, 1978).

Yoshimoto (1975) described species of *Archaeromma, Triadomerus, Carpenteriana, Protooctonus*, and *Macalpinia* from Canadian amber. The new genus *Galloromma bezonnaisensis* was described from French Cretaceous amber by Schlüter (1978A).

Tertiary amber deposits containing Mymaridae are also extensive. Representatives of *Anaphes, Arescon, Gonatocerus* (= *Lymaenon*), *Litus, Malfattia, Stethynium, Mymar, Ooctonus* and *Palaeomymar* occur in Baltic amber, although Spahr (1987) placed the latter three genera in the family Serphitidae.

Doutt (1973B) reviewed the fossil Mymaridae and examined representatives in Mexican amber. He identified two extant species in these deposits, *Alaptus psocidivorus* Gahan and *A. globosicornis* Girault, as well as the new species *Litus mexicanus* and undetermined species of *Palaeomymar* and *Anaphes*. Doutt (1973A) also described *Polynemoidea mexicana* from the same source.

The family is also represented in Dominican (Appendix B), Sicilian (Doutt, 1973B), and Arkansas amber (Saunders et al., 1974). Representatives previously placed in the family Mymarommatidae are included here in the Mymaridae.

Perilampidae and Pteromalidae. Members of these families are small, often stoutish wasps that frequently display metallic coloration. Many of the Perilampidae are hyperparasites on other fly and wasp parasites. Their presence in amber is restricted to undescribed reports of *Perilampies* and *Pteromalus* in Baltic amber (Spahr, 1987). Pteromalids are also represented in Mexican and Dominican amber (Appendices A, B) and in Siberian amber (Spahr, 1987).

Tetracampidae. This small family, which comprises 11 genera and 27 species worldwide, shows characteristics intermediate between those of the Eulophidae and Pteromalidae. The great number of fossils in Canadian amber indicates that these wasps were fairly common in North American forests during the Cretaceous. Extant forms parasitize the eggs of sawflies. Yoshimoto (1975) studied the Canadian amber forms, which comprise six species distributed among the genera *Baeomorpha, Bouceklytus*, and *Distylopus*. He suggested that these three fossil genera are the early descendants of a primitive eurytomid-torymid-like ancestor.

Thysanidae. The family Thysanidae, comprising small chalcids that attack Homoptera or parasites of Homoptera, are represented only in Mexican amber (Appendix A).

Torymidae. Representatives of this family are small, elongate chalcids with large hind coxae and a long ovipositor. They include both insect and plant parasites. The family is represented by *Monodontomerus primaevus* Brues (1923) in Baltic amber and by *Neopalachia bouceki* Grissell (1980) and the new genus *Zophodetus woodruffi* Grissell (1980) in Dominican amber.

Trichogrammatidae. Members of this family resemble the Mymaridae in being tiny (0.3–1.0 millimeters in length), possessing hairs on the hind wings, and attacking insect eggs. They differ from the latter and all other Chalcidoidea in that both sexes possess three-segmented tarsi. In contrast to the abundance of the Mymaridae, only the single genus and species *Enneagmus pristinus* Yoshimoto (1975) has been described from Canadian deposits. The family is also represented in Baltic (Larsson, 1978) and Mexican amber (Appendix A).

Superfamily Cynipoidea (gall wasps and others). Most of the representatives in this group are small, black forms with filiform antennae, a shiny somewhat compressed abdomen, and an ovipositor that issues anterior to the tip of the abdomen. Wing venation is also reduced in this superfamily.

Cynipidae. This family includes the well-known gall wasps that make galls on various plants and act as parasites and hyperparasites of insects. Cynipids are uncommon in amber, which is somewhat surprising because many extant species attack oak (*Quercus*) and such trees were present in the Baltic amber forest. The earliest known cynipid is *Protimaspis costalis* Kinsey (1937) from Canadian amber. From Baltic amber were described *Aulacidea succinea* Kinsey, *Cynips succinea* Presl, and a *Diastrophus* sp. (Keilbach, 1982). The family is also represented in Dominican amber (Appendix B).

Ibaliidae. Representatives of this family are relatively large (7–16 millimeters in length) and parasitize horntails. Extant forms are uncommon and a single report of this family in Siberian deposits (Spahr, 1987) furnishes the only occurrence of the group in amber.

Superfamily Evanioidea. Members of this superfamily possess filiform antennae and two-segmented trochanters and have the abdomen attached high above the hind coxae. All parasitize insects. Only a single record of a Gasteruptiidae in amber occurs from Lebanese deposits (Whalley, 1981).

Aulacidae. These wasps are parasitic in larvae of wood-boring beetles and wood wasps. Species of *Aulacus* (= *Micraulacinus*) and *Pristaulacus* (= *Oleisoprister*) were described by Brues (1923, 1933A) from Baltic amber. Cockerell (1917B, 1917F) erected three new genera and species from Burmese amber; these three genera—*Electrofoenus, Protofoenus,* and *Hyptiogastrites*—were listed in the Evaniidae by Keilbach (1982).

Evaniidae (ensign wasps). These stout-bodied wasps can be recognized by the shape of the abdomen, which is small and oval, and by its peculiar attachment to the thorax, which is high on the propodeum through a slender stalk. The abdomen and stalk give the appearance of a flag, and suggest the common name. All representatives parasitize the egg cases of cockroaches. The only species of this family described from amber were placed in the genus *Evania* by Brues (1933A). In addition, Spahr (1987) cites a *Brachygaster* from Baltic amber, and the family is also represented in Mexican and Dominican amber (Appendices A, B).

Superfamily Pelecinoidea: Pelecinopteridae. This family was erected by Brues (1933A) to accommodate the unusual species *Pelecinopteron tubuliforme* from Baltic amber. A similar form, *P. dubium,* possibly belonging to the above species, was reported by Zherichin (1978) from Siberian amber (Spahr, 1987).

Superfamily Proctotrupoidea. Most individuals of this group are small, black wasps that resemble chalcids but are distinguished from the latter by the triangular-shaped pronotum and the ovipositor arising from the abdomen tip, not anterior to the tip.

Ceraphronidae (including the Megaspilidae). Members of this group are parasites or hyperparasites of other wasps that attack aphids or scales. Species of *Ceraphron* (= *Calliceras*), *Lagynodes,* and *Conostigmus* occur in Baltic amber (Spahr, 1987; Szabó and Oehlke, 1986). Muesebeck (1963) described *Allocotidus bruesi* from Alaskan amber (some authors place this species in the family Stigmaphronidae), and Brues (1937) described *Lygocerus dubitatus* from Canadian amber. *Prolagynodes penniger* Alekseyev and Rasnitsyn and *Conostigmus dolicharthrus* Alekseyev and Rasnitsyn were described from Siberian amber (Alekseyev and Rasnitsyn, 1981). Representatives of this group also occur in Mexican amber (Appendix A).

Diapriidae. Diaprids are small parasites of Diptera that can be recognized by the placement of the antennae on a shelflike protuberance on the face. Some attack fungus gnats (Mycetophilidae) which are moderately common in amber. Representatives of the genera *Ambositra*, *Belyta*, *Cinetus*, *Pantolyta*, *Paramesius*, and *Psilus* have been described or reported from Baltic amber. The family is also represented in Mexican amber (Appendix A). Because a Baltic amber *Ambositra* closely resembles the African *A. famosa* Masner, Masner (1969B) considers *Ambositra* to be a genus of southern origin that spread from Africa to Europe in the Mesozoic.

Maimetshidae. This family was erected by Rasnitsyn (1975) to accommodate *Maimetsha arctica* Rasnitsyn from Siberian amber.

Platygasteridae. These small black insects have reduced wing venation and, normally, 10-segmented antennae situated low on the face, just above the clypeus (a part of the insect head below the frons). Most are parasites of flies belonging to the family Cecidomyiidae. No species has been described from amber, although representatives have been noted in Baltic, Mexican, and Dominican deposits (Spahr, 1987) (Appendices A, B).

Proctotrupidae. These small wasps, 3–6 millimeters in length, can be recognized by the large stigma and small marginal vein in the fore wing. They are insect parasites, and Cretaceous representatives of this family include unidentified forms in Lebanese amber (Whalley, 1981). Species in the genera *Cryptoserphus* and *Proctotrupes* were described from Baltic amber (Spahr, 1987), and representatives of this family also occur in Mexican and Dominican amber (Appendices A, B).

Scelionidae. These small wasps parasitize the eggs of insects and spiders. They are represented in Cretaceous Canadian amber by *Baryconus fulleri* Brues and *Proteroscelio antennalis* Brues (1937) and by *Cenomanoscelio pulcher* Schlüter (1978A) in French amber. The family is also reported from Siberian amber (Zherichin and Sukacheva, 1973). Species in the genera *Aneurobaeus*, *Archaeoscelio*, *Baryconus*, *Brachyscelio*, *Ceratobaeoides*, *Ceratoteleia*, *Chromoteleia*, *Dissolcus*, *Electroteleia*, *Gryon* (= *Hadronotus*), *Hadronotoides*, *Hoploteleia*, *Idris*, *Macroteleia*, *Microtelenomus*, *Parabaeus*, *Proplatyscelio*, *Pseudobaeus*, *Sembilanocera*, *Sparasion*, *Trachelopteron*, and *Uroteleia* have been described from Baltic amber (Brues, 1940; Spahr, 1987).

Masner (1969A) described *Palaeogryon muesebecki* from Mexican amber and from Recent specimens in Mexico. If the specimens are indeed conspecific, the survival of a species for some 30 million years would be indicated. Scelionids also occur in Dominican amber (Appendix B).

Serphitidae. This family was erected by Brues (1937) to account for a fossil in Canadian amber that could not be placed in any existing family. This species, *Serphites paradoxus*, was regarded by Brues as a primitive, degenerate form that had the closest affinity with certain Serphoidea. Later workers have described other fossil material in this family, including species of *Aposerphites*, *Microserphites*, *Palaeomymar*, and *Serphites* (Color Plate 4) from Siberian amber (Kozlov and Rasnitsyn, 1979) and species of *Archaeromma* and *Distylopus* from Canadian amber (Yoshimoto, 1975). Reference to the Serphitidae in Tertiary Baltic amber is vague (Spahr, 1987).

Stigmaphronidae. This family was erected by Kozlov (1975) for three new genera and species—*Elasmomorpha melpomene* Kozlov, *Stigmaphron orphne* Kozlov, and *Hippocoon evadne* Kozlov—from Siberian amber. Some authors place *Allocotidus bruesi* Muesebeck (1963) from Alaskan amber in this family (Spahr, 1987).

Trupochalcididae. The Trupochalcididae was erected by Kozlov (1975) to accommodate the species *Trupochalcis inops* Kozlov from Siberian amber.

Vanhorniidae. This rare family is mentioned only as possibly occurring in Canadian amber (McAlpine and Martin, 1969).

Superfamily Bethyloidea. These wasps are generally larger than those of the preceding groups. The larvae are endoparasites or ectoparasites of other insects.

Bethylidae. The females of many Bethylidae are wingless and antlike in shape. Winged females overlap in size with small bees and sometimes sting. All members parasitize larvae of Coleoptera and Lepidoptera. Evans (1973) described *Archaepyris minutus* and *Celonophamia taimyria* from Siberian amber, and Cockerell (1917A, 1920B) described species of *Apenesia*, *Bethylitella*, *Epyris*, and *Sclerodermus* (= *Scheroderma*) from Burmese amber. All of the remaining amber species of Bethylidae were described by Brues (1939) from Baltic amber. These include species of *Artiepyris*, *Bethylopteron*, *Calyoza*, *Ctenobethylus*, *Epyris*, *Eupsenella*, *Holepyris* (= *Misepyris*), *Homoglenus*, *Isobrachium*, *Laelius*, *Palaeobethyloides*, *Palaeobethylus*, *Parapristocera*, *Parasierola* (= *Perisierola*), *Pristapenesia*, *Prosierola*, *Protopristocera*, *Rhabdepyris*, *Sierola*, and *Uromesitius*.

Of the nine extant genera recorded by Brues (1939) from Baltic amber, the present-day distributions include two species that are cosmopolitan, six that are Neotropical, five that are Palaearctic, and three each that occur in the Ethiopian, Indomalayan, and Australian regions. The most dominant genus in Baltic amber, *Eupsenella*, is now restricted to Australia. The family is also represented in Mexican and Dominican amber (Appendices A, B). L. Caltagirone identified a representative of the genus *Parasierola* from the latter deposits.

Chrysididae (cuckoo wasps). These medium-sized wasps are frequently metallic green or blue in color, with various patterns of sculpture over the body. They differ from bees in having no closed cells in the hind wing. Some forms are split off as a separate family, the Cleptidae, by certain authors. Evans described *Hypocleptes rasnitsyni* and *Protamisega khatanga* from Siberian amber (Evans, 1973) and *Procleptes carpenteri* from Canadian amber (Evans, 1969).

Species or representatives of *Chrysis, Cleptes, Omalus,* and *Protochrysidis* (= *Protochrysis*) occur in Baltic amber (Spahr, 1987), and Krombein (1986) recently described *Palaeochrum diversum* and *Protadelphe aenea* from the same source. Brues (1939) stated that the species *Protochrysidis succinalis* (Bischoff) from Baltic amber "cannot be placed in any family with assurance."

Dryinidae. In the females of this family, the front tarsi are often modified for grasping their insect host during oviposition (Fig. 102). The antennae are 10-segmented, and the females may be wingless and resemble ants. Members of the Homoptera are parasitized, and as the parasitic larva develops, it protrudes from the host in a saclike structure (Fig. 140). The small family Embolemidae is often included here.

N. G. Ponomarenko described *Cretodryinus zherichini* and *Laberius antiquus* from Siberian amber and *Avodryinus canadensis* from Canadian amber (Spahr, 1987). Species or representatives of *Ampulicomorpha, Chelogynus, Dryinus, Electrodryinus, Embolemus, Harpactosphecion, Lestodryinus, Neodryinus,* and *Thaumatodryinus* occur in Baltic amber (Brues, 1933A; Spahr, 1987) and *Dryinus palaeodominicanus* has been described from Dominican amber (Currado and Olmi, 1983).

Trigonalidae. These medium-sized, often brightly colored wasps have long, many-segmented antennae. A species from Siberian amber, *Cretogonalys taimyricus* Rasnitsyn (1977A), and one from Burmese amber, *Trigonalys pervetus* Cockerell (1917E), constitute the only records of this family in amber (Spahr, 1987).

Aculeate Hymenoptera. Representatives of this group include the ants, bees, and true wasps. They are distinguished by usually having 12

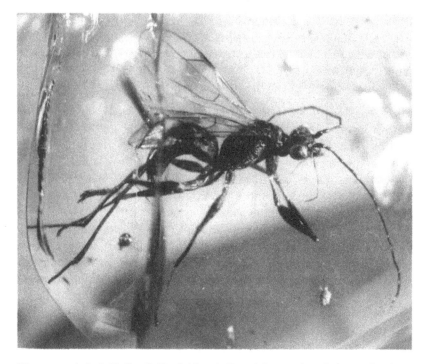

Figure 102. A dryinid (family Dryinidae) in Dominican amber (Poinar collection).

antennal segments in the female and 13 in the male, a single trochanter, a costal cell in the wing, and the ovipositor (often used for stinging) emerging from the terminal abdominal segment. The food habits of aculeates are extremely varied, and foods consist of various plant and animal products. The aculeates probably underwent their principal radiation in the Jurassic and Early Cretaceous.

Superfamily Formicoidea: Formicidae (ants). Ants are often considered to be the most successful of all insects. They occupy most terrestrial habitats and outnumber in individuals all other arthropods. Ants have the first abdominal segment extended into a nodelike stalk (pedicel), possess mostly elbowed antennae with an extended first segment, and have a quadrate pronotum.

All ants are social, and most colonies consist of queens, males, and various types of workers (sterile wingless females). Only the males and fertile females are winged. At certain times of the year, winged males and queens are produced and engage in mating flights.

Ant colonies occur in a variety of habitats, often in trees, which may account in part for the large representation of ants, both winged forms and workers, in amber (Color Plate 5). Although subfamilies and families of ants are recognized, all will be treated here under the family Formicidae.

Ants have been described from Cretaceous amber and these finds represent the earliest fossil record for the aculeates (Fig. 103). The first described Cretaceous ant was *Sphecomyrma freyi* Wilson et al. (1967) based on two workers found in New Jersey amber (Color Plate 1). The describing authors placed *S. freyi* in a separate subfamily, the Sphecomyrminae and concluded that the fossil formed a near-perfect link between certain nonsocial tiphiid wasps and the most primitive myrmecioid ants. But *Sphecomyrma freyi* is not universally accepted as being an ant. On the basis of its shortened basal antennal (scape) segments and the apparent absence of a metapleural (thoracic) gland, Baroni Urbani (1988) concludes it is probably a nonsocial wasp.

Dlussky (1975, 1983) described ants belonging to the genera *Archaeopone, Armania, Armaniella, Cretomyrma, Cretopone, Dolichomyrma, Paleomyrmex, Petropone, Poneropterus,* and *Pseudarmania* from Siberian am-

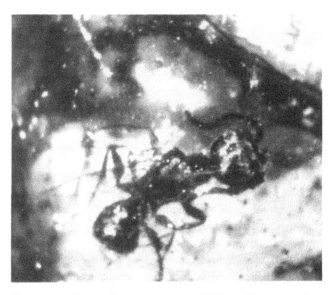

Figure 103. The earliest known ant? This still-to-be-studied specimen occurs in Lebanese amber (A. Acra collection of Lebanese amber; photo courtesy of A. Acra).

ber. These genera were placed in the separate family Armaniidae by Dlussky; however, Wilson (1985D) reported that those of the above that were well-enough preserved to disclose subfamily-level diagnostic characters seemed to fall within the Sphecomyrminae. Acceptance of Wilson's view would establish that the most primitive group of ants, the Sphecomyrminae, was distributed over a large portion of the Northern Hemisphere during the Late Cretaceous. The known range of this group was further extended by the discovery and description of *Sphecomyrma canadensis* Wilson (1985D) from Canadian amber. An "ant" discovered by A. Acra in Lebanese amber (Fig. 103) still needs to be carefully studied by experts before its placement can be established. If it is indeed an ant, then it would be the earliest known fossil of the family, roughly equivalent in age to *Cretaceoformica* from Australia (Jell and Duncan, 1986). The latter species lacks the critical characters necessary for it to be a true ant (Baroni Urbani, personal communication).

Tertiary ants in amber have been well studied, at least in Baltic and Dominican amber. Most of the Baltic amber ants were described by Mayr (1868) and Wheeler (1915) and include species or representatives in the genera *Agroecomyrmex*, *Anomma*, *Aphaenogaster*, *Asymphylomyrmex*, *Bothriomyrmex*, *Bradoponera*, *Camponotus*, *Cataglyphis*, *Dolichoderus*, *Drymomyrmex*, *Ectatomma*, *Electromyrmex*, *Electroponera*, *Enneamerus*, *Formica*, *Gesomyrmex* (= *Dimorphomyrmex*), *Glaphyromyrmex*, *Gnampto-genys*, *Hypoclinea*, *Iridomyrmex*, *Lasius*, *Leptothorax*, *Liometopum*, *Mono-morium* (= *Lampromyrmex*), *Monoriscus*, *Myrmica*, *Nothomyrmica*, *Oecophylla*, *Oligomyrmex* (= *Aeromyrma*, *Erebomyrma*, *Pheidologeton*), *Pachycondyla*, *Parameranoplus*, *Paraneuretus*, *Pityomyrmex*, *Plagiolepis*, *Platythyrea*, *Polyrhachis*, *Ponera*, *Prenolepis*, *Prionomyrmex*, *Procerapachys*, *Prodimorphomyrmex*, *Pronolepis*, *Protaneuretus*, *Pseudolasius*, *Rhopalo-myrmex*, *Sicilomyrmex*, *Sima*, *Stenamma*, *Stigmomyrmex*, *Stiphromyrmex*, and *Vollenhovia* (= *Propodomyrma*).

Two of the above species are of particular interest because of their morphological similarity to extant species. The Baltic amber *Formica flori* Mayr closely resembles the Holarctic species *Formica fusca* Linnaeus found today in southern Europe and North America. This similarity was used to suggest that ant evolution since the Oligocene has been stagnant. However, Baroni Urbani and Graeser (1987) compared *F. fusca* with a pyritized *F. flori* in Baltic amber and found significant differences at the level of the scanning electron microscope (Fig. 104). These fine differences may also apply to *Lasius schiefferdeckeri* Mayr, which is very similar to *Lasius niger* (Linnaeus), which has a Holarctic distribution today.

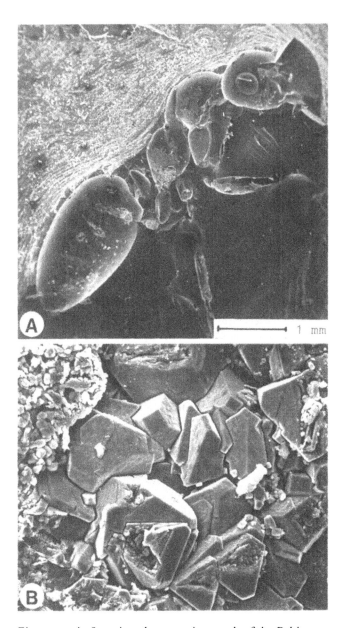

Figure 104. A. Scanning electron micrograph of the Baltic amber ant *Formica flori* Mayr, showing patches of pyrite crystals on its abdomen. B. Close up of the pyrite crystals. Such micrographs reveal useful differences between fossil and closely related living forms. (Photos courtesy of State Museum of Natural History, Stuttgart; from Baroni Urbani and Graeser, 1987.)

Wheeler (1915) noted that the Baltic amber ants comprised two basic groups, one associated with subtropical conditions (thermophilous) and the other with more temperate climates (boreal). Wheeler's explanation for this was that the amber was initially deposited when the climate was subtropical and resin production continued during a period of cooling in that part of the world (during the Oligocene), at which time the boreal fauna was introduced. This hypothesis is discussed by Larsson (1978).

Emery (1890) studied the Sicilian amber ants. He and Guérin (1838) published descriptions of Sicilian amber species belonging to the genera *Cataulacus, Crematogaster, Ectatomma, Gesomyrmex, Gnamptogenys, Hypopomyrmex, Leptomyrmula* (= *Leptomyrmex*), *Oecophylla, Oligomyrmex* (= *Aeromyrmar*), *Plagiolepis, Podomyrma, Ponera, Pseudomyrmex, Tapinoma*, and *Technomyrmex*.

Baroni Urbani (1980A–D) was the first to describe species of ants in Dominican amber. Among these was *Trachymyrmex primaevus*, the first certain fossil garden ants, well known for their habit of cultivating fungi on pieces of leaves carried by foraging workers back to the nest (Baroni Urbani, 1980A). Also described was the first known fossil ant of the subtribe Odontomachiti, *Anochetus corayi* (Baroni Urbani, 1980B). Members of this subtribe normally live in small colonies in the soil (but sometimes in trees) and forage in the litter on the ground. Closely related extant forms, among them *A. mayri*, still occur in the Caribbean. Baroni Urbani's (1980C) description of *Leptomyrmex neotropicus* was especially interesting because the present-day distribution of this genus is restricted to New Guinea, New Caledonia, and the eastern coastal area of Australia. Apparently, *Leptomyrmex* species do not excavate a nest but use the abandoned burrows of lizards or small marsupials. They walk with the gaster (abdomen) bent forward over the trunk, and the queens are apterous.

Baroni Urbani (1980C) provided evidence that the Neotropical region might be the potential center of origin for the tribe Leptomyrmicini. The description of *Leptomyrmex neotropicus* brings additional evidence for a faunal relationship between South America and Australia. However, Baroni Urbani and Saunders (1980) later suggested that a former Tertiary cosmopolitan distribution of *Leptomyrmex* combined with a post-Miocene contraction might explain the present-day distribution. This view is supported by the report of a *Leptomyrmex*-like representative, *Leptomyrmula maravignae* (Emery), in Sicilian amber. Wilson (1985C) first considered *Leptomyrmex neotropicus* to be a *Camponotus*, stating that the specimen closely resembled *C. branneri* of Brazil, but later he

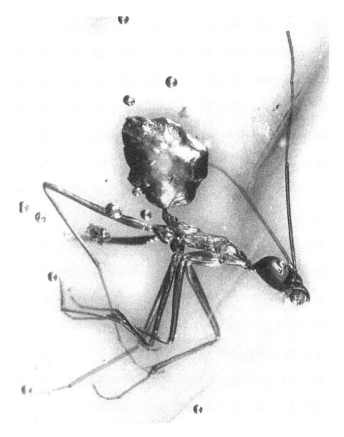

Figure 105. A swollen-abdomen worker-subcaste ant of the genus *Leptomyrmex* (family Formicidae) in Dominican amber (Poinar collection).

recognized it as a true *Leptomyrmex* (Baroni Urbani and Wilson, 1987). The discovery in Dominican amber of a worker subcaste of *Leptomyrmex* with a swollen abdomen is interesting (Fig. 105). This specialized worker subcaste arose in several ant genera as an adaptation to the ingestion of liquid food in dry climates and suggests that the amber forest had at least an extended dry season.

Another genus of ants found in Dominican amber that today has a Neotropical and Indo-Malayan distribution is *Gnamptogenys*, represented by *G. levinates* Baroni Urbani (1980D) and *G. pristina* Baroni Urbani. The latter species is similar to recent Central American and Antillean representatives of the genus, but the former species is morphologically distinct enough to possibly represent an extinct phyletic line.

Wilson published a series of papers (1985A, B, C, E) on the Dominican amber ants and emphasized how this group offered an excellent opportunity to study dispersal and evolution in a Tertiary West Indian fauna. Wilson (1985E) summarized his data regarding the fossil amber and present-day genera of ants on Hispaniola. Two genera, represented by *Ilemomyrmex caecus* Wilson and *Oxyidris antillana* Wilson, are now ex-

Figure 106. A worker *Paracryptocerus* (= *Zacryptocerus*) ant (family Formicidae) in Dominican amber (Poinar collection).

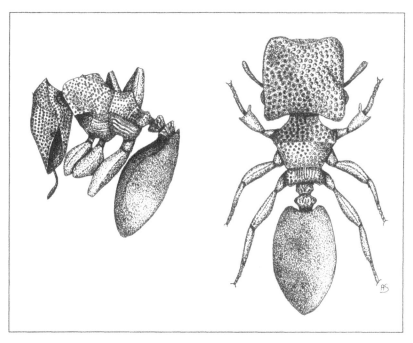

Figure 107. Two *Paracryptocerus* worker ants (family Formicidae) drawn from specimens in Dominican amber (drawing by A. Ann Sorensen).

tinct worldwide (Wilson, 1985A). Both were small and eyeless, and possessed narrow, sharp-toothed mandibles. Species that are extinct in the Greater Antilles but have generic ties elsewhere in the Neotropical area are *Azteca alpha* Wilson, *A. eumeces* Wilson, *Dolichoderus dibolia* Wilson, *Monacis prolaminata* Wilson, *M. caribbaea* Wilson, *Hypoclinea primitiva* Wilson, *Neivamyrmex ectopeis* Wilson, and representatives of the genera *Erebomyrma*, *Leptothorax*, *Paraponera*, and *Prionopelta* (Wilson, 1985B, 1985C, 1985E).

Now extinct on Hispaniola but present elsewhere in the Greater Antilles are representatives of *Cylindrymyrmex*, *Octostruma*, and *Prenolepis* (Wilson, 1985E). Those amber fossils that represent genera still present on Hispaniola include *Anochetus*, *Aphaenogaster*, *Camponotus*, *Crematogaster*, *Cyphomyrmex*, *Gnamptogenys*, *Hypoponera*, *Iridomyrmex* (*I. hispaniola* Wilson [1985C] and *I. humiloides* Wilson [1985C]), *Leptothorax*, *Odontomachus*, *Pachycondyla*, *Paracryptocerus*, (= *Zacryptocerus*) (Figs. 106–108), *Paratrechina*, *Pheidole* (*P. tethepa* Wilson [1985A]), *Platythyrea*, *Pseudomyrmex*, *Smithistruma*, *Solenopsis*, *Tapinoma* (*T. trochis* Wilson [1985C])

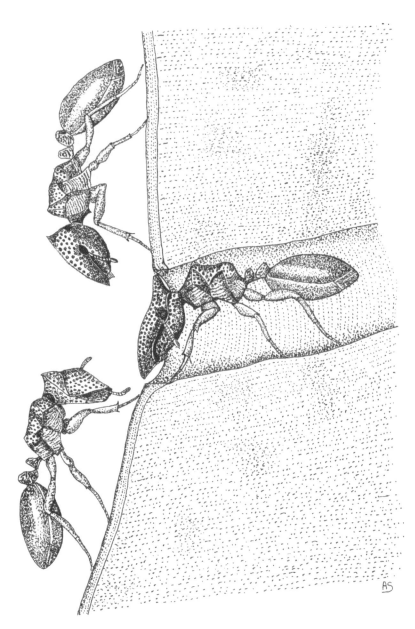

Figure 108. Reconstruction of a habitat setting with three worker ants of *Paracryptocerus* (family Formicidae) from Dominican amber. The broad head of a major worker is used to plug the entrance hole of the nest to protect against enemies (drawing by A. Ann Sorensen).

and *Trachymyrmex* (*T. primaevus* Baroni Urbani [1980A]) (Wilson, 1985E). In discussing the subject of survival in the West Indian ant fauna, Wilson (1985E) concluded that a higher extinction rate occurred in genera and subgenera that were highly specialized or possessed less colonizing ability.

Also described from Tertiary Eocene amber is *Eomyrmex guchengziensis* Hong et al. (1974) from Chinese amber and *Eocenidris crassa* Wilson, *Iridomyrmex mapesi* Wilson, and *Protrechina carpenteri* Wilson from Arkansas amber (Wilson, 1985D). The latter three species represent the first fossils of the most advanced subfamilies of ants, the Myrmicinae, Dolichoderinae, and Formicinae, respectively. The Mexican amber ants have not been critically studied. Brown (1973) commented that representatives of the genera *Azteca*, *Camponotus*, ?*Crematogaster*, *Dorymyrmex*, ?*Lasius*, *Mycetosoritis*, *Pachycondyla*, *Pheidole*, and *Stenamma* occurred in the Chiapas deposits (Fig. 109). From Australian Tertiary resin was described *Hypoponera scitula* (Clark) (Hills, 1957).

Superfamily Scolioidea (scolioidid wasps). Members of this superfamily are medium to large, more or less hairy, stocky, and relatively short-

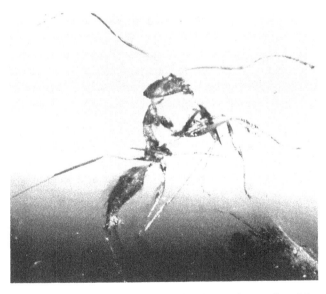

Figure 109. An ant (family Formicidae) in Mexican amber (Museum of Paleontology collection, University of California, Berkeley; photo by P. Hurd).

legged. The females, which often are wingless, deposit their eggs on various arthropod hosts. Upon hatching, the larvae feed externally on the host.

Mutillidae (velvet ants). The females in this family are wingless and antlike, but are covered with a dense pubescence. The adults usually have characteristic "felt lines" on the second abdominal segment. Mutillids occur generally in open, dry areas where females search out the larvae and pupae of wasps, bees, beetles, and flies. Three species of *Mutilla* and a few unidentified specimens occur in Baltic amber (Brischke, 1886; Larsson, 1978; Menge, 1856). *Dasymutilla dominica* from Dominican amber is the first fossil mutillid from the New World (Manley and Poinar, 1991).

Sapygidae. Members of this small family parasitize bees and wasps. Reference to a *Sapyga* sp. and other unidentified forms in Baltic amber are the only reports of this group in amber (Spahr, 1987).

Scoliidae. The adults of these wasps are large and hairy, and the larvae are external parasites of scarabaeid beetles. Reference to a *Scolia* sp. and other unidentified forms from Baltic amber are the only reports of this group in amber (Spahr, 1987).

Tiphiidae. These are medium-sized, darkish, somewhat hairy wasps that parasitize beetles, bees, and wasps. Representatives of this family have been reported from Siberian amber (Zherichin and Sukacheva, 1973), and Bischoff (1916) described seven species of *Protomutilla* (sometimes placed in the Myrmosidae) from Baltic amber. A possible *Tiphia* sp. from Baltic amber was reported by Brischke (1886). The family is also represented in Dominican amber (Appendix B).

Superfamily Vespoidea. This superfamily includes the paper wasps, yellow jackets, hornets, mud daubers, and mason and potter wasps. They tend to have the posterior margin of the pronotum strongly U-shaped in dorsal view.

Vespidae (paper wasps, yellow jackets, and hornets). These are social wasps, with colonies composed of queens, workers, and males. The nest is constructed out of a papery material produced by the workers from plant material. Two species, *Palaeovespa baltica* Cockerell and *Vespa dasypodia* Menge, and representatives of *Polistes* and *Vespa* have been reported from Baltic amber (Spahr, 1987). These are the only reports of this group from amber.

Superfamily Sphecoidea: Sphecidae. These solitary wasps are medium-sized to large and the posterior margin of the pronotum is straight. They nest in burrows in the ground or in plant material, and some construct mud nests. They are sometimes called digger wasps. All of the Sphecoidea are included here under the family Sphecidae, but the subfamilies are sometimes treated as separate families.

Cretaceous amber sphecids are represented by *Pittoecus pauper* Evans (1973) from Siberian amber, *Lisponema singularis* Evans (1969) from Canadian amber, and *Gallosphex cretaceus* Schlüter (1978A) from French amber. *Taimyrisphex pristinus* Evans (1973) and *Falsiformica cretacea* Rasnitsyn (1975) from Siberian amber have been placed in a separate extinct family, the Falsiformicidae, by Rasnitsyn (1975). Also from Siberian amber is *Cretabythus sibiricus* Evans (1973), which is placed in the family Scolebythidae.

Tertiary representatives of this group have been described from Baltic amber. These include species or individuals of the following genera: *Cerceris, Crabro, Crossocerus, Dolichurus, Pison, Mellinus, Gorytes, Mimesa, Passaloecus, Pemphredon,* and *Psen* (Spahr, 1987). Antropov and Pulawski (1989) described a Baltic amber *Pison* (subgenus *Entomopison*) whose extant members exist only in the Neotropical region.

Representatives of *Crabro, Crossocerus,* and *Pison* have been identified from Dominican amber by R. Coville (Fig. 110).

Pompilidae (spider wasps). These graceful long-legged wasps are normally dark colored, with smoky or yellowish wings. The females can be found on the ground or on plants searching for spiders, which are used as an oviposition site. A single species, *Pompilus scelerosus* Meunier, and a *Pepsis* have been recorded from Baltic amber (Spahr, 1987; Larsson, 1978).

Superfamily Apoidea (bees). Bees differ from wasps in possessing plumose or branched body hairs. On most, the hind basitarsus (first tarsal segment) is flattened, and the pollen collectors possess scopae (brushes of hairs on hind legs or abdomen) or corbiculae (flattened surface of the hind tibia). The living habits of bees range from nonsocial to social and their feeding habits include saprophagy (feeding on dead animals), parasitism, and nectar and pollen gathering. This diversity has resulted in the erection of several separate families for this group.

Andrenidae. Andrenids often nest in groups in ground burrows although they are considered solitary bees. Andrenids carry pollen on scopae, and because of their short tongues they tend to take nectar from

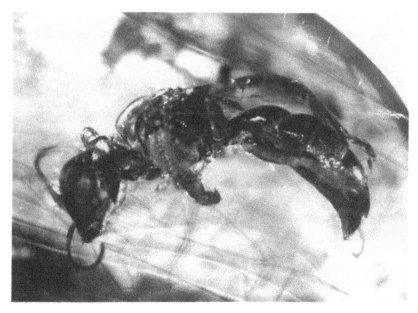

Figure 110. A sphecid wasp of the genus *Crossocerus* (family Sphecidae) in Dominican amber (Poinar collection).

exposed nectaries or flowers they can bodily enter. A single amber species, *Andrena wrisleyi* Salt (1931), has been described from Baltic deposits.

Anthophoridae. This family includes the stout and hairy digger bees, most of which collect pollen and nest in ground burrows. Representatives of *Anthophora* sp. in Baltic amber have been reported by several authors (Spahr, 1987). This family also includes the carpenter bees, which are dark blue-black forms with a bare and shiny abdomen that nest in wood or plant stems. Mention of a *Xylocopa* sp. in Baltic amber is the only report of this group in amber (Bachofen-Echt, 1949).

The Anthophoridae also contains the cuckoo bees, all of which are parasites in the nests of other bees. There are, however, no reports of this group in amber.

Apidae. Most of the Apidae are social, and the workers possess corbiculae for carrying pollen on their hind legs. They include the honeybees (Apinae), well known to us by our common bee, *Apis mellifera* Linnaeus, bumblebees (Bombinae), euglossine bees (Euglossinae), and the stingless bees (Meliponinae). Kelner-Pillault (1969) reported 173 fossil representatives of the superfamily Apoidea. These comprise 112 species, of which 41 have been described from amber, including 34 from Baltic amber.

The earliest bee in the fossil record, and the only Cretaceous record of any bee, is *Trigona prisca*, described from New Jersey amber (Michener and Grimaldi, 1988) (Color Plate 1). The next earliest bees are members of the extinct genus *Electrapis*. They are considered to be social, as evidenced by the presence of many specimens together in Baltic amber and the presence of pollen collecting apparatus. Whether *Electrapis* is directly ancestral to our present-day *Apis* or whether it represents an evolutionary sideline is still an open question (Burnham, 1978).

The following species of *Electrapis* Cockerell have been described from Baltic amber: *E. apoides* Manning, *E. bombusoides* Kelner-Pillault, *E. indecisus* Cockerell, *E. meliponoides* (Buttell-Reepen), *E. minuta* Kelner-Pillault, *E. palmnickenensis* (Roussy), *E. proava* (Menge), *E. tornquisti* Cockerell, and *E. tristellus* (Cockerell) (Spahr, 1987). Also from Baltic amber are the following species of *Bombus*: *B. carbonarius* Menge, *B. muscorum* Roussy, and *B. pusillus* Menge. Cockerell (1909) also described species of the extinct genera *Chalcobombus*, *Sophrobombus*, and *Protobombus* from Baltic amber, and together with *Bombusoides mengei* Motschulsky these indicate the presence of bumblebee-like forms in the early Tertiary.

Stingless bees are well represented in amber. *Melipona* sp. and *Kelneriapis eocenica* (Kelner-Pillault) occur in Baltic amber, *Meliponorytes sicula* Tosi and *Meliponorytes succini* Tosi are from Sicilian amber, and *Tetragonula devicta* (Cockerell) was described from Burmese amber (Spahr, 1987). The latter species may have been in younger deposits, since it resembles the Recent *T. laeviceps* Smith (Cockerell, 1922). *Trigona (Nogueirapis) silacea* Wille (1959, 1977) (now in the genus *Plebeia* [see Michener, 1990]) was described from Mexican amber. The latter closely resembles the rare extant species *T. butteli* Friese, which occurs in Peru, Bolivia, the upper Amazon, and the Territory of Rio Negro, and *T. mirandula* from Costa Rica.

Wille and Chandler (1964) described *Trigona (Liotrigona) dominicana* from Dominican amber and felt that it was related to the *Liotrigona* group, which is now represented by three African species. At present no indigenous member of the tribe Meliponini is known from the Greater Antilles. After examining specimens of *T. dominicana*, Michener (1982) decided that it belonged to a new subgenus, *Proplebeia*, and later raised *Proplebeia* to generic level. Thus the name of the Dominican amber trigonid is now *Proplebeia dominicana* (Wille and Chandler) (Michener, 1990), thus removing a biogeographical anomaly by aligning *P. dominicana* with Neotropical forms. I feel that this question is still open and hopefully will be resolved by future workers. The Dominican trigonid is relatively common in amber, presumably because individuals became entangled when they attempted to collect resin balls for nest construction (Color Plate 5). Such resin balls can still be seen on the corbiculae of fossilized individuals (Fig. 111). The lack of pollen masses on the corbiculae suggest that the amber forms had the specialized task of collecting only resin. The presence of phoretic mites and triungulin larvae on *P. dominicana* establishes early symbiotic relationships (see Chapter 5).

Halictidae. These small to moderate-sized bees include the sweat bees (*Lasioglossum* spp.) that are attracted to perspiration. Most collect pollen but some are parasitic on other bees. Mention of a *Halictus* sp. in Baltic amber is the only record of this family in amber (Bachofen-Echt, 1949).

Megachilidae (leafcutting bees). The stoutish females of this family carry pollen in a scopa on the ventral side of the abdomen. Many species

Figure 111. A worker stingless bee, *Proplebeia dominicana* (Wille and Chandler) (family Apidae), in Dominican amber. Note the resin balls still attached to the corbiculae of the hind legs (Poinar collection).

The oldest known fossil bee, *Trigona prisca* Michener & Grimaldi (family Apidae), in New Jersey (Kinkora) amber. (American Museum of Natural History collection; photo by D. A. Grimaldi)

The Cretaceous ant *Sphecomyrma freyi* Wilson, Carpenter & Brown in New Jersey amber. (Harvard Museum collection; photo by F. M. Carpenter)

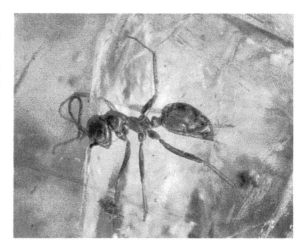

Left Pollen falling from an anther of the extinct tree *Hymenaea protera* (family Leguminoseae), which produced much of the Dominican amber. (Poinar collection)   *Right* A petal of *Hymenaea protera* in Dominican amber. (Poinar collection)

A frog of the genus *Eleutherodactylus* (family Leptodactylidae) in Dominican amber. (Work collection)

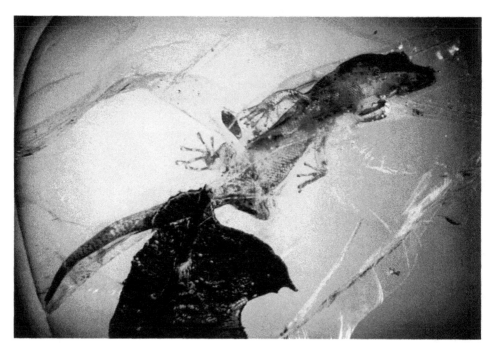

A lizard of the genus *Anolis* (family Iguanidae) in Dominican amber.

Color Plate 2

A feather in Dominican amber.
(Smithsonian Institution collection)

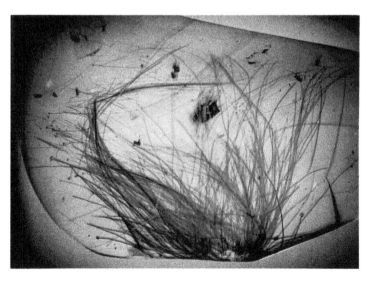

Mammalian hair in Dominican amber. (Costa collection,
Puerto Plata, Dominican Republic)

A walking stick (family Phasmatidae) in Dominican amber.

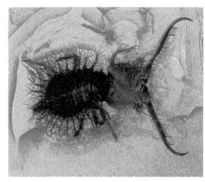

A larva of the owlfly *Neadelphus protae* MacLeod (family Ascalaphidae) in Baltic amber. (Harvard Museum collection; photo by F. M. Carpenter)

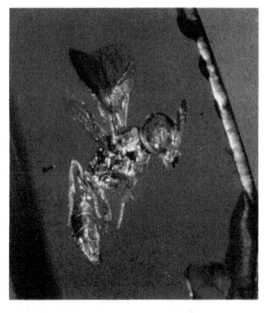

A serphitid wasp, *Serphites dux* Kozlov (family Serphitidae), in Siberian amber (Yantardak site). The Serphitidae became extinct in the Paleocene. (Photo by A. P. Rasnitsyn; courtesy of V. V. Zherichin)

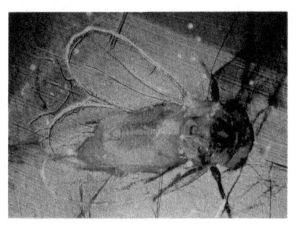

A primitive whitefly, *Bernaea neocomica* Schlee, in Lebanese amber. (Stuttgart Natural History Museum collection)

Left to right A winged termite (order Isoptera) in Dominican amber; a worker ant carrying an ant pupa (family Fomicidae) in Dominican amber (Poinar collection); a fungus gnat (family Mycetophilidae) in Dominican amber.

The stingless bee *Proplebeia dominicana* (Wille & Chandler) in Dominican amber. (Poinar collection)

A parasitic wasp (superfamily Chalcidoidea) in Dominican amber.

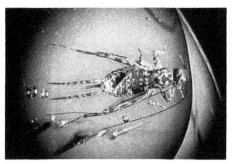

A platypodid beetle (family Platypodidae) in Dominican amber. (Poinar collection)

A cricket (family Gryllidae) in Dominican amber.

A millipede (order Diplopoda) in Dominican amber. (Poinar collection)

A winged ant (family Formicidae) in Dominican amber.

Color Plate 5. Arthropods commonly occurring in tertiary amber

A wedge-shaped beetle (family Rhipiphoridae) in Dominican amber.

An ant bug, *Praecoris dominicana* Poinar, in Dominican amber. (Poinar collection)

A soldier fly (family Stratiomyidae) in Dominican amber. (Poinar collection)

A cockroach in Dominican amber.

A pseudoscorpion (order Pseudoscorpiones) in Dominican amber.

A fungus-like organism in Triassic amber (Carnian stage, ca. 225 Ma) from Germany. (Courtesy of Ulf-Chr. Bauer)

A butterfly (family Lycaenidae or Riodinidae) in Dominican amber.

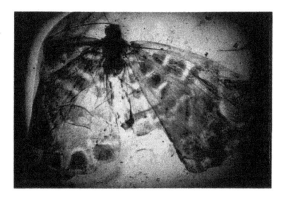

A mushroom, *Coprinites dominicana* Poinar & Singer, in Dominican amber. (Poinar collection)

A weevil (family Curculionidae) in Dominican amber. (Poinar collection)

A pair of water striders (family Gerridae) in Dominican amber, the first example of mate-guarding in the fossil record. (Poinar collection)

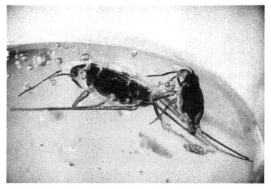

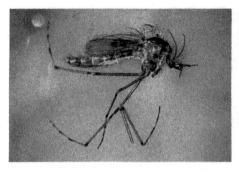

A sand fly (family Psychodidae) in Dominican amber. (Poinar collection)

A parasitic wasp (family Ichneumonidae) in Dominican amber.

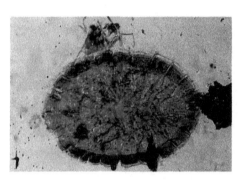

A termite bug (family Termitaphididae) in Mexican amber. (Poinar collection)

A dusty wing (family Coniopterygidae) in Dominican amber.

A *Strobilops* snail (class Gastropoda) in Dominican amber. (Brodzinsky collection)

A plasmodium stage of a myxomycete (order Physarales) in Dominican amber. (Poinar collection)

line their brood cells with pieces of leaves. Records of *Megachila* sp. and *Osmia* sp. from Baltic amber are the only records of this family in amber (Spahr, 1987).

Melittidae. These small, dark bees nest in the soil. Species or representatives of the genera *Ctenoplectrella*, *Dasypoda*, and *Glyptapis* occur in Baltic amber (Cockerell, 1909; Spahr, 1987), but other amber forms are unknown.

Kingdom Animalia: Arthropoda: Arachnida

The arachnids are generally considered a class in the subphylum Chelicerata of the phylum Arthropoda. The Chelicerata lack antennae and have six pairs of appendages. The first pair of appendages are called chelicerae and, in the Arachnida, the second pair are known as the pedipalps. The body is divided into an anterior prosoma (cephalothorax) and a posterior opisthosoma (abdomen). Eleven different orders are generally recognized in the Arachnida, and most have representatives in amber.

Scorpiones (Scorpions)

This order is recognizable by their pincers (chelate pedipalps) and the exposed stinger, borne on the narrow terminal portion of the opisthosoma known as the metasoma. The anterior portion of the abdomen is called the mesosoma. On the ventral side of the second abdominal segment lie a pair of comblike structures, the pectines. The function of these unique organs is not certainly known, but they are considered by some to be tactile sensory organs. Scorpions were at first aquatic and later became terrestrial. They have been described from Baltic and Dominican amber. From the former source was described *Tityus eogenus* Menge (1869), the first fossil record of the scorpion family Buthidae. According to Schawaller (1979B), the assignment of this juvenile scorpion to *Tityus* is questionable, but the family assignment is probably correct. Unfortunately, the specimen is now lost.

Still earlier appeared the description of *Scorpio schweiggeri* Holl (1829) from Baltic amber, first mentioned by Schweigger in 1819. Apparently the specimen is lost, and Schawaller (1979B) considered it a nomen nudum but thought that it was probably a representative of the Buthidae.

Three fossil scorpion species have been described from Dominican amber. The first was *Centruroides beynai* Schawaller (1979B) (Buthidae), a description based on a single juvenile. The second was *Microtityus am-*

Figure 112. Holotype of the scorpion *Tityus geratus* Santiago-Blay and Poinar in Dominican amber.

barensis (Schawaller, 1982A; Santiago-Blay et al., 1990) (Buthidae), based on two juvenile specimens and a tail of a larger animal. The third, *Tityus geratus* Santiago-Blay and Poinar (1988), was based on a single second-stage nymph (Fig. 112). These three fossil scorpions are the only known Tertiary scorpions from the Americas. Two extant *Tityus* species, *T. quis-queyanus* Armas and *T. crassimanus* (Thorell), occur on Hispaniola today, but they are clearly distinct from the fossil form.

Pseudoscorpiones (False or Moss Scorpions)

Members of this order resemble small scorpions possessing large chelate pedipalps (pincers); however the abdomen is not drawn out into a tail with a stinger. Their relative abundance in amber is due to their habit of living under bark and other concealed habitats where they prey on small insects. Most species have venom glands on their pedipalps, and pseudoscorpions also emit silk from their chelicerae, used in forming overwintering cocoons.

Pseudoscorpions appear first in the fossil record in the Devonian (Shear et al., 1984). They are not identifiable to genera and species, however. The Baltic amber pseudoscorpions have been studied by Beier (1937, 1947, 1955) and recently by Schawaller (1978). These include species of the genus *Chelignathus* in the family Dithidae; *Chthonius*, *Lechytia*, and *Pseudochthonius* in the family Chthoniidae; *Microcreagris*, *Neobisium*, and *Roncus* in the Neobisiidae; *Garypinus* and *Geogarypus* of the Garypidae; *Pseudogarypus* of the Pseudogarypidae; *Cheiridium* of the Cheiridiidae; *Oligochernes* of the Chernetidae; *Progonatemnus* of the Atemnidae; *Chelifer*, *Electrochelifer*, *Oligochelifer*, *Pycnochelifer*, and *Oligowithius* of the Cheliferidae (Keilbach, 1982). Larsson (1978) comments that a relatively large number of specimens of *Pseudogarypus* were found in Baltic amber, whereas the genus is presently restricted to the southern parts of North America. On the basis of Recent forms, the habitats of the Baltic amber pseudoscorpions included the moist subterranean, the upper litter, and warm, dry bark zones. Filaments of silk in the amber with *Microcreagris koellneri* Schawaller indicate that the ability to spin was already established by the Late Eocene (Schawaller, 1978).

Pseudoscorpions have also been identified or described from Dominican amber (Color Plate 6). These include *Pachychernes effossus* Schawaller (one tritonymph), *Pachychernes* sp. (one protonymph), and *Americhernes* sp. (one tritonymph), all of the family Chernetidae (Schawaller, 1980A); *Pseudochthonius squamosus* Schawaller (one female) and *Lechytia tertiaria* Schawaller (one male), both of the family Chthoniidae (Schawaller, 1980B); *Cryptocheiridium antiquum* Schawaller (one male and three females) in the family Cheiridiidae (Schawaller, 1981A); and two specimens of a *Parawithius* sp. in the family Cheliferidae (Schawaller, 1981B). The two specimens of *Parawithius* sp. were still attached to platypodid beetles, indicating that the phoretic association between pseudoscorpions and insects was well established in the Tertiary (see

Figure 113. A pseudoscorpion holding onto the abdomen of a platypodid beetle. This type of phoretic behavior illustrates how some pseudoscorpions move from one site to another. Note also the dauer nematodes adjacent to the beetle (Poinar collection).

Fig. 113). These descriptions are the first records of fossil pseudoscorpions from the Western Hemisphere.

A member of the Chernetidae was reported by Schawaller (1982B) from Mexican amber. It was represented by a single protonymph and could not be assigned to either species or genus. Protescu (1937) also reported a pseudoscorpion of the genus *Cheiridium* (Cheiridiidae) from Romanian amber. Two other species, *Electrobisium acutum* Cockerell (1917F) of the Neobisiidae and *Garypus burmiticus* Cockerell (1920C) of the Garypidae, were described from Burmese amber.

Amblypygi (Tail-less Whipscorpions)

Previous authors have placed the interesting arthropods of this order in the families Tarantulidae and Phrynidae of the order Phrynichida. Their long, spiny pedipalps are used for capturing prey, and the first pair of legs, which are long and slender, are probably used as sensory probes. The abdomen is distinctly segmented and lacks a terminal appendage. Fossil records of this group extend back to the Carboniferous (Crowson et al., 1967).

The first report of this group in amber was the description, from a partial specimen in Mexican amber, of *Electrophyrnus mirus* Petrunkevitch (1971) in the new family Electrophrynidae. This genus differed from Recent genera in having weak, elongate chelicerae and a specialized tibia in the pedipalpus.

A second amber specimen was described by Schawaller (1979A) from Dominican amber as *Tarantula resinae* Schawaller on the basis of a single juvenile specimen. Later, after examining four additional specimens, Schawaller (1982C) transferred the species to the genus *Phrynus* (Phrynidae) and commented on how closely the fossil *P. resinae* resembled the extant Hispaniolan species *P. marginemaculatus* C. L. Koch. An exuvia of an amblypigid from Dominican amber was identified as a male *Phrynus* by D. Quintero Arias (Fig. 114).

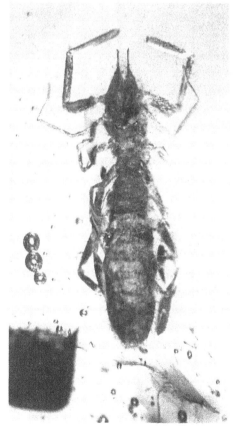

Figure 114. A tail-less whipscorpion, *Phrynus* (order Amblypygi), in Dominican amber (Poinar collection; identified by D. Quintero Arias).

Figure 115. Holotype of the windscorpion (or solpugid) *Happlodontus proterus* Poinar and Santiago-Blay in Dominican amber (Truman collection). (Published in *J. New York Entomol. Soc.*)

Solpugida (= Solifugae) (Windscorpions)

This order is characterized by large forward-directed chelicerae, stout pedipalps with a blunt end bearing an adhesive organ, slender first legs that are tactile in function, distinct segmentation of the abdomen, and the presence of racquet organs (T-shaped structures) on the fourth pair of legs. The only solpugid known from amber is *Happlodontus proterus* Poinar and Santiago-Blay (1989), a member of the Ammotrechidae found in Dominican deposits (Fig. 115). This is the best preserved of the two known fossil solpugids. The other fossil, earlier and not from amber, is a single specimen of *Protosolpuga carbonaria* Petrunkevitch from Pennsylvanian deposits in North America.

Opiliones (Harvestmen, Daddy Longlegs)

In members of this order, the cephalothorax and abdomen are broadly joined, giving the body a roundish appearance. Members of the largest suborder, Palpatores, characteristically possess long legs. Most opilionids are predaceous on other arthropods; however, some are saprophagous and others utilize plant juices as a source of nourishment. Their fossil record extends back to the Carboniferous (Crowson et al., 1967). An undetermined representative of this group has been reported from French Cretaceous amber (Schlüter, 1978A).

Most descriptions of opilionids are from Baltic amber and include species in the genera *Acantholophus, Cheiromachus, Dicranopalpus, Leiobunum, Opilio, Phalangopus,* and *Platybunus* of the Phalangiidae; *Nemastoma* of the Nemastomatidae; and *Sabacon* of the Ischyropsalidae (Keilbach, 1982) (Fig. 116). Larsson (1978) comments that the isolated legs of many Opiliones have been discovered in Baltic amber but identification of these is impossible. Thus, the presence of harvestmen in the amber forest was probably much greater than indicated by the number of entire specimens, simply because many were "saved" from entrapment by possessing legs that easily autonomize at the coxal joint. Cokendolpher (1986) described the Dominican amber *Pellobunus proavus*, the first fossil representative of the suborder Lanistores from the New World, and Cokendolpher and Poinar (1992) described *Philacarus hispaniolensis* and *Kimula* sp. from Dominican deposits.

Acari (Mites and Ticks)

Representatives of this order are small to minute, oval forms with little differentiation between the cephalothorax and abdomen. Newly

Figure 116. A harvestman (order Opiliones) in Baltic amber (photo by G. Brovad). (Published in *Baltic Amber: A Palaeobiological Study*, by S. Larsson. Scandinavian Science Press Ltd.)

hatched young (larvae) possess only three pairs of legs and acquire the fourth pair after the first molt. The stages between the larva and adult are called nymphs. This large order contains groups with a considerable amount of biological diversity, for example forms that are free-living, plant-parasitic, animal-parasitic, or aquatic. As a consequence, the order is commonly broken down into seven or eight suborders.

Both hard ticks (Ixodidae), characterized by a hard dorsal plate or scutum and anteriorly protruding mouthparts, and soft ticks (Argasidae), which are soft-bodied, lack a scutum, and have ventrally located mouthparts, have been recorded from amber. One soft tick considered to be an *Ornithodoros* sp. was recorded from Dominican amber but still awaits description (R. Woodruff, personal communication). Two hard ticks have been reported, one female from Baltic amber described as *Ixodes succineus* Weidner (1964) and a male from Dominican amber described as a representative of *Amblyomma* (Lane and Poinar, 1986) (Figs. 117, 118).

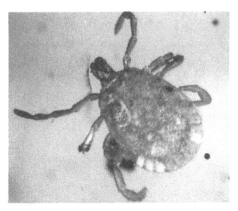

Figure 117. A larval tick, *Amblyomma* sp. (suborder Ixodida) in Dominican amber (Poinar collection).

Figure 118. An adult female tick, *Amblyomma* sp. (suborder Ixodida), in Dominican amber (Work collection). (Published in *International Journal of Acarology*, Vol. 12, No. 2, Indira Publishing House, P.O.B. 37256, Oak Park, Michigan 48237.)

Unfortunately, because the host range of present-day species of *Amblyomma* is extensive (occurring on birds, reptiles, and mammals), the presence of a specific host group cannot be inferred on the basis of these finds.

Many mites have been described from various ambers, mostly Baltic (Keilbach, 1982). At least two have been described from Cretaceous deposits: *Eocamisia sukatshevae* Bulanova-Zachvatkins (Camisiidae) from Siberian amber and *Bdella vetusta* Ewing (Bdellidae) from Canadian amber. Unidentified mites have also been reported from Lebanese (Schlee, 1972) and French (Schlüter, 1978A) amber, and representatives of the Erythraeidae, Gymnodamoeidae, and Oribatei occur in Canadian amber (McAlpine and Martin, 1969).

Baltic amber mites include species in the following genera and families: *Bdella* and *Scirus* in the Bdellidae; *Camisia* in the Camisiidae; *Cheyletus* in the Cheyletidae; *Acarus* and *Erythraeus* in the Erythraeidae; *Banksia, Brachychthonius, Caleremaeus, Carabodes, Cepheus, Ceratoppia, Chamobates, Cultroribula, Damaeus, Dameosoma, Embolacarus, Eporibatula, Eremaeus, Euzetes, Galumna, Gradidorsum, Gymnodamaeus, Hermanniella, Hoploderma, Licneremaeus, Liebstadia, Lucoppia, Melanozetes, Micreremus, Neoliodes, Neoribates, Notaspis, Nothrus, Odontocephalus, Otocepheus, Oribatus, Oribatella, Oribotritia, Oripoda, Pelops, Plategeocranus,*

Platyliodes, Protoribates, Punctoribates, Scapheremaeus, Scheloribates, Scutoribates, Spaeozetes, Strieremaeus, Suctobelba, Tectocepheus, Tectocymba, Tectoribates, and *Trhypochthonius* of the Oribatidae; *Sejus* of the Parasitidae; *Tetranychus* of the Tetranychidae; and *Actineda, Arytaena, Penthaleus, Rhyncholophus*, and *Trombidium* of the Trombidiidae.

On the basis of this list, the Oribatidae are by far the most diverse and common group found in Baltic amber. Present-day forms occur in leaf litter, under bark and stones, and in soil, where they are scavengers, feeding mostly on plant matter (algae) and fungi. Sellnick (1931), who has been the most recent to examine and revise the Baltic amber mites, mentioned how close many of the amber forms are to present-day species, indicating that little or no evolution has occurred in these groups during the past 35–45 million years.

Mites have been described also from Mexican amber. Türk (1963) was the first to describe two deutonymphs (second-stage nymphs) of *Amphicalvolia hurdi* Turk (Tyroglyphidae) from this source. This was followed by descriptions of a male of *Dendrolaelaps fossilis* Hirschmann (1971) (Digamasellidae), *Damaeus mexicanus* Woolley (1971), and *D. setiger* Woolley (1971) of the Damaeidae (or Oribatidae); *Hydrozetes smithi* Woolley (1971) of the Hydrozetidae; *Exoripoda chiapasensis* Woolley (1971) of the Oripodidae; *Liebstadia durhami* Woolley (1971) of the Oribatulidae; *Scapheremaeus brevitarsus* Woolley (1971) of the Cymberemaeidae; *Eremaeus denaius* Woolley (1971) of the Eremaeidae (or Oribatidae); and *Oppia hurdi* Woolley (1971) of the Oppiidae.

A single member of the Trombidiidae, *Cheyletus burmiticus* Cockerell (1917A), has been described from Burmese amber, and Womersley (1957) described *Acronothrus ramus* Womersley (Camisiidae) from Australian amber. Mites of the suborders Gamasida (Gamasidae), Oribatida (Oribotritiidae, Licaridae, Carabodidae), Actinedida (Trombidioidea), Prostigmata (Bdellidae, Cunaxidae, Erythraeidae), and Cryptostigmata (Galumnidae, Oppiidae, Ceratozetoidea, Plateremaeoidea) have been identified by J. Krantz and A. Moldenke in Dominican amber (Figs. 119–122). The first fossil fur mites (Listrophoridae) associated with mammalian hair occur in Dominican amber (Poinar, 1988A) (Fig. 123). Parasitic erythraeid mites still attached to lepidopteran hosts in Dominican amber have also been reported (Poinar et al., 1991).

Araneae (Spiders)

The body of a spider is composed of an anterior cephalothorax and a posterior, normally unsegmented abdomen, attached to the former by a

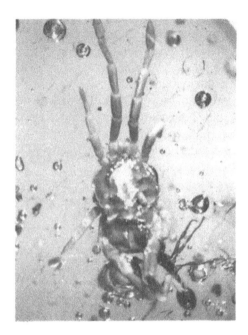

Figure 119. A red mite in Dominican amber (Poinar collection).

Figure 120. A penknife mite (family Orbibotritiidae) in Dominican amber (Poinar collection).

Figure 121. A mite near the genus *Fallopia* (family Erythraeidae) in Dominican amber (Poinar collection).

Figure 122. A predaceous mite (family Bdellidae) in Dominican amber (Poinar collection).

Figure 123. A listrophorid fur mite in Dominican amber (Costa collection).

short pedicel. The mouthparts, eyes, and legs are attached to the cephalothorax whereas the abdomen bears the spiracles, spinnerets, anus, and genital structures. Most spiders have eight simple eyes, two segmented chelicerae containing openings of the poison glands, leglike pedipalps containing the male copulatory organ, and book lungs. Silk, produced in glands, issues from an udderlike structure called the spinnerets. The silk, its use, and the structure of the produced web vary among species of spiders. For example, the silk of orb weavers is often covered with drops of a gluey substance that helps to entrap the prey. Spiders represent an extremely common order that occurs in a variety of habitats.

All spiders are predaceous, and prey are generally killed by poison introduced with the first spider bite. Spiders are normally divided taxonomically into two to four suborders and two or more sections. Fossil records definitely establish the presence of spiders in the Carboniferous and possibly even earlier, in the Devonian (Crowson et al., 1967).

Spiders are moderately common in amber, and appear to be present in all amber deposits that contain insects, including Cretaceous Canadian (Araneidae, Linyphiidae, Theridiidae) (McAlpine and Martin, 1969), Siberian (Zherichin and Sukacheva, 1973), French (Schlüter, 1978A), and possibly Lebanese (Schlee, 1972). No formal description of a Cretaceous amber species has appeared, however.

Petrunkevitch (1942, 1950, 1958, 1963, 1971) has described, discussed, and written more on amber spiders than any other person. Aside from completely examining and classifying the Baltic amber spiders (1942,

1950, 1958), he spent considerable effort on the Mexican amber spiders (1963, 1971). Recent workers on amber spiders include Schawaller (1981C, 1981D, 1982D, 1984), Ono (1981), and Wunderlich (1986, 1988).

In the following account of the Tertiary amber spiders, no attempt will be made to utilize suborders or sections, because spider classification is not consistent. Rather, all families and genera will be listed alphabetically under the order Araneae.

Species of Baltic amber spiders have been described in the following genera and families: *Adjunctor, Adjutor,* and *Admissor* of the Adjutoridae; *Agelena, Eocryphoeca, Myro, Tegenaria, Textrix,* and *Thyelia* of the Agelenidae; *Amaurobius* and *Auximus* of the Amaurobiidae; *Anyphaena* of the Anyphaenidae; *Acrometa, Araneus, Cyclosoma, Elucus, Eometa, Epeira, Eustaloides, Gea, Graea, Memoratrix, Priscometa, Siga, Singa, Theridiometa,* and *Zilla* of the Araneidae; *Archaea* and *Entomocephalus* of the Archaeidae; *Arthrodictyna* of the Arthrodictynidae; *Ablator, Abligurtor, Clubiona, Concursator, Cryptoplanus, Eodoter, Eomazax, Eophrurolithus, Erithus, Heteromma, Laccolithus, Macaria, Machilla, Massula, Mizalia, Phrurolithus,* and *Sosybius* of the Clubionidae; *Eolathys* of the Dictynidae; *Clostes* of the Dipluridae; *Dasumia, Dysdera, Harpactes,* and *Thereola* of the Dysderidae; *Ephalmator* of the Ephalmatoridae; *Eresus* of the Eresidae; *Eogonatium, Erigone,* and *Micryphantes* of the Erigonidae; *Adulatrix, Caduceator, Collacteus, Eoprychia, Eostaianus, Eostasina, Ocypete,* and *Zachria* of the Eusparassidae; *Captrix, Eomactator, Idmonia, Melanophora, Pythonissa,* and *Spheconia* of the Gnaphosidae; *Eohahnia* of the Hahniidae; *Gerdia* and *Hersilia* of the Hersiliidae; *Inceptor* of the Inceptoridae; *Insecutor* of the Insecutoridae; *Antopia, Custodela, Eopopino, Impulsor, Linyphia, Liticen, Malleator, Meditrina, Mystagogus, Obnisus,* and *Viocurus* of the Linyphiidae; *Linoptes* of the Lycosidae; *Ero* of the Mimetidae; *Orchestina* of the Oonopidae; *Oxyopes* of the Oxyopidae; *Micropholcus* of the Pholcidae; *Eopisaurella* and *Esuritor* of the Pisauridae; *Eomatachia* of the Psechridae; *Almolinus, Cenattus, Eolinus, Evophrys, Gorgopsina, Paralinus, Parevophrys, Phidippus, Prolinus, Propetes,* and *Steneattus* of the Salticidae; *Segestria* of the Segestriidae; *Adorator* and *Spatiator* of the Spatiatoridae; *Astodipoena, Clya, Clythia, Cornitis, Dipoena, Eodipoena, Eomysmena, Euryopis, Flegia, Lithyphantes, Municeps, Mysmena, Nactodipoena, Steatoda,* and *Theridion* of the Theridiidae; *Anatone, Athera, Artamus, Eothanatus, Facundia, Fiducia, Filiolella, Medela, Misumena, Opisthophylax, Philodromus,* and *Syphax* of the Thomisidae; *Androgeus* of the Uloboridae; *Paruroctea* of the Urocteidae; *Anniculus, Di-*

elecata, and *Eocydrele* of the Zodariidae; and *Adamator* of the Zoropsidae (Petrunkevitch, 1958).

To this list of Baltic amber spiders can be added the following genera recently described by Wunderlich (1986): *Balticonesticus* and *Heteronesticus* of the Nesticidae and *Cornuanandrus* and *Pseudoacrometa* of the Acrometidae.

The families Adjutoridae, Arthrodictynidae, Ephalmatoridae, Inceptoridae, Insecutoridae, and Spatiatoridae are known only from species in Baltic amber (Schawaller 1981C). Petrunkevitch (1942, 1950, 1958) mentions that of the 140 genera and over 260 species of Baltic amber spiders, including six extinct families, none of the species and only a few genera have extant representatives. The majority of the spider genera in Baltic amber are poorly represented in the Holarctic today, but are related to tropical and subtropical forms in South Africa, Malaysia, Australia, and South America. Wunderlich (1986) argues, however, that although Petrunkevitch (1942) considered, for example, *Adamator* to be closely related to the Central American genus *Zorocrates*, there is no convincing indication for this or other generic affinities of the Baltic amber spiders with extant South American or Neotropical fauna. Wunderlich (1986) considers that the Baltic amber families Archaeidae and Cyatholipidae had a broad distribution in the Tertiary, whereas the former family is now restricted to South Africa, Madagascar, and Australia and the latter family to South Africa, Australia, New Zealand, and Jamaica. A pair of mating *Orchestina* in Baltic amber is interesting from the aspect of behavior (Wunderlich, 1982) (Fig. 124).

Mexican amber spiders were studied by Petrunkevitch (1963, 1971) and include species or representatives of the following genera and families: *Aranea* and *Mirometa* of the Araneidae; *Chiapasona*, *Mimeutychurus*, and *Prosocer* of the Clubionidae; *Mistura* of the Dysderidae; *Veterator* of the Eusparassidae; *Fictotama*, *Perturbator*, *Priscotama*, and *Prototama* (Fig. 125) of the Hersiliidae; *Malepellis* of the Linyphiidae; *Orchestina* of the Oonopidae; *Plantoxyopes* of the Oxyopidae; *Propago* of the Pisauridae; *Eomysmena*, *Municeps*, *Mysmena*, and *Pronepos* of the Theridiidae; and representatives of the Mimetidae and Salticidae. According to Roth (1965) the genus *Propago* should belong in the family Zodariidae.

The Dominican amber spiders have been studied by Ono (1981), Schawaller (1981C, 1981D, 1982D, 1984), and Wunderlich (1986, 1988) (Figs. 126–129). Included are representatives or species of the following genera: *Palaeoanapis* of the Anapidae; *Aysha*, *Teudis*, and *Wulfila* of the

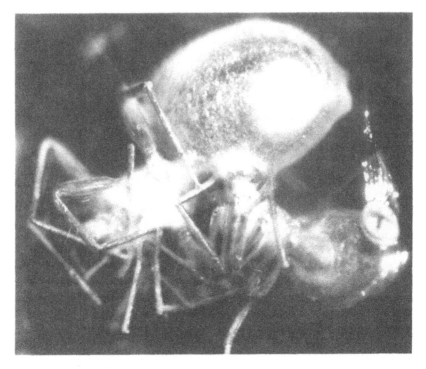

Figure 124. A pair of mating spiders, *Orchestina* sp. (family Oonopidae), in Baltic amber. The smaller male has placed his pedipalps into the genital opening of the female (photo by J. Wunderlich).

Anyphaenidae; *Araneometa, Araneus, Cyclosa, Fossilaraneus,* and *Pycnosinga* of the Araneidae; *Psalistops* of the Barychelidae; *Nops* of the Caponiidae; *Clubionoides* and *Strotarchus* of the Clubionidae; *Nanoctenus* of the Ctenidae; *Bolostromus* of the Ctenizidae; *Hispaniolyna, Palaeodictyna, Palaeolathys,* and *Succinyna* of the Dictynidae; *Ischnothele, Masteria,* and *Microsteria* of the Dipluridae; *Drassyllinus* of the Gnaphosidae; *Tama* of the Hersiliidae; *Tentabuna* of the Heteropodidae; *Agyneta, Lepthyphantes,* and *Palaeolinyphia* of the Linyphiidae; *Mimetus* of the Mimetidae; *Castianeira, Chemmisomma, Corinna, Megalostrata,* and *Veterator* of the Myrmeciidae; *Hispanonesticus* of the Nesticidae; *Arachnolithulus* of the Ochyroceratidae; *Oecobius* of the Oecobiidae; *Fossilopaea, Gamasomorpha, Heteroonops, Opopaea,* and *Orchestina* of the Oonopidae; *Oxyopes* of the Oxyopidae; *Otiothops* of the Palpimanidae; *Modisimus, Pholcophora,* and *Serratochorus* of the Pholcidae; *Corythalia, Descangeles, Descanso, Lysso-*

Figure 125. Holotype of the hersilid spider *Prototama succinea* Petrunkevitch in Mexican amber. The insect clinging to the spider's legs has not been identified, but is perhaps a fungus gnat (family Sciaridae) (Museum of Paleontology collection, University of California, Berkeley; photo by P. Hurd).

Figure 126. Paratype (male) of the diplurid spider *Microsteria sexoculata* Wunderlich (family Dipluridae) in Dominican amber (Poinar collection).

Figure 127. A spider, *Mimetus biturberculatus* Wunderlich (family Mimetidae), in Dominican amber (Poinar collection).

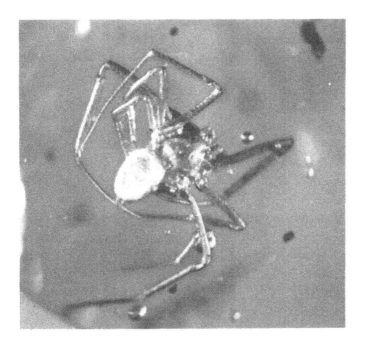

Figure 128. Paratype (male) of the spider *Chrosiothes curvispinosus* Wunderlich (family Theridiidae) in Dominican amber (Poinar collection).

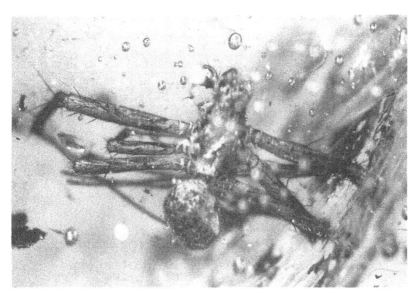

Figure 129. A spider, *Nephila dommeli* Wunderlich (family Tetragnathidae), in Dominican amber (Poinar collection).

manes, Nebridia, Pensacolatus, Phlegrata, and *Thiodina* of the Salticidae; *Scytodes* of the Scytodidae; *Ariadna* of the Segestriidae; *Selenops* of the Selenopidae; *Loxosceles* of the Sicariidae; *Monoblemma* of the Tetrablemmidae; *Azilia, Cyrtognatha, Homalometa, Leucauge, Nephila,* and *Tetragnatha* of the Tetragnathidae; *Ischnocolinopsis* of the Theraphosidae; *Achaearanea, Anelosimus, Argyrodes, Chrosiothes, Chrysso, Cornutidion, Craspedisia, Dipoenata, Episinus, Lasaeola, Spintharus, Stemmops, Styposis,* and *Theridion* of the Theridiidae; *Palaeoepeirotypus* and *Theridiosoma* of the Theridiosomatidae; *Heterotmarus* and *Komisumena* of the Thomisidae; and *Miagrammopes* of the Uloboridae. Also included are representatives of the following families: Agelenidae, Amaurobiidae, Hahniidae, Lioncranidae, Microstigmatidae, Philodromidae, Pisauridae, Pycnothelidae, and Symphytognathidae.

Wunderlich (1986) has made some comparisons between Dominican and Baltic amber spiders. Some 31 of the 45 spider families known to occur in either Dominican or Baltic amber occur in both sources. All of the described species from the combined amber sources are extinct, and about 75 percent of the Baltic amber spider genera are extinct in contrast to about 17 percent of the Dominican amber genera. With respect to the present-day geographical range of amber spiders, all of the families and subfamilies and five or six genera of the Dominican amber spiders still exist today in the Neotropics, whereas five of the Baltic amber spider families are either extinct or do not presently occur in Europe.

Wunderlich (1986) found that the Dominican amber spiders show more affinities to the Recent fauna of the area than do the Baltic forms. This he attributes to two causes, the greater climate change in the Baltic region since the early Tertiary (from subtropical to temperate) and the greater age of the Baltic deposits. The latter aspect may not be crucial, however, because new studies on Dominican amber show a possible age range from Late Miocene to Early Eocene (Lambert et al., 1985), overlapping with that of Baltic amber.

All spider families reported from Mexican amber were also found in Dominican amber (Wunderlich, 1986). The genus *Veterator*, recovered from both amber sources, has not been found among the Recent fauna in either region.

Spider webs have been found in amber, sometimes still containing prey or the shed spider skin or even egg masses, rarely in the process of hatching. Wunderlich (1986) discussed some of the likely enemies of Baltic amber spiders that were themselves trapped in amber. These include spiders of the families Archaeidae, Mimetidae, and Theridiidae; robber flies (Asilidae) of the genera *Asilus* and *Helopogon*; representatives of the dipterous families Chloropidae and Acroceridae; and wasps (*Pimpla, Crabro*), mites, and possibly fungi. What was determined to be a first-stage larva belonging to a tribe of wasps parasitic on spiders, the Polysphinctini (Ichneumonidae), was reported adjacent to the body of a clubionid spider in Dominican amber (Poinar, 1987) (Fig. 141). Spider inquilines in ant nests include representatives of the minute genus *Mastigusa* (Agelenidae; the funnel spiders) in Baltic amber (Wunderlich, 1986).

Kingdom Animalia: Vertebrata

The find of a vertebrate in amber, or even a trace of one, is always a cause for excitement. Unfortunately, because of the rarity of such finds and, consequently, their high commercial and scientific value, forgeries abound. Larsson (1978) commented that all amphibians and fishes reported in Baltic amber have been forgeries, as have been most lizards. I have personally examined many specimens, particularly lizards, that were intentionally placed in amber substitutes and represented as authentic. The typical medium for these fakes is copal, which can be melted in an oven and then resolidified after the animal is embedded. Commonly in the forgery process, a cavity in a large piece of natural copal is melted or hollowed out, the animal is introduced, and melted copal is poured over the animal to fill the cavity. Choice embalming materials are New Zealand Kauri gum (from *Agathis australis*) or South American copal (from trees of the genera *Hymenaea* and *Copaifera*). One such forgery was a Recent New Zealand gecko (*Naultinus elegans*) embedded in Kauri gum.

Many such forgeries were made around the turn of the century and some are quite good. Nevertheless, the surface of the copal, because of its softness, is subject to rapid oxidation, resulting in numerous cracks and fissures that can be detected. Thus before publicizing a vertebrate in amber, or for that matter any important find in amber, the find should be verified as representing a naturally occurring event.

Vertebrates or portions of vertebrates in amber include representatives of four classes: Amphibia, Reptilia, Aves, and Mammalia.

Amphibia (Frogs)

This class, with some 2,400 living species, includes toads, caecilians, and salamanders, as well as frogs. They are characterized by a moist, glandular, smooth skin lacking scales. They are distinguished from lizards by

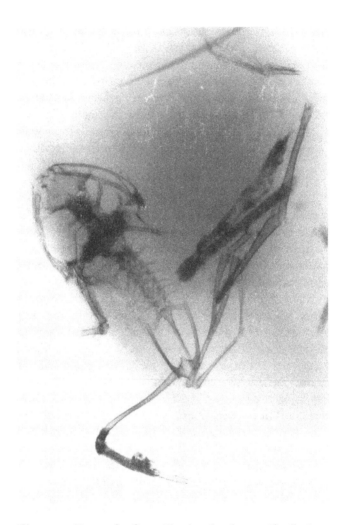

Figure 130. X-ray of a frog, *Eleutherodactylus* sp. (family Leptodactylidae), in Dominican amber. The left hind and right front legs are broken, and the bones of other decomposed frogs are also present, suggesting the activities of a predator (Work collection).

this feature and by the absence of claws and no more than four toes on the fore limb (lizards always have five toes on the fore limb).

As Larsson (1978) has mentioned, all reports of frogs in Baltic amber were later discovered to be forgeries. Some illustrations of frogs embedded in "Baltic amber" are presented by Andrée (1939). The first authentic frog in amber was reported from Dominican amber (Poinar and Cannatella, 1987). This was an adult *Eleutherodactylus* (Leptodactylidae) that was well preserved (Fig. 130, Color Plate 2). Because the left hind and right front legs of the fossil frog were broken, and bones of another *Eleutherodactylus* were in the same piece of amber, it was supposed that the frog was probably captured by a predator (bird?) and brought back to a nest in a cavity of a resin-producing tree. The frog fell into the resin before being eaten and a portion of the predator's nest was also covered by the resin. Whatever the circumstances leading to entombment, this report constitutes the first described frog in amber, the oldest known fossil of the genus *Eleutherodactylus*, and the oldest amphibian fossil from Mesoamerica (Mexico, Central America, and the Antilles).

While in the Dominican Republic, I confirmed instances of other frogs from Dominican amber. One specimen is now in the Stuttgart Natural History Museum collection (see fig. 12 in Schlee, 1984). Another is in a private collection in Switzerland (Siber & Siber, personal communication); this is apparently the specimen figured by Case (1982: fig. 26–26). Two other "frogs" were sold in Italy, a sixth is owned by an amber dealer in the United States, and a seventh is owned by an amber dealer in the Dominican Republic. Unfortunately, none of the above specimens has been verified for authenticity or described.

There are no records of toads, salamanders, or caecilians in amber.

Reptilia (Lizards)

This class includes lizards, snakes, crocodiles, and turtles, among others, but only lizards have been found in amber. They can be recognized by their scaly, dry skin, the presence of horny claws, and the five toes on the fore limbs. Lizards are easy to obtain and numerous forgeries exist.

Several reports of lizards in Baltic amber exist, but all are fakes involving recent lizards placed in semi-fossilized (copal) resin. Unfortunately, none of these can be located today. The first was *Platydactylus minutus*, reported by Giebel in 1862. The second was *Hemidactylus viscatus*, reported by Vaillant in 1873. Both of these were apparently recent specimens of African lizards that had been embedded in East Indian copal, also known as Zanzibar copal, which is the semi-fossilized resin of *Hymenaea* trees in East Africa. The third "Baltic amber" lizard (originally from the Königsberg Museum) was reported by Klebs (1910) as belonging to the extant species *Nucras tesselata* (Smith). It had earlier been assigned to the genus *Cnemidophorus* by Boettger (in Klebs, 1910), but in 1917, Boulenger renamed it *Nucras succinea*. Larsson (1978) regarded *Nucras* as being authentic, but Loveridge (1942) had already shown that it was embedded in copal, not amber. Loveridge (1942) also commented on two other lizards that had been placed in African copal.

One of these was *Holaspis guentheri* (Smith) Gray, which was part of the Hagen collection that originally came from Königsberg. Although the posterior portions of the hind limbs and tail were missing, the lizard was indistinguishable from the extant forms of *H. guentheri* that extend across tropical Africa. The second was the gecko *Lygodactylus grotei* Sternfeld, which originated from a collection of animals entombed in Zanzibar copal that O'Swald had brought back from Africa. This gecko is common on the coast and islands neighboring Zanzibar.

Besides being placed in African copal, lizards have also been entombed in South American and New Zealand copal. I was recently presented with a large *Anolis* in what was considered "Baltic amber." The specimen was suspicious, since *Anolis* is a New World lizard with no records (either extant or extinct) from the Old World. An analysis of the embedding medium showed it to be South American copal. Most of this material comes from Colombia, near the villages of Velez and Girón in the District of Santander.

Another lizard that I examined in "Baltic amber," which was being offered for sale (twenty thousand dollars), was identified as a New Zealand green gecko (*Naultinus elegans* Gray) that had been entombed in kauri gum (copal from the conifer *Agathis australis*). However, the possibility of lizards being preserved in Baltic amber exists, especially since Katinas (1983) figured a shed lizard skin in amber from Lithuania.

The first lizard in authentic amber was *Anolis electrum* Lazell (1965) (Iguanidae) in Mexican amber. The description was based on two partial juvenile specimens possibly representing a single specimen. Lazell (1965) considered *A. electrum* to be an arboricolous form and mentioned that its closest living relatives occurred in northern South America, although *A. limifrons*, closely comparable to *A. electrum*, occurs in southern Mexico today.

Rieppel (1980) described a specimen of *Anolis dominicanus* from Dominican amber. This individual, considered a juvenile or small adult, belongs to the alpha anoles and is closely related to the green anole group that lives in the crowns of trees today in Hispaniola. Representatives of the alpha *Anolis* group lack transversal processes on the posterior tail vertebrae. This primitive character is shared today by forms living in the southeastern United States and the Antilles.

In 1984, Böhme described the gecko *Sphaerodactylus dommeli* (Gekkonidae) from Dominican amber. The description was based on one entire specimen (designated the holotype), now deposited at the Museum A. Koenig in Bonn, and a second specimen lacking the tail (designated a paratype) now deposited at the Natural History Museum in Stuttgart. Böhme commented that the fossil did not closely resemble any extant *Sphaerodactylus* on Hispaniola.

From my discussions with miners and amber dealers, I learned that at least II lizards (all representatives of *Anolis* or *Sphaerodactylus*) have been found in Dominican amber. The first, *A. dominicanus*, is now in Switzerland; the second, the holotype of *S. dommeli*, is now in Bonn, Germany; the third, a paratype of *S. dommeli*, is in the Stuttgart collection; the fourth, a *Sphaerodactylus* sp., is in the Amber Museum in Puerto Plata; the fifth, an *Anolis* from the La Toca mine, and the sixth, an *Anolis* from El Valle, are in my collection; the seventh, an *Anolis* from La Toca, is in a private collection in the Dominican Republic; the eighth, a partial *Sphaerodactylus* is in a collection in the Dominican Republic; the ninth, a *Sphaerodactylus* (Fig. 131), is now in Japan; the tenth, a *Sphaerodactylus*, is in the Amber Museum in Puerto Plata; and the eleventh, another *Anolis*, was recently sold to a private collector in North America (Color Plate 2).

Aves (Birds)

The presence of birds is reflected in amber by feathers or portions of feathers. One of the more interesting reports is the discovery of a feather in Lebanese amber (Schlee, 1973), which as Schlee and Glöckner (1978) point out, represents the oldest known feather. The structure of this feather is reminiscent of those of certain species of present-day waterfowl. A color photograph of this rarity is presented by Schlee and Glöckner (1978).

Feathers also occur in Baltic amber, the earliest record being that of Meyer (1887). Other finds have been summarized by Bachofen-Echt (1949). He mentions feathers resembling those of present-day sparrows, titmice (*Parus* spp.), nuthatches (*Sitta* spp.), woodpeckers (*Picus* spp.), and motmot (*Momotus* spp.). The latter find is especially interesting because *Momotus* (Momotidae) is a genus of nonmigratory Neotropical

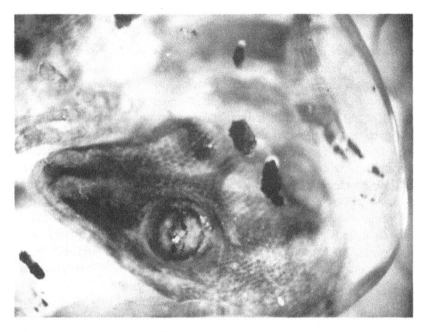

Figure 131. Head of a gecko (*Sphaerodactylus* sp.) in Dominican amber.

birds that range from Mexico to Argentina. Although a motmot could possibly have occurred in Europe some 35–40 Ma, convergence in a completely separate, now extinct group could also account for the similarities noted between the fossil specimen and present-day motmots (C. Sibley, personal communication). Photographs of Baltic amber feathers are provided by Bachofen-Echt (1949) and Weitschat (1980).

Feathers have also been recovered from Dominican amber. The first description of a fossil feather from New World amber, and the oldest known bird fossil from the West Indies, was reported by Poinar et al. (1985) on the basis of a portion of a feather barb (Fig. 132). Since then, several complete feathers have been recovered from Dominican amber. The specimen shown in Color Plate 3 is being studied by Dr. R. Layborne at the Smithsonian Institution, who concludes that it belongs to a woodpecker or relative (Picidae).

Mammalia (Mammals)

Mammalian remains in amber are represented by one impression and a variety of hairs. A single footprint of a small mammal tentatively

Figure 132. Portion of a feather barb in Dominican amber.

identified as a marsupial or carnivore has been found in Baltic amber (Bachofen-Echt, 1949: fig. 188, p. 188). Mammalian hair occurs in amber from at least four sources. The oldest known fossil hair occurs in Siberian amber from Yantardak (Zherichin and Sukacheva, 1973), although no identification of this material has been made.

Numerous pieces of Baltic amber containing mammalian hair have been reported. The first of these reports was made by Eckstein (1890), who examined 17 pieces of amber with hair inclusions and identified, among them, the hair of a dormouse (*Glis*, = *Myoxus*) and a squirrel (*Sciurus* sp.). Lühe (1904) identified hairs in two pieces of Baltic amber as belonging to members of the marsupial genus *Phascogale* (= *Phascologale*), shrewlike, carnivorous animals today limited to Australia. Bachofen-Echt (1944) also described mammalian hair from Baltic amber. His findings included hair morphologically similar to that of a doormouse (*Glis glis*) and certain bats (Vespertilionoidea).

Voigt (1952) found louse eggs attached to squirrel(?) hair in Baltic amber. The presence of such ectoparasites can be of use in determining the identity of hair. In reporting on mammalian hair in Dominican amber (Color Plate 3), Poinar (1987) partly based his conclusion that the hair belonged to a rodent on the presence of two ectoparasites that occur mainly on rodents today: a listrophorid fur mite and a parasitic staphylinoid beetle larva.

Hair has also been found in Mexican amber, but no identifications have been reported (Poinar, unpublished data). Although mammalian hair has certain characteristics (size, scale pattern, cortex configuration) that can be used to identify its bearer, even to species, the inevitable process of deterioration, coupled with the fact that scale patterns are not visible on hair embedded in amber, make identifications difficult. Some hairs, such as those of certain bats, have characteristic jutting scales that aid in identification. Sloth hairs are also distinctive from those of other mammals, and hair of the now-extinct ground sloth *Neomylodon lighai*, has typical sloth hair that most closely resembles that of the present-day *Bradypus* (a sloth). Because of the arboreal habit of sloths, their hair might be expected in amber deposits.

CHAPTER FIVE

Paleosymbiosis

Examples of demonstrated symbiosis in the fossil record (paleosymbiosis) are rare, and it is probably accurate to say that the majority of such cases are found in amber. This is because the rapidity of entombment often catches the symbionts in situ. External symbionts are still attached to their carrier and internal symbionts are sometimes trapped as they attempt to leave their host.

The traditional concept of symbiosis as outlined by Ahmadjian and Paracer (1986) will be used here. Under symbiosis (defined as the living together of different organisms) are included *commensalism* (where one organism benefits and the other is neither harmed nor benefited), *mutualism* (where both organisms benefit), and *parasitism* (where one organism receives nutrients at the expense of the other). A fourth category, *inquilinism* (where two organisms live in the same niche, but neither organism benefits from nor is harmed by the other), is sometimes included as a separate entity. In practice, distinguishing between inquilinism, commensalism, and mutualism is often difficult, because benefit or harm may not be directly evident or may occur in only some stages of the life cycle.

Commensalism

Perhaps the most common type of relationship observed in amber is that of phoresy (phoresis) where one organism is carried, but is not parasitic, on or inside another organism. The one being carried is benefited, and the bearer is neither benefited nor harmed. In some instances the bearer may feed on its cargo, as with macrochelid mites feeding on phoretic nematodes, and in these instances the association would be considered mutualism. The cases of phoresy observed in amber will be discussed under the category of the organism being transported.

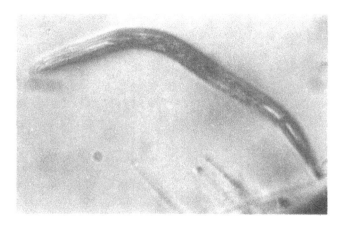

Figure 133. Dauer stage of a rhabditid nematode associated with an ant in Dominican amber (Poinar collection).

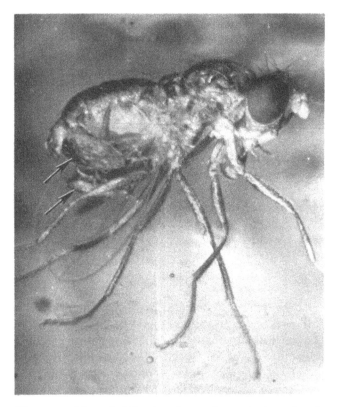

Figure 134. Macrochelid mites (arrows) attached to the abdomen of the drosophilid fly *Protochymomyza miocena* Grimaldi in Dominican amber (Poinar collection). (Published in *J. New York Entomol. Soc.*)

Both external and internal phoretic associations occur between nematodes and arthropods (Poinar, 1983) but only external phoretic associations have been observed in amber. In general, it is the "dauer" or non-feeding third-stage juvenile nematodes that are carried by arthropods. These stages have been found associated with the intersegmental membranes of worker ants (Poinar, 1981) (Fig. 133) and under the elytra of platypodid beetles in Dominican amber. The latter association is known today, but the former, although possibly existing, has not been reported. After the carrier reaches a suitable habitat, the dauer nematodes leave their hosts, mature to the adult stage, and reproduce in the environment of the arthropod. Formation of dauer juveniles is often associated with a reduction of food or the onset of adverse environmental conditions, especially drought.

Mites, too, are known to have phoretic associations with arthropods, and again the immature forms are normally the transported stage. Three macrochelid mites, which are relatively large, were attached to the abdomen of the drosophilid fly *Protochymomyza miocena* Grimaldi (1987) (Poinar and Grimaldi, 1990) (Fig. 134). Other phoretic mite stages can be very small, as are for example the phoretic deutonymphs (hypopi) of the suborder Astigmata that were associated with the bee *Proplebeia domini-*

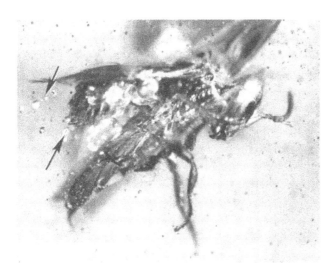

Figure 135. Phoretic deutonymphs (hypopi) of the suborder Astigmata (arrows) on a worker stingless bee, *Proplebeia dominicana* (Wille and Chandler), in Dominican amber (Poinar collection).

cana (Fig. 135). I have also observed phoretic mites on adult platypodid beetles, adult chironomid midges, adult tipulid flies, and termites in Dominican amber. An example of a water mite being carried by an adult midge in Baltic amber is illustrated by Schlee and Glöckner (1978), and Bachofen-Echt (1949) mentions finding mite larvae associated with Trichoptera in Baltic amber.

Pseudoscorpions have the characteristic of holding on to insects with the chela of their pedipalps, although they may occasionally wedge themselves under the elytra of beetles. Examples of the former behavior in Baltic amber include the pseudoscorpions *Oligochernes bachofeni* Beier and *Pycnochelifer kleemanni* Koch and Berendt on adult wasps of the Braconidae (Beier, 1937) (see photograph in Bachofen-Echt, 1949), *Oligochelifer berendtii* (Menge) on an adult wasp of the Ichneumonidae (Menge, 1855) and an adult caddis fly (Trichoptera) (Beier, 1948), and an undetermined representative of the Chernetidae on an adult crane fly (Tipulidae) (Lemoniinae) (Schlee and Glöckner, 1978).

Schawaller (1981B) described two examples of phoretic pseudoscorpions in Dominican amber. In both cases, a species of *Parawithius* (Cheliferidae) was carried by a female of the beetle *Mitosoma rhinoceroide* Schawaller (Platypodidae). One specimen was under the beetle's elytra,

Figure 136. Triungulin (arrow), a modified larva of a meloid beetle (family Meloidae), on the "neck" of a worker stingless bee, *Proplebeia dominicana* (Wille and Chandler), in Dominican amber (Poinar collection).

and the other was attached to both hind tarsi. These two examples are illustrated in Schlee (1980). A third example from Dominican amber, which is in my collection, shows the pseudoscorpion attached to the beetle's abdominal tergites (Fig. 113). Similar phoretic associations between pseudoscorpions and platypodid beetles occur today (Muchmore, 1971). Pseudoscorpions living in the same habitat as the beetles use the insects to reach a new habitat after the old one has deteriorated.

An association that involves both phoresy at the initial stage and parasitism at a later stage of the life history occurs in certain bee-attacking beetles of the family Meloidae. The first instar of these blister beetles is an active, legged form called a triungulin. This stage climbs on a flower and waits for a bee to arrive. When it does, the triungulin climbs on the bee and is carried back to the nest. This is the phoretic portion of the life cycle. After reaching the bee's nest, the triungulin feeds on the young stages of the bees and molts into a legless larva. The presence of such an association in the early Tertiary is shown by the presence of a triungulin on the neck region of a worker of *Proplebeia dominicana* (Fig. 136).

Mutualism

Examples of mutualism in amber are difficult to find. There are individuals of *Proplebeia dominicana* with pollen on their corbicules, if one wishes to consider pollination as an example of mutualism. Also, many of the soft-bodied worker termites are associated with air bubbles that arise from their abdomen and extend into the amber. Such bubbles probably represent the production of gases by mutualistic, cellulose-digesting microorganisms within the termite's gut.

Parasitism and Disease

Examples of parasitism and disease also occur in amber. A parasite is defined as an organism that lives in or on and obtains nourishment from another organism. Parasites are sometimes separated into microparasites, which include microorganisms and other microscopic forms of life (e.g., some fungi), and macroparasites, which include nematodes and arthropods. Although mosquitoes, ticks, and braconid wasps are also examples of macroparasites occurring in amber, in this section we examine only those relationships in which the parasite and host are physically together, and thus provide direct evidence of a parasitic association. When parasitism is caused by a microorganism that causes a disease—that is, a disturbance of function or structure of a tissue, organ, or entire organism over a period of time—then the agent is called a pathogen.

Fungi are commonly found in amber. The great majority of these are saprophytic forms that obtained their nourishment from partially exposed insects entrapped in the resin. Eventually, additional resin covered up the entire insect along with the associated fungal colonies (Thomas and Poinar, 1988). Reports of insect-parasitic fungi in amber involve an Entomophthorales growing on the body of an adult termite (Poinar and Thomas, 1982) and a *Beauveria* growing on a worker ant (Poinar and Thomas, 1984), both in Dominican amber. In addition, Bachofen-Echt (1949) has commented that of four walking-stick (Phasmidia) eggs in Baltic amber, three had developing embryos but the fourth was covered with a *Mucor*-like fungus. There is no way of knowing whether the fungus was a pathogen or parasite.

Fossil nematophagous fungi in Mexican amber were reported by Jansson and Poinar (1986). In this instance, fungal spores and mycelia were associated with the aphelenchoidid nematode *Oligaphelenchoides atrebora* Poinar (Fig. 24). Leaves with blotches resembling leaf spot fungi are sometimes found in Dominican amber, but these have not yet been characterized.

Aside from the strictly phoretic mites that derive no nourishment from their transporter, there are also parasitic mites that penetrate the cuticle of their host and feed on the hemolymph. Examples include parasitic water mites (Hydracarina) attached to adult caddis flies (Trichoptera) (Bachofen-Echt, 1949) and adult midges (Chironomidae) (Schlee and Glöckner, 1978) in Baltic amber, and four parasitic mites attached, in feeding position, to the ventral abdominal segments of an adult midge in Dominican amber (Poinar, 1985A) (Fig. 137). The first fossil evidence of moths parasitized by mites was reported from Dominican amber (Poinar et al., 1991). Two adult moths (families Gracillaridae and Tineidae) each contained a pair of larval erythraeid (Erythraeidae) mites attached to their bodies. Parasitic fur mites of the family Listrophoridae have been found associated with mammalian hair in Dominican amber (Poinar, 1988A) (Fig. 123).

There is fossil evidence in amber of two groups of insect-parasitic nematodes. The first group is the Mermithidae (Mermithida), a worldwide family that parasitizes a variety of aquatic and terrestrial insects. The normal life cycle of these parasites involves an infective second-stage juvenile that enters the body cavity of the larval stage of the host. The juvenile develops within the host and leaves when mature, which may oc-

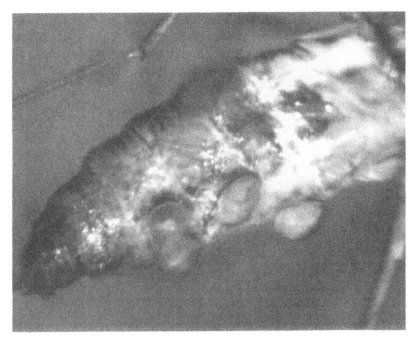

Figure 137. Four parasitic mites attached to the abdomen of an adult midge (family Chironomidae) in Dominican amber (Poinar collection). (Published in *International Journal of Acarology*, Vol. 11, No. 1, Indira Publishing House, P.O.B. 37256, Oak Park, Michigan 48237.)

cur while the host is a juvenile or adult. The nematodes develop to the adult stage and mate outside the host (Poinar, 1983). In many dipterous hosts such as mosquitoes (Culicidae) and midges (Chironomidae) the nematodes are carried around by the adult host for several weeks. If these parasitized adults fly into resin, the nematodes may be mature enough to emerge and then become entrapped themselves, often before they have completely exited from their host (Fig. 138).

Records of mermithids associated with insects in Baltic amber have been summarized by Poinar (1984C) and include two species of the fossil mermithid genus *Heydenius* Taylor and three undescribed specimens. One of the latter, which is figured in Schlee and Glöckner (1978), is emerging from an adult midge (Chironomidae). The other two undescribed forms are in the Copenhagen amber collection. The holotypes of the two described species of *Heydenius*, *H. matutina* Menge (1863)

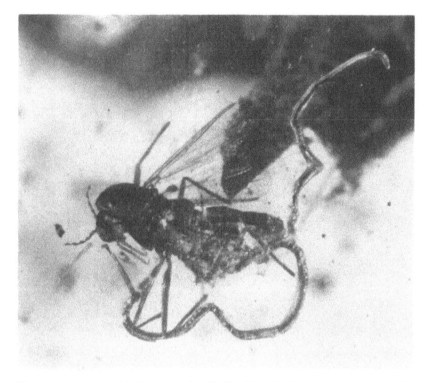

Figure 138. A mermithid nematode (family Mermithidae) emerging from its midge host (family Chironomidae) in Dominican amber (Poinar collection).

from an adult midge and *H. quadristriata* Menge (1872) from an unknown host, have not been located.

Two examples of mermithids associated with insect hosts occur in Dominican amber. One of these consists of two parasitic juveniles of *Heydenius dominicus* that are free in the amber. Close by are an adult crane fly and an adult mosquito, and the parasites are thought to have emerged from the mosquito host (Poinar, 1984B). A second example is a mermithid partly emerged from its adult midge host (Fig. 138). These records clearly establish the family Mermithidae, complete with characters as known today, in both Old World temperate-subtropical and New World tropical zones during the early Tertiary.

The second group of insect parasitic nematodes found in amber, the Allantonematidae and Iotonchiidae (Tylenchida), have been found only in Dominican deposits (Poinar, 1984A; Poinar and Brodzinsky, 1986;

Poinar, 1991C). The life cycle of this group, which parasitizes only insects and mites, is initiated by a fertilized female that enters the body cavity of an immature stage of the host (Poinar, 1983). The female then enlarges and produces eggs that hatch inside the host's body. The juveniles normally develop to the fourth stage before they leave the host, after which they mature to the adult stage and mate. Some iotonchiids have an alternating insect-parasitic and fungal-parasitic life cycle (Poinar, 1991C). As with the mermithids, these nematodes were usually carried into amber by the adult stage of the insect. When parasitized hosts became entangled in resin, the fourth-stage juveniles attempted to emerge. In a case involving the drosophilid *Chymomyza primaeva* Grimaldi (1987), 120 juvenile allantonematids were trapped around a parasitized adult female. The juveniles closely resemble juvenile forms of *Parasitylenchus diplogenus* Welch, an allantonematid parasite of present-day drosophilid flies (Poinar, 1984A) (Fig. 139). A second example of allantonematid parasitism of insects involves an unidentified staphylinid beetle partly surrounded by 44 juvenile nematodes that closely resemble *Proparasitylenchus platystethi* Wachek, a

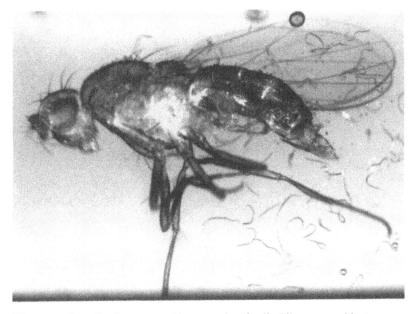

Figure 139. Juvenile allantonematid nematodes (family Allantonematidae) emerging from an infected female drosophilid fly, *Chymomyza primaeva* Grimaldi (family Drosophilidae), in Dominican amber (Poinar collection). (Published in *Journal of Parasitology.*)

Figure 140. A parasitic dryinid wasp larva (family Dryinidae) inside a sac protruding from the abdomen of its homopteran host (Poinar collection).

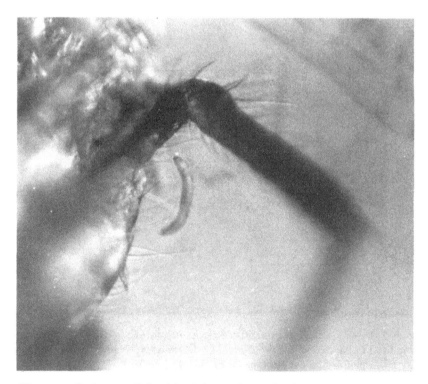

Figure 141. Early stage Polysphinctini wasp larva (family Ichneumonidae) adjacent to the abdomen of a spider (family Clubionidae) in Dominican amber (Poinar collection).

present-day allantonematid parasite of a staphylinid beetle (Poinar and Brodzinsky, 1986). These records include the only reported fossil evidence of the Iotonchiidae and Allantonematidae and document that insect parasitism by members of the Tylenchida was well established in the Tertiary.

The adults of many parasitic insects (e.g., Hymenoptera, Diptera) are found in amber and thus, on the basis of our knowledge of closely related extant forms, parasitic relationships must have existed. Finds of developing parasites in their hosts are rare, however. I have found five pieces of Dominican amber containing adults or nymphs of planthoppers (Fulgoroidea), each insect with a developing dryinid larva (Dryinidae) inside. Their presence can be determined from saclike structures that contain larvae and protrude from the body of the host (Fig. 140). Adult Dryinidae also occur in Dominican amber (Fig. 102).

A second example of insect parasitism in amber involves an early stage larva of a Polysphinctini wasp (Ichneumonidae: Hymenoptera) adjacent to the base of the abdomen of a spider (Clubionidae) (Fig. 141) (Poinar, 1987). Contemporary species of this wasp tribe sting a spider into submission, then deposit their egg at the base of the spider's abdomen. The hatched larva sits exposed on the spider's abdomen where it periodically punctures the latter's integument and feeds on the released hemolymph.

CHAPTER SIX

Implications for Biological Science

An analysis of the fossil flora and fauna found in amber can be useful in many respects. Most important, amber fossils allow us to compare, in detail, past life forms with their present-day descendants. Important geological information on the origin, continuation, or extinction of species, genera, families, or even orders can be obtained from amber inclusions. Variability in a taxonomic line during the past 100 million years aids our study of macro- and microevolution. In addition, the distribution of groups of terrestrial organisms at various periods in the past can be established, thus elucidating our understanding of their biogeography. Finally, the detailed study of amber inclusions from a particular site can give us clues to the environment and ecosystem structure during the period of resin production.

Evolution and Extinction

Amber fossils are important for studies of microevolution (species level and below) because their excellent degree of preservation allows a detailed comparison with extant forms. Amber fossils are also important with respect to macroevolution (from the genus level upward), and can assist in establishing higher taxa. Unfortunately, however, well-preserved fossiliferous amber does not exist in deposits older than the Cretaceous, and thus it is not possible to learn much about the origin of most insect families or orders from amber. Nevertheless, in amber are the only confirmed fossils of the insect orders Zoraptera and Siphonaptera and the earliest confirmed fossils of the insect orders Strepsiptera, Embioptera, Thysanura, Isoptera, and Thysanoptera. In Cretaceous amber are the earliest known records of social insects (bees, ants, and termites).

Amber provides us with a means to measure the longevity of a species. As Eldredge (1985) and others have claimed, extinction is more obvious than origin. Eldredge also states that "most species living in the past, probably including all those which were alive more than about 10 million years ago, are extinct". Ten million years is probably a conservative figure for the time span of terrestrial species, although the length of their survival depends on their reproductivity rate and the homogeneity of their habitat and ecosystem. For most insects, 2–3 million years is an approximate period for species survival. What can the amber deposits tell us about the duration and change of arthropod species?

The great majority of life forms described from amber have been considered extinct at least at the specific level. There are possible exceptions:

1. The ant *Formica flora* Mayr was described from Baltic amber and considered to be indistinguishable from the extant *Formica fusca* Linnaeus. However, later studies (Baroni Urbani and Graeser, 1987) using scanning electron microscopy on the fossil *F. flora* and the extant *F. fusca* showed striking differences in the structure of the integument.

2. Another possible case of an extended species line was Hennig's (1966) report of the fly *Fannia scalaris* Linnaeus in Baltic amber. According to Frank Carpenter (personal communication), there was some question about whether the fossilized resin was copal or amber. In this case, it will be necessary to reexamine and test the matrix material, because if it was copal, then the specimen could be as young as 50 years.

3. The extant beetle *Micromalthus debilis* LeConte was described from Mexican amber by Rozen (1971), but again it would be prudent to reexamine the piece for authenticity. If the fossil was of an extant species, it would indicate a species existence of about 25 million years, which would be exceptional for insects.

4. Rodendorf and Zherichin (1974) reported finding the wasp *Serphites paradoxus* Brues in Siberian amber. If correct, then this species could have existed some 8 million years, because it was originally described from younger Canadian amber. The family Serphitidae was present during the Cretaceous but disappeared completely by the end of that period.

5. Mockford (1972) restudied psocoptids of the genus *Belaphotroctes* and concluded that the Mexican amber *B. similis* should be a synonym of the extant *B. ghesquierei*, thus establishing a minimum age of 22 million years for this species.

6. Also in Mexican amber is the scelionid wasp *Palaeogryon muesebecki*, which is unchanged from Recent Mexican specimens (Masner, 1969A).

If such specimens are indeed authentic, then it would appear that certain insect species have a survival line much longer than the postulated norm of 2–3 million years.

Species become extinct for different reasons. The climate may change, a competitor, predator, or pathogen may appear, or the food source may change or disappear. As a response, the species can either adapt to the changing conditions or search out more favorable ones. The Baltic amber species were faced with a climatic change and subsequent habitat change, and those that perished were unable to adapt or to relocate their preferred habitat (migrate). On the other hand, some may have moved south, because many descendents of Baltic amber organisms occur today in the tropics and subtropics throughout the world. For Dominican amber forms, which were not subjected to any drastic climatic change, competition may have been the major factor responsible for extinction.

An analysis of the Baltic, Dominican, and Mexican amber flora and fauna and the present geographical range of their lineages led me to offer a hypothesis I call "genetic confinement." This hypothesis proposes that a species is genetically adapted only to the climatic environment in which it originally evolved. From all available evidence, it appears that the Baltic amber insects were heterogeneous in climatic preference. Some must have evolved under tropical-subtropical conditions, while others appear to have evolved under warm to cool-temperate conditions. Those of the former group could tolerate only minor changes in the physical environment, and when the climate in northern Europe cooled, many Baltic amber lineages perished or were forced to migrate to geographical areas resembling those in which they evolved. Those Baltic amber species that evolved under more temperate conditions were able to survive under conditions found at the extremes of a temperate climate. These forms produced lineages that survive today in the general area of the original Baltic amber forest.

Descendants of the tropical Mexican and Dominican amber plants and animals are rarely found outside tropical or subtropical zones today, but it was not major climatic changes that drove their lineages to extinction (extinctions occurred, but for other reasons). Since new evidence presented by Čepek in Schlee (1990) indicates that at least some of the amber mines in the Dominican Republic overlap in age with the Baltic deposits, a comparison of existing lineages in these two deposits may help us to understand the causes of extinctions during the past 40 million years. Some comparative studies on Baltic and Dominican amber extinctions have already been conducted.

A comparison between Baltic and Dominican amber spiders was presented by Wunderlich (1986). All of the species described from Baltic or Dominican amber are extinct. But on the generic level, some 75 percent of the Baltic amber spiders are now extinct, whereas only 17 percent of the Dominican amber spider genera are extinct. In addition, several Baltic amber spider tribes and two or three families are also extinct, whereas no taxa at these levels in Dominican amber are extinct. Wunderlich (1986) considered Dominican amber as Miocene in age; however, at present, there are indications that much of this amber dates from the Oligocene-Eocene period and is therefore roughly equivalent in age to Baltic amber. Thus, the different extinction rates can be attributed to climatic change rather than an age difference. Furthermore, the affinities of the Dominican amber spiders to the Recent Neotropical fauna are much closer than those of the Baltic amber spiders to the extant northern European fauna. All of the Dominican amber spider families and subfamilies, as well as five or six genera, still occur in the Neotropics while surprisingly few genera of Baltic amber spiders are distributed in the Palaearctic. Many of the closest relatives of the Baltic amber forms occur in the Oriental, Ethiopian, and Australian regions. Wunderlich (1986) provides eight cases of transition extinctions (those where close relatives of the fossil spiders exist today) in Dominican amber spiders but only a single case in Baltic amber forms.

A similar pattern exists with the ant fauna in Baltic and Dominican amber (Wilson, 1985E). Some 44 percent of the Baltic amber ant genera are extinct, in contrast to 8 percent of the Dominican amber ant genera. Wilson reported that the Baltic amber extinction rate of ant genera is roughly equivalent to that of the Florissant shales (40 percent) and both of these relatively high extinction rates, in contrast to the lower rates in Dominican amber, were attributed to the greater age of the Baltic and Florissant deposits. These relatively high extinction rates might better be attributed to the widespread climatic changes in the Northern Hemisphere that resulted in extinctions and forced migrations. In contrast, selection pressures in the tropical habitats affecting Dominican amber insects would have resulted from competition, predation, and diseases possibly correlated with changes in microhabitat.

Extinctions in general are shown in the data of Rodendorf and

Zherichin (1974), who listed the percentage of extinct insect families, genera, and species as recorded from copal and other deposits (Rodendorf and Zherichin, 1974) (Figs. 142, 143). It is clear that in earlier deposits, more extinct species, genera, and families occur. It is also obvious that species become extinct at a faster rate than genera, which in turn become extinct at a faster rate than families. The sudden increase in extinct families at about 65 Ma is in part due to the higher rate of families containing a single or very few species (Fig. 143).

In the fossil record, extinctions are much easier to document than are origins. Groups just seem to appear and disappear, and transition forms are exasperatingly few. These observations resulted in several hypotheses regarding speciation, the most recent of these being punctuated equilibrium, a hypothesis suggesting that once a species evolves, it will remain stable for a relatively long period and, then, when change occurs,

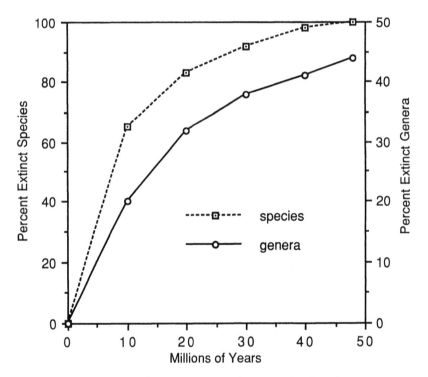

Figure 142. Percentages of insect genera and species that have become extinct over the past 50 million years, based in part on amber inclusions. (Modified from Rodendorf and Zherichin, 1974.)

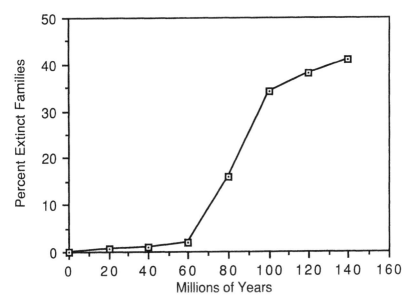

Figure 143. Percentages of insect families that became extinct over the past 140 million years. (Modified from Rodendorf and Zherichin, 1974.)

it does so in a rapid spurt (in a geological time frame). If such a system were to exist, it would help explain why so few transitional forms are in the fossil record. There are a few, however. From amber deposits come several extinct families, especially in the Hymenoptera, that could be interpreted as bridging the gap between two groups. One of these is *Sphecomyrma freyi* Wilson et al. (1967) which was the basis for a new subfamily, the Sphecomyrminae. This taxon is thought to have bridged the gap between aculeate wasps (*Sphecomyrma* has characters resembling present-day members of the Tiphiidae) and the ants.

No clear lineage showing a gradual transition from one insect family to another has been demonstrated in amber or the fossil record in general. In his study of behavioral evolution, Boucot (1990) states, "It is of general interest that the evolution of behavior appears to have been sudden, coming in with the first appearance of the taxon in question together with the taxon's unique morphologic features and community group's character, and then long persisting in essentially unchanging form."

Coevolution has several definitions but, in the broad sense, it refers to the simultaneous evolution of two or more interacting species that favors the survival of both. It has traditionally been applied to herbivorous insect-plant interactions, but few studies have attempted to follow patterns of coevolution through the fossil record. As Futuyma and Slatkin (1983) state, "The ideal paleontological evidence would be a continuous deposit of strata in which each of two species shows gradual change in characters that reflect their interaction." Interactions of two or more species occurring in amber have been discussed under the section of Paleosymbiosis, and these provide no clear evidence of change over the past 40 million years. Patterns of phoresy and parasitism observed in Tertiary deposits are very similar to present-day associations, and it will be necessary to go back further in the fossil record to discover how these symbiotic relationships evolved.

Biogeography

Amber inclusions assist in the determination of animal and plant biogeography. Many examples show that the closest living relatives of fossil insects in amber live some considerable distance from the amber location. Living beetles of the family Mordellidae similar to the genus *Ctenidia* found in Siberian amber occur today as relicts in South Africa (Rodendorf and Zherichin, 1974). A micropterigid moth from Lebanese amber, *Parasabatinca aftimacrai* Whalley (1978), has Recent relatives in the Southern Hemisphere (Australia, New Zealand, and Africa). The spider families Pycnothelidae and Coponiidae (Nopinae) in Baltic amber are now restricted to North and South America, predominately with tropical distributions. Spiders of the subfamily Urocteinae (Oecobiidae) and the family Archaeidae from Dominican amber occur today in the Palearctic, Ethiopian, Oriental, and Ethiopian-Australian regions, respectively (Wunderlich, 1986). Numerous other examples demonstrate possible dispersal (or survival if the original distribution of the species was worldwide) of northern amber insects into the tropics, as they apparently gravitated toward their preferred physical environment as the climate cooled.

Thus, the shift toward a cooler climate during the early Tertiary played an important and probably dominant role in determining distributional patterns of amber insects. One surprising find in Dominican amber was ants of the genus *Leptomyrmex*, which today occur in eastern Australia, New Caledonia, New Guinea, the Aru Islands, and the Moluccas (Baroni Urbani and Wilson, 1987). Because the majority of Dominican amber finds represent species with present-day Neotropical and even

Caribbean descendants, one can assume that little migrational activity occurred during the past 30–40 million years, undoubtedly the result of a relatively stable climate during this period. In the unusual case of *Leptomyrmex* (and also the related Sicilian amber fossil *Leptomyrmula*), it seems likely that this ant group was well distributed in the New and Old Worlds and that for unknown reasons (possibly competition) the group underwent a drastic retreat during the late Tertiary. There are other, similar cases. One of the most primitive termites, *Mastotermes* sp., which was recovered in Mexican amber (Krishna and Emerson, 1983), now occurs only in Australia. Almost half of all Dominican amber caddis flies belong to the species *Antillopsyche oliveri* Wichard of the Polycentropodidae, but although the genus is endemic today in the Greater Antilles, the subfamily is represented today mainly in the Ethiopian and Oriental regions (Wichard, 1987).

Dominican amber finds are significant in explaining the present-day distribution of plants and animals not only within the Greater Antilles but also the pantropics. One line of evidence for a vicariance event (splitting up continuous populations by separation of landmasses through continental drift) regarding the South American and African continents is the lineage of the tree that produced the Dominican amber, namely *Hymenaea protera* of the caesalpinioid division of the Leguminosae (sometimes elevated to family level Caesalpinioidae) (see also pp. 78–81). On the basis of flower morphology (floral parts of *Hymenaea protera* occur in Dominican amber), infrared spectra, and nuclear magnetic resonance spectra, the Dominican amber tree appears to be most closely related to *H. verrucosa* on the east coast and adjoining islands of tropical Africa and *H. oblongifolia* var. *latifolia* along the east coast of Brazil. Hueber and Langenheim (1986) and Langenheim and Lee (1974) suggest oceanic dispersal of seed pods to explain this amphi-Atlantic distribution. They hypothesize that the ancestral stock of *Hymenaea* (of which *H. verrucosa* is a relict) originated in west Africa and that ancestors of the South American species *H. oblongifolia* were carried from west Africa to the Amazonian region via the South Equatorial Current and ancestors of *H. oblongifolia* var. *latifolia* were brought to the east coast of Brazil via the Brazil current. Fruit of African ancestral stock could have been transported via the South Equatorial Current to the West Indies during the early Tertiary (Langenheim and Lee, 1974).

I suggest an alternative hypothesis based on the age of Dominican amber. It is possible that the genus *Hymenaea* expanded and speciated during the late Mesozoic from a distributional center located along the equator. Diversification resulted in the genus extending from Mexico across South America and into the then adjoining African continent. When the African continent drifted away from South America, it took with it populations of *Hymenaea*, now represented by the relict species *H. verrucosa* along the eastern coast. Our fossil knowledge of the Leguminoseae and especially the Caesalpiniaceae is incomplete. However, the discovery of pollen of the tropical caesalpinioid genera *Crudia* from the Paleocene and *Sindora* from the Maestrichtian (Late Cretaceous) (Muller, 1970, 1984) together with finds of caesalpinioid and other leguminous vegetative parts in the Green River Eocene formations (Grande, 1984) give evidence that the Caesalpiniaceae was well established by the early Tertiary. Thus, a complicated voyage of seed pods via various ocean currents need not be proposed to explain the presence of Caribbean *Hymenaea* from African stock during or before the Late Eocene.

A more specialized biogeographical topic relates to the origins of the present-day flora and fauna on the Greater Antilles. The Dominican amber finds clearly represent the earliest known fossils of all flora and fauna in the Greater Antilles and probably of all Mesoamerica (Caribbean plus Central America and Mexico), although some of the Dominican deposits overlap in age with the Chiapas deposits in Mexico. These amber finds can provide an answer to the question of whether the present biota on the Greater Antilles originated as a result of vicariance or dispersal (Poinar, 1988C). The vicariance model considers that ancestors of the Recent biota were carried to their present position on land masses originating from a "Proto-Antillean Archipelago" that existed in the last Mesozoic between North and South America (Rosen, 1975). The more classical, dispersalist model argues that the Antilles were colonized from the mainland after having reached their present position (Pregill, 1981).

Within the vicariance view, there is still some degree of speculation regarding the physical location and structure of the Proto-Antilles and when it began to drift toward its present location. Suggested dates for the northeastward movement vary from Late Jurassic (Rosen, 1975) to Late Cretaceous–early Tertiary (Coney, 1982), Eocene (Malfait and Dinkelman, 1972), and Late Eocene–Early Oligocene (Buskirk, 1985). From the fauna and flora of Dominican amber, it appears that a well-established forest habitat was on the land mass before the drifting began. Indeed, trees of *H. protera* were probably thriving when the islands were moving toward their present position because a Miocene date for their arrival has been generally accepted (Donnelly, 1985). Of course we must assume that during this interim period, although the coastline certainly fluctuated, Hispaniola was never completely submerged and the climate never changed drastically enough to exterminate a large portion of the biota.

There are other instances in which a vicariance model of biogeography is supported by finds in Dominican amber. One example of these involves the caddis flies (Trichoptera). Caddis flies are aquatic or semi-aquatic in fresh water during their larval stage, and the adults are small to medium-sized mothlike insects. The adults feed principally on liquid foods, are rather weak fliers, and survive normally a month or so. They would not be considered good dispersalists, in contrast to groups of strong fliers like the dragonflies (Odonata), and no caddis flies are considered endemic to the Hawaiian Islands for example (Zimmerman, 1948). Trichoptera are not common in Dominican amber (less than 0.1 percent of all inclusions) but 8 out of 12 Recent families and 12 out of 26 Recent genera from the Greater Antilles have been identified (Wichard, 1987). The most common species, *Antillopsyche oliveri* Wichard, accounts for almost half of the Dominican amber caddis flies. Although *Antillopsyche* is endemic in the Greater Antilles today, the genus belongs to a subfamily (Pseudoneuroclipsinae) whose present distribution (aside from *Antillopsyche*) is restricted to the Ethiopian and Oriental regions. Wichard (1987) suggested that *Antillopsyche* is a relict of the Mesozoic that occurred throughout the South American–African landmass. Continental movements coupled with extinctions could account for its present distribution.

After *Antillopsyche*, the next largest group of Dominican amber caddis flies is related to South American forms, whereas only about 1 percent of the amber forms belong to the Nearctic element (Wichard, 1987). This pattern is similar to the present-day distribution of caddis flies in the Dominican Republic (Flint, 1978), which consists of a minor North American element, a large Neotropical element, and a very small southern African element. The present-day West Indian Trichoptera are highly endemic with over 80 percent of the species known from a single island. Many have speciated to such a degree that they are not clearly related to any mainland form. Such a condition suggests an original colonization of the island landmass in the early Tertiary or Cretaceous (Flint, 1978), before the islands reached their present locations. Flint also noted that the smaller-sized caddis flies had a wider distribution than the average-sized caddis flies and suggested that the former might be carried by the wind. If so, this might establish dispersal as a secondary pattern of distribution for the West Indian caddis flies.

Another example of vicariance from Dominican amber involves the ant *Leptomyrmex neotropicus* Baroni Urbani. As mentioned earlier, this genus today occurs in eastern Australia, New Caledonia, New Guinea, the Aru Islands, and the Moluccas (Baroni Urbani, 1980C; Baroni Urbani and Wilson, 1987). Aside from the related *Leptomyrmula* in Sicilian amber, there are no other fossil remains of this group. Because the queens of *Leptomyrmex*, as well as the workers, are large-bodied and do not fly, dispersal would be limited to walking or rafting. Thus, it is probable that *Leptomyrmex* had a wide distribution at the beginning of the Tertiary, including the Old and New World tropics, and gradually disappeared as a result of competition.

In his study of Dominican amber drosophilids, Grimaldi (1987) described *Neotanygastrella wheeleri* Grimaldi (1987) from Mexican amber. Today this genus occurs in South America (six species), west Africa (five species), the Indo-Pacific (four species), Australia (one species), and Jamaica (one species) (Grimaldi, 1988). The find in Mexican amber together with the clearly Gondwanan distribution of this genus suggests a vicariant event.

The finds of *Anolis*, *Sphaerodactylus*, *Eleutherodactylus*, and mammalian hair in Dominican amber is further evidence of a substantial and varied population of terrestrial fauna on the Greater Antillean landmasses when the resin was being produced. This is, at the least, evidence of interisland vicariance and, together with the amount of endemism found among the present-day flora and fauna, indicates that the "Proto-Antillean Archipelago" consisted of a thriving community.

Reconstructing Ancient Landscapes

Reconstructing from amber inclusions the ecosystem that existed during the actual period of resin flow is definitely a challenge. Of all the fossiliferous amber deposits in the world, only the Baltic has been studied sufficiently as regards the flora and fauna to allow some reconstruction of the original ecosystem (Larsson, 1978). The most interesting aspect of the Baltic amber deposits is the climatic range expressed by the flora and fauna. This is especially demonstrated by representatives of the Angiosperms, which include families that today are placed in temperate, subtropic, and even tropical climates (both New and Old World). The subtropical-tropical aspect is understandable in light of other paleobotanical data (Dorf, 1964) showing that the climate in western Europe has gone through several shifts, ranging from subtropical in the Paleocene to tropical in the Early Eocene to subtropical in the Late Eocene and Early Oligocene to warm temperate in the Miocene to cool temperate in the Late Pliocene. Thus when the Baltic amber was forming (Eocene-Oligocene), the climate varied from subtropical to tropical in western Europe. The rapid cooling of the climate in the Northern Hemisphere during the Oligocene could have accounted for the disappearance of the amber-bearing forests.

As regards moisture in the amber-bearing forests, the fauna in Baltic amber includes many groups (Plecoptera, Ephemeroptera, Trichoptera, Simuliidae) that breed in fast-flowing streams as well as others (Chironomidae, Culicidae) that occur in still water. The closeness of standing water to the trees is surmised by discoveries of the aquatic larvae of caddis flies, mayflies, water bugs, and beetles in the amber. As Larsson (1978) has pointed out, the Baltic amber forest was immense, covering large areas of what is now Scandinavia and northern Europe almost to the Urals. It could easily have encompassed a wide range of biotypes and climatic zones, thus explaining the tropical, subtropical, and boreal elements. A variation in altitude (the forest certainly included hills and possibly small mountains) could explain some of the temperate forms in the amber. One might imagine a dense, humid, subtropical rain-forest habitat comprising at least a significant portion of the Baltic amber forest.

While making reconstructions of the original forest habitat, the researcher should be aware of certain biases regarding amber inclusions. Organisms found in amber do not constitute a random sample of life present in the original habitat (Brues, 1933B; Skalski, 1975). Small organisms (under 5 millimeters in length) are much more numerous in amber than are medium-sized (5–10 millimeters in length) or large (over 1 centimeter in length) organisms. Larger organisms can more easily escape from the resin, although they often leave their appendages as evidence. Organisms that live in the bark microhabitat are more likely to come into contact with the resin and will be more numerous than forms living in the soil or aquatic habitats. Smaller insects lacking strong powers of flight or light plant parts are likely to be blown against resin deposits more frequently than are strong fliers or heavier parts. Organisms attracted to the resin as a source of nesting material (*Proplebeia* bees) or food will more frequently be entrapped than others. Finally, there will always be some (usually rare) large forms that fall into the resin by accident.

It should be obvious from the above that it is difficult to quantify the abundance of or even to qualify the presence of all groups of organisms in an ecosystem on the basis of a limited number of amber inclusions. A closer correlation between amber fauna and the actual population may be achieved if only organisms from a small, select habitat, such as the bark microhabitat, were selected for study. Whatever the biases, it is still possible to reconstruct, if only partially, the original ecosystem, provided that a large enough sample of specimens is available. Because most amber deposits have biological inclusions in only 1–3 percent of the pieces, thousands of amber pieces from a particular site need to be studied in order to approach a representative, qualitative sample of the original flora and fauna.

Tissue Preservation

The high degree of preservation of entire organisms or their parts in amber has been one of the main attractions in studying life in amber. But how well preserved in amber are the tissues and cells of these organisms (Fig. 144)?

In 1903, Kornilovich described well-preserved striated muscles in Baltic amber insects and, later, Petrunkevitch (1950) noted that in some Baltic amber spiders, the esophagus, gizzard, heart, chylenteron (liver), and hypodermis were preserved. Mierzejewski (1976A, 1976B) used the scanning electron microscope to examine exposed surfaces of insects, spiders, and plant remains. He found the remains of book lungs, spinning glands, liver, muscles, and blood cells in spiders and commented that the arachnids in Baltic amber appeared to have better tissue preservation than the insects.

Using the transmission electron microscope, Poinar and Hess (1982) examined the ultrastructure of preserved tissue in a Baltic amber gnat (Mycetophilidae). Recognizable muscle fibers, cell nuclei, ribosomes, lipid droplets, endoplasmic reticulum, and mitochrondria were identified (Figs. 145, 146). Similar cellular inclusions were discovered in the tissues of an adult braconid wasp in Canadian Cretaceous amber (70–80 Ma) (Fig. 147).

Tissue preservation in amber represents an extreme form of mummification resulting from inert dehydration and natural fixation by compounds in the original tree resin. These preservatives, as well as the antibiotic qualities of resin that destroy or retard bacterial and fungal growth, have been known since antiquity (Poinar and Hess, 1985). Tincture of myrrh, an oleoresin from the *Commiphora* (Burseraceae) plant, has been used to treat an assortment of ailments, including sores in the mouth. The Egyptians used resin to coat the wrappings of their mum-

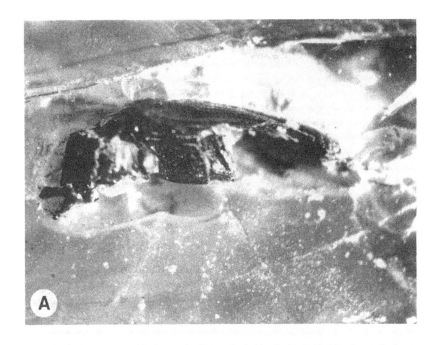

Figure 144. Amber insects are not represented only by the outer cuticle, contrary to popular opinion. The Baltic amber beetle shown in A was the source of the removed internal tissue shown in B. (Specimen courtesy of L. Brost.)

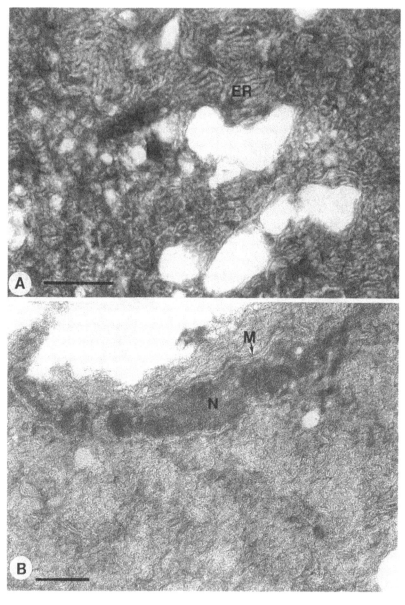

Figure 145. Electron micrographs of tissue removed from the abdomen of a fossilized fungus gnat (family Mycetophilidae) in Baltic amber about 40 million
years old. A. Here are seen profiles of the principal membranes in the cytoplasm,
namely curvilinear arrays spaced 200–400 A° apart that may represent smooth
endoplasmic reticulum (ER). B. The electron-dense nucleus (N) still possesses a
well-defined membrane (M) probably representing the nuclear envelope (bars =
0.25 μm). (Photos by R. Hess-Poinar.)

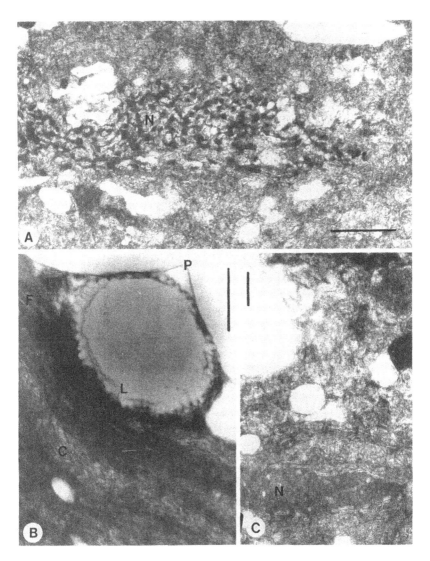

Figure 146. Electron micrographs of tissue removed from the abdomen of a fossilized fungus gnat (family Mycetophilidae) in Baltic amber about 40 million years old. A. A distinct nucleus (N, at left center) shows electron-dense areas believed to represent chromatin interspersed with lighter areas of nucleoplasm (bar = 0.75 μm). B. Cross section through a trachea shows preservation of the outer plasma membrane (P) and the inner lipoprotein cuticulin layer (L). Note also adjacent muscle fibrils (F) and the cristae (C) of a mitochondrion (bar = 0.32 μm). C. Portion of a nucleus (N) shows less electron-dense patches and adjacent vacuoles (bar = 0.25 μm). (Photos by R. Hess-Poinar; from G. Poinar, *Science* 215 [3/5/82]: 1242, copyright 1982 by the AAAS.)

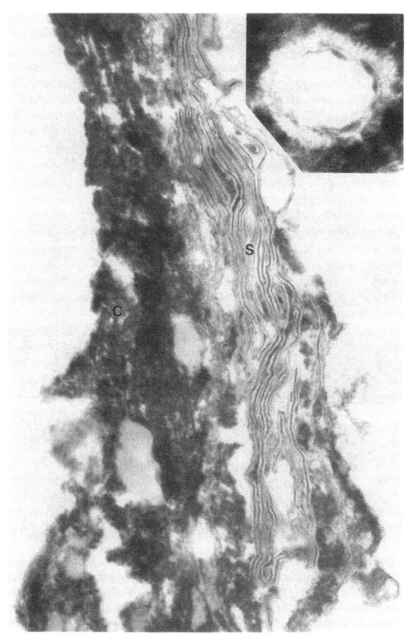

Figure 147. Electron micrographs of tissue removed from the abdomen of a fossilized wasp (family Braconidae) from Canadian amber about 70–80 million years old. Note laminated membraneous structures (S) adjacent to the vacuolated cytoplasm (C). Insert shows a cross section through a trachea. (Specimen courtesy of Frank Carpenter; photos by R. Hess-Poinar.)

mies, which undoubtedly helped preserve the skin and underlying tissues. Resins have been placed on wounds to prevent secondary infections and added to wines to prevent spoilage. In the case of the Greek retsina wines, the resin taste was not only tolerated but later desired and continued as a tradition.

There has been interest in the possibility of recovering DNA from cells preserved in amber. Such DNA would have to survive not only 4 to 120 million years of time, but also fluctuations in temperature, in pressure, and in the chemical, physical, and microbial environment. If amber were a true sealant and preserved cells in an airtight tomb, as recent researchers have suggested (Berner and Landis, 1987, 1988), then the DNA would be maintained in an anaerobic environment. However, it now appears that the amber matrix is porous and can absorb gases, which it does at diffusivities and solubilities common to glassy polymers (Hopfenberg et al., 1988). Consequently, amber not buried in an anaerobic environment would absorb oxygen, which could cause a change in pH and damage to the remaining DNA.

Our knowledge of the ability of extant organisms to survive long periods of dormancy is not great. Sussman (1966) listed the longevity of spores and vegetative cells, and in this list the maximum longevity of bacterial spores was given as 118 years, for a thermophilic bacillus in canned meat (Wilson and Shipp, 1938). However, pseudomonads and other bacteria have been claimed to have been isolated from Paleozoic salt from the Kali and Zechstein deposits in Germany (Dombrowski 1960, 1961). Also, bacteria were reported by Galippe (1920) to be in Baltic amber. These accounts would suggest that bacterial DNA could remain viable for millions of years, but for lack of completeness these reports have not been accepted by the scientific community.

Seeds of the lotus *Nelumbium nucifera* from an ancient dry lake in Manchuria were successfully germinated after having remained dormant for possibly 50,000 years (Libby and Arnold, 1951; Chaney, 1951). This is the longevity record for a vascular plant.

For metazoan animals, the record is 40 years, set by larvae of the cerambycid beetle *Eburia quadrigeminata* (Jaques, 1953). Long periods of dormancy have also been reported for dried juveniles of the plant-parasitic nematodes *Anguina tritici* Steinbuch (28 years) (Fillding, 1950) and *Tylenchus polyhypnus* Steiner and Albin (39 years) (Goodey, 1923.)

Reviving metazoan eucaryotic cells from amber has always been a dream. Recovery of DNA from insects in amber was attempted by Higuchi and Wilson (1984), who had already successfully cloned DNA from dried museum specimens of the quagga (*Equus quagga*, a zebra-like animal that became extinct in 1883) (Higuchi et al., 1984) and from a 40,000-year-old mammoth carcass found frozen at Magadan, Siberia, in 1977 (Higuchi and Wilson, 1984). In the amber studies, tissue was removed from the body cavities of eight insects in Dominican amber. In the cases of two insects, the extracts made from the remains seemed to contain templates capable of directing the synthesis of radioactive complementary DNA in the presence of exogenous primers, purified DNA polymerase 1 from *Escherichia coli*, and P^{32}-labelled nucleoside triphosphates (Higuchi and Wilson, 1984). These results indicate that viable DNA was present in the samples. Tests to determine whether these responses were from insect tissue in amber, and not contamination, were not performed.

A recent study demonstrated the successful extraction and evaluation of genomic DNA from stingless bees (*Proplebeia dominicana*) preserved in Dominican amber (Cano et al., 1992) and opens the door to the molecular study of extinct organisms in amber.

Conclusions and Prospects

The catalog this book has offered portrays the wide spectrum of organisms that occur in amber. Some of the smaller, more fragile forms of life, such as protozoa, nematodes, and rotifers, attest to the delicate, protective nature of fossilization in amber, because representatives of these groups appear nowhere else in the fossil record. Insects abound, but the largest complete organisms in amber are lizards, some of which are as remarkably well preserved as the insects. Parts of larger animals, both birds and mammals, are represented directly by feathers or hairs, and indirectly by small parasites (fleas, ticks, mites) that normally live on warm-blooded hosts. The plant world is also represented in amber: spores, seeds, fruits, flowers, leaves, and portions of stems, bark, and even roots reveal glimpses of the plant world extant when the amber was being formed.

One of the most difficult tasks facing someone interested in the spectrum of life in amber is finding experts who can recognize and identify the many different inclusions. First, not all scientists are interested in life forms of the past, even when the fossils belong to the group they are now studying. Second, many scientists, especially those from lands where there is no tradition of amber collecting and study, regard biological inclusions in amber as curiosities, or they mistrust them as possible fakes.

Entomologists tend to shunt amber inclusions to paleontologists, who, for the most part, are concerned with studying groups in the fossil record that have no direct descendants today. Those who can contribute most to the study of life in amber are specialists with a worldwide knowledge of a group of organisms. But such scientists are becoming more and more difficult to find, because there has been a striking decline in the number of organismic biologists, especially taxonomic systematists, during the past two decades. And as the amount of scientific literature in any particular subject has increased with time, biologists have had to specialize further in order to survive, always learning more and more about less and less. One frequent result of narrowing a field to a manageable size has been geographical limitations. Thus, the days of the generalist are fading, if not gone.

The study of amber insects has given biologists new records for the first appearance of certain groups in the fossil record, new dates for the longevity of symbiotic associations, a chance to study microevolution in certain groups, and new insights into speciation, extinction, and biogeography.

Amber also provides an opportunity to study the paleontological history of biological diversity, as revealed by the number of species, their relative abundance, and their relationship with other species in a particular habitat. We can compare the organisms and ecology of a 40-million-year-old forest with the present flora and fauna of that region, or with any known habitat in the world. Obviously, the kinds and numbers of species living when the resin was being deposited were quite different from those living today. But, what brought about these changes, what are the causes of natural extinctions, and how do new species evolve?

Mass extinctions of life on earth caused by past catastrophic events have been well documented, especially for marine invertebrates. Two of the mass extinctions (the Late Permian and Cretaceous extinctions) were known to affect terrestrial life as well. For example, of the 21 insect orders known to have existed prior to or in the Permian, only 13 survived into the Mesozoic (data based on Carpenter and Burnham, 1985). Surprisingly, however, the effect on insects of the Cretaceous extinction, which recent evidence suggests arose when a massive asteroid smashed into the earth about 65 Ma, appears to be minimal. Although much discussion has focused on the two orders of dinosaurs that became extinct at this time, none of the 17 orders of insects occurring in the Cretaceous became extinct. As more investigations of Cretaceous amber insects are completed, extinction levels at the family rank can perhaps be documented. At present, however, only the family of parasitic wasps known as the Serphitidae, which are found in Canadian and Siberian amber, seem to have disappeared at the end of the Cretaceous.

Many lines of research on amber are being followed today. New amber deposits are being found, and the amber is being examined for biological inclusions. Methods for directly dating amber are being explored. Although infrared, mass, and nuclear magnetic resonance spectroscopy are important techniques for identifying the source of amber samples, they have not yet been very helpful in providing information about the age of samples. One possible exception is the use of nuclear magnetic resonance spectroscopy in measuring the exo-methane peaks of fossilized resins; these peaks are inversely proportional to age. If this method is shown to be consistent and effective, it would settle much speculation regarding the age of various amber deposits. The normal method of dating amber, by determining the age range of associated marine foraminifera, normally provides a minimum date, because such amber has been redeposited.

The several kinds of spectroscopy and other tests previously mentioned ("hot point" and solvent tests) can be used to identify many amber forgeries. Nothing but a sharp eye, however, can detect insects placed inside pieces of real amber that have been cut and glued back together. Such forgeries, many with beautiful inclusions, do appear from time to time, and what the unwary scientist or collector may believe or claim to be a spectacular find may be someone's misdeed.

A technique allowing scientists to view the inside of objects in amber would assist identifications and morphological studies. Objects in amber can be x-rayed without damaging the amber or the specimen, and this technique has been used for studies of frog and lizard skeletons. It would also be helpful and interesting to see the structure of seeds inside a fruit or an ovary within a partially closed flower. For ecological studies, being able to see food items in the alimentary tracts of various invertebrates would also be helpful.

Being able to remove and examine spores and pollen in amber would add a considerable amount of information about plant communities in the original amber forest. Unfortunately, all attempts to remove such items (including insects) have been unsuccessful. Most larger arthropods, such as insects and spiders, "float apart" during the dissolving process, and the solvents often destroy pollen and spores. Under the right physical conditions and with the correct solvents, it might be possible to remove both minute and large objects from amber.

The scanning electron microscope has been used successfully on arthropods in amber in a procedure wherein the amber is chipped away and the organism exposed. Chipping is a risky procedure, however, because it may damage the specimen. If some type of image of objects just beneath the amber surface could be obtained, then the fine details of amber inclusions could be examined, allowing exact comparisons with extant descendants.

Preserving amber fossils is a problem for museum curators and researchers. Although protection from air and light greatly diminishes surface oxidation, it does not completely stop the slow weathering process. Older methods that involved placing the amber in glycerin in sealed containers aided preservation but were not practical for large numbers of specimens, and if the amber had cracks that touched the enclosed organism, the glycerin would reach the organism and darken it, hindering observation. Embedding amber in a plastic matrix is possible but time-consuming, and it makes viewing from all sides more difficult. Hopefully, some type of liquid coating will be developed that could cover an amber piece with a hard, nearly airtight, and completely transparent deposit.

Finally, the idea of cloning DNA from amber inclusions might not be a topic only for fictional stories. This exciting possibility may eventually become a reality as better techniques are developed for removing organisms from amber, for repairing damaged nucleic acids, and for replicating existing nucleotides. It may then be possible to "revive" single-celled organisms such as bacteria, fungi, and protozoa, and to begin the very ambitious task of analyzing multicellular organisms, possibly even vertebrates.

Appendices

Arthropod Classes, Orders, and Families Reported from Mexican Amber

Species and higher taxa of arthropods reported from Mexican amber are listed here. Each species name is followed by the name of the describing author, which in most cases serves also as a bibliographic reference. In a few cases, the name of the describing author is followed by a separate bibliographic reference that serves as the source for the record. A species that has not yet been formally described is indicated by "sp." Records were compiled from various sources, including Hurd et al. (1962) and records in the Museum of Paleontology, University of California, Berkeley.

Myriapoda
  Henicopidae

Arachnida
  Acari
    Cymberemaeidae
      Scapheremaeus brevitarsus Woolley, 1971
    Damaeidae
      Damaeus mexicanus Woolley, 1971
      Damaeus setiger Woolley, 1971
    Digamasellidae
      Dendrolaelaps fossilis Hirschmann, 1971
    Hydrozetidae
      Hydrozetes smithi Woolley, 1971
    Oppiidae
      Oppia hurdi Woolley, 1971
    Oribatidae
      Eremaeus denaius Woolley, 1971
    Oribatulidae
      Liebstadia durhami Woolley, 1971
    Oripodidae
      Exoripoda chiapasensis Woolley, 1971
    Tyroglyphidae
      Amphicalvolia hurdi Türk, 1963

Other acarid families represented
  Aceoseiidae
  Belbidae
  Carabodidae
  Ceratozetidae
  Eremaeidae
  Liodidae
  Neoliodidae
  Uropodidae
Amblypygi
  Electrophyrnidae
    Electrophyrnus mirus Petrunkevitch, 1971
Araneae
  Araneidae
    Aranea exusta Petrunkevitch, 1963
    Mirometa valdespinosa Petrunkevitch, 1963
  Clubionidae
    Chiapasona defuncta Petrunkevitch, 1963
    Mimeutychurus paradoxus Petrunkevitch, 1963
    Prosocer mollis Petrunkevitch, 1963
  Dysderidae
    Mistura perplexa Petrunkevitch, 1971

Eusparassidae
  Veterator extinctus Petrunkevitch,
    1963
Hersiliidae
  Perturbator corniger
    Petrunkevitch, 1971
  Fictotama extincta Petrunkevitch,
    1963
  Priscotama antiqua
    Petrunkevitch, 1971
  Prototama succinea
    Petrunkevitch, 1971
Linyphiidae
  Malepellis extincta Petrunkevitch,
    1971
Oonopidae
  Orchestina mortua
    Petrunkevitch, 1971
Oxyopidae
  Planoxyopes eximius
    Petrunkevitch, 1963
Pisauridae
  Propago debilis Petrunkevitch,
    1963
Theridiidae
  Eomysmena asta Petrunkevitch,
    1971
  Pronepos exilis Petrunkevitch,
    1963
  Pronepos fossilis Petrunkevitch,
    1963
  Municeps chiapasanus
    Petrunkevitch, 1971
  Mysmena fossilis Petrunkevitch,
    1971
Other families of Araneae
    represented
  Mimetidae
  Salticidae
Pseudoscorpiones
  Chernetidae: Schawaller, 1982B

Insecta
  Coleoptera
  Alleculidae
    Hymenorus chiapasensis
      Campbell, 1963
  Anobiidae
    Cryptorama sp.: Spilman, 1971
    Stictoptychus mexambrus Spilman,
      1971
  Cantharidae
    Silis chiapasensis Wittmer, 1963
  Carabidae
    Polyderis antiqua Erwin, 1971

Chrysomelidae
  Crepidodera antiqua Gresset,
    1971
  Profidia nitida Gresset, 1963
Curculionidae
  Cryptorhynchus hurdi
    Zimmerman, 1971
  Cryptorhynchus sp. 1:
    Zimmerman, 1971
  Cryptorhynchus sp. 2:
    Zimmerman, 1971
  Zygops durhami Zimmerman,
    1971
Dermestidae
  Cryptorhopalum eleckron Beal,
    1972
Elateridae
  Agriotes succiniferus Becker, 1963
  Glyphonyx sp.: Becker, 1963
  Mionelater planatus Becker, 1963
Micromalthidae
  Micromalthus debilis LeConte:
    Rozen, 1971
Platypodidae
  Cenocephalus hurdi Schedl, 1962
  Cenocephalus succinicaptus
    Schedl, 1962
  Cenocephalus quadrilobus Schedl,
    1962
Staphylinidae
  Oxypoda binodosa Seevers, 1971
  Palaeopsenius mexicanus Seevers,
    1971
  Palaminus sp.: Seevers, 1971
  Paracyptus minutissima Seevers,
    1971
Other beetle families represented
  Anthicidae
  Anthribidae
  Bostrichidae
  Buprestidae
  Cistelidae
  Clambidae
  Coccinellidae
  Colydiidae
  Cossonidae
  Cucujidae
  Dermestidae
  Euglenidae
  Helodidae
  Histeridae
  Hydrophilidae
  Limulodidae
  Melandryidae
  Mordellidae

Ostomatidae
Phalacridae
Pselaphidae
Ptiliidae
Ptilodactylidae
Scaphidiidae
Scolytidae
Scydmaenidae
Collembola
  Entomobryidae
    Drepanura sp.: Christiansen,
      1971
    Entomobrya decora Nicolet?:
      Christiansen, 1971
    Entomobrya litigiosa Denis?:
      Christiansen, 1971
    Entomobrya sp.: Christiansen,
      1971
    Entomobrya trifasciata
      Handschin?: Christiansen,
      1971
    Isotoma sp.: Christiansen, 1971
    Isotomorus retardus Folsom:
      Christiansen, 1971
    Lepidocyrtinus frater Bonet?:
      Christiansen, 1971
    Lepidocyrtus geayi Denis:
      Christiansen, 1971
    Lepidocyrtus sp.: Christiansen,
      1971
    Paronella sp.: Christiansen, 1971
    Salina tristani Denis?:
      Christiansen, 1971
  Isotomidae
    Isotomurus retardatus Folsom:
      Christiansen, 1971
Dermaptera
Diplura
  Japygidae
Diptera
  Bibionidae
    Plecia pristina Hardy, 1971
  Cecidomyiidae
    Bremia sp.: Gagné, 1973
    Clinodiplosis sp.: Gagné, 1973
    Contarinia sp.: Gagné, 1973
    Henria sp.: Gagné, 1973
    Lestodiplosis sp.: Gagné, 1973
    Monardia sp.: Gagné, 1973
    Phaenolauthia sp.: Gagné, 1973
  Drosophilidae
    Neotanygastrella wheeleri
      Grimaldi, 1987
  Milichiidae
    Phyllomyza hurdi Sabrosky, 1963

Mycetophilidae
  Manota sp.: Gagné, 1980
Periscelidae
  Periscelis annectans Sturtevant,
    1963
Phoridae
  Metopina sp.: Grimaldi, 1989
Psychodidae
  Brunettia hurdi Quate, 1961
  Philosepedon labecula Quate, 1963
  Phlebotomus paternus Quate, 1963
  Phlosepedoin mexicana Quate,
    1963
  Psychoda sp.: Quate, 1963
  Psychoda usitata Quate, 1963
  Telmatoscopus hurdi Quate, 1963
  Trichomyia antiquria Quate, 1961
  Trichomyia declivivena Quate,
    1963
  Trichomyia discalis Quate, 1963
  Trichomyia glomerosa Quate, 1963
  Trichomyia mecoceria Quate, 1963
  Trichomyia smithi Quate, 1963
  Trichomyia sp.: Quate, 1963
Scatopsidae
  Scatopse bilaminata Cook, 1971
  Procolobostema hurdi Cook, 1971
  Procolobostema incisa Cook, 1971
  Procolobostema longicorne Cook,
    1971
  Procolobostema obscurum Cook,
    1971
  Scatopse primula Cook, 1971
  Swammerdamella prima Cook,
    1971
Sciaridae
  Bradysia sp.: Gagné, 1980
Stratiomyiidae
  Pachygaster antiqua James, 1971
Other dipteran families represented
Agromyzidae
Anisopodidae
Asilidae
Calobatidae
Ceratopogonidae
Chironomidae
Chloropidae
Culicidae
Dolichopodidae
Empididae
Ephydridae
Phyllomyzidae
Sphaeroceridae
Tethinidae
Tipulidae

Embioptera
Hemiptera
Cydnidae
Amnestus guapinolinus Thomas,
1988
Dipsocoridae
Ceratocombus hurdi
Wygodzinsky, 1959
Saldidae
Leptosalda chiapensis Cobben,
1971
Other hemipteran families
represented
Aradidae
Discolomidae
Hebridae
Isometopidae
Reduviidae
Schizopteridae
Termitaphididae
Homoptera
Cixiidae
Mnemosyne sp.: Fennah, 1963
Oeclixius amphion Fennah, 1963
Other homopteran families
represented
Aleyrodidae
Cicadellidae
Dactylopiidae
Delphacidae
Derbidae
Diaspididae
Eriococcidae
Flatidae
Fulgoridae
Membracidae
Phylloxeridae
Pseudococcidae
Psyllidae
Hymenoptera
Apidae
Trigona silacea Wille, 1959
Braconidae
Ecphylus oculatus Muesbeck,
1960
Formicidae
Azteca sp.: Brown, 1973
Camponotus sp.: Brown, 1973
?*Crematogaster* sp.: Brown, 1973
Dorymyrmex sp.: Brown, 1973
?*Lasius* sp.: Brown, 1973
Mycetosoritis sp.: Brown, 1973
Pachycondyla sp.: Brown, 1973
Pheidole sp.: Brown, 1973
Stenamma sp.: Brown, 1973

Mymaridae
Alaptus globosicornis Girault:
Doutt, 1973B
Alaptus psocidivorus Gahan:
Doutt, 1973B
Anaphes sp.: Doutt, 1973B
Litus mexicanus Doutt, 1973B
Palaeomymar sp.: Doutt, 1973B
Polynemoidea mexicana Doutt,
1973A
Scelionidae
Palaeogryon muesebecki Masner,
1969
Other hymenopteran families
represented:
Bethylidae
Chalcididae
Ceraphronidae
Diapriidae
Dryinidae
Encyrtidae
Eulophidae
Eurytomidae
Evaniidae
Ichneumonidae
Platygasteridae
Proctotrupidae
Pteromalidae
Thysanidae
Trichogrammatidae
Isoptera
Kalotermitidae
Calcaritermes vetus Emerson,
1969
Incisitermes krishnai Emerson,
1969
Mastotermidae
Mastotermes electomexicus
Krishna and Emerson,
1983
Rhinotermitidae
Coptotermes sucineus Emerson,
1971
Heterotermes primaerus Snyder,
1960
Lepidoptera
Oecophoridae
Tineidae
Mecoptera
?Panorpidae
Neuroptera
Coniopterygidae
Orthoptera
Blattidae
Gryllidae

Gryllotalpidae
Phasmatidae
Tettigoniidae
Psocoptera
    Amphientomidae
      Amphientomum elongatum
        Mockford, 1969
    Archipsocidae
      Archipsocus (Archipsocopsis)
        antiguus Mockford, 1969
    Ectopsocidae
      Ectopsocus sp.: Mockford, 1969
    Epipsocidae
      Epipsocus clarus Mockford, 1969
    Liposcelidae
      Belaphotroctes similis Mockford,
        1969
      Liposcelis sp.: Mockford, 1969
    Myopsocidae: Mockford, 1969

    Psyllipsocidae
      Psyllipsocus sp.: Mockford, 1969
    Trichopsocidae
      Trichopsocus maculosus Mockford,
        1969
    Other psocopteran family
      represented
    Pseudocaeciliidae
Thysanoptera
    Merothripidae
    Phlaeothripidae
Thysanura
    Machilidae
      Neomachilellus sp.: Wygodzinsky,
        1971
Trichoptera
    Hydroptilidae
    Sericostomatidae

Arthropod Classes, Orders, and Families Reported from Dominican Amber

This list of arthropod classes, orders, and families reported from Dominican amber was compiled from information supplied by Jake Brodzinsky, the U.S. National Museum, Schlee and Glöckner (1978), and Wunderlich (1986, 1988). I also gratefully acknowledge the following for identifications of Dominican amber specimens: Orthoptera, L. M. Roth; Embioptera, E. S. Ross; Psocoptera, E. L. Mockford; Ephemeroptera, W. L. Peters; Neuroptera, N. Penny; Trichoptera, W. Wichard; Hemiptera, G. Stonedahl and J. Maldonado Capriles; Coleoptera, J. Doyen, J. F. Lawrence, and J. Chemsak; Diptera, D. Grimaldi, J. Baxter, and R. Lane; Lepidoptera, J. Heppner, J. Powell, and R. Hodges; Hymenoptera, L. Caltagirone, R. Coville, C. Baroni Urbani, M. Prentice, and E. O. Wilson; Crustacea, H. Schmalfuss; Acari, J. Krantz, R. Norton; Araneae, J. Wunderlich; and Scorpiones, J. Santiago-Blay.

Crustacea
  Isopoda (pillbugs, sowbugs)
    Philosciidae
    Platyarthridae
    Pseudarmadillidae
    Sphaeroniscidae

Diplopoda (millipedes)
  Chelodesmidae
  Chytodesmidae
  Glomeridesmidae
  Lophoproctidae
  Polydesmidae
  Polyxenidae
  Pseudononnolenidae
  Pyrgodesmidae
  Siphonophoridae
  Stemmiulidae

Chilopoda (centipedes)
  Cryptopidae
  Scolopendridae

Arachnida
  Acari (mites, ticks)

Bdellidae
Carabodidae
Ceratozetoidea
Cunaxidae
Erythraeidae
Galumnidae
Gamasidae
Ixodidae
Liacaridae
Listrophoridae
Oppiidae
Oribatidae
Oribotritiidae
Plateremaeoidea
Trombidioidea
Amblypygi (tail-less whipscorpions)
  Phrynidae
Araneae (spiders)
  Agelenidae (grass spiders)
  Amaurobiidae (white-eyed spiders)
  Anapidae
  Anyphaenidae
  Araneidae (orb spiders)
  Barychelidae
  Caponiidae (caponiid spiders)

Clubionidae (hunting spiders)
Ctenidae (wandering spiders)
Ctenizidae (trap-door spiders)
Dictynidae (hackled-band weavers)
Dipluridae (funnel-web tarantulas)
Gnaphosidae (hunting spiders)
Hahniidae
Hersiliidae (hersiliid spiders)
Heteropodidae (giant crab spiders)
Linyphiidae (sheet-web spiders)
Liocranidae
Lycosidae (wolf spiders)
Microstigmatidae
Mimetidae (spider-hunting
　spiders)
Myrmeciidae
Nesticidae
Ochyroceratidae
Oecobiidae (oecobiid spiders)
Oonopidae (minute jumping
　spiders)
Oxyopidae (lynx spiders)
Palpimanidae
Philodromidae
Pholcidae (long-legged spiders)
Pisauridae (nursery web spiders)
Pycnothelidae
Salticidae (jumping spiders)
Scytodidae (spitting spiders)
Segestriidae
Selenopidae (selenopid crab
　spiders)
Sicariidae
Symphytognathidae
Tetrablemmidae
Tetragnathidae (long-jawed orb
　spiders)
Theraphosidae (tarantulas)
Theridiidae (comb-footed spiders)
Theridiosomatidae (ray spiders)
Thomisidae (crab spiders)
Uloboridae
Opiliones (harvestmen)
　Phalangodidae
Pseudoscorpiones (pseudoscorpions)
　Cheiridiidae
　Cheliferidae
　Chernetidae
　Chthoniidae
Scorpiones (scorpions)
　Buthidae
Solpugida (windscorpions)
　Ammotrechidae

Insecta
　Coleoptera (beetles)
　　Alleculidae (comb-clawed beetles)
　　Anobiidae (anobiid beetles)
　　Anthicidae (antlike flower beetles)
　　Biphyliidae (fungus beetles)
　　Bostrichidae (twig borers)
　　Brachypsectridae
　　Brentidae (straight-snouted
　　　weevils)
　　Bruchidae (seed beetles)
　　Buprestidae (metallic wood-boring
　　　beetles)
　　Cantharidae (soldier beetles)
　　Carabidae (ground beetles)
　　Cerambycidae (long-horned
　　　beetles)
　　Cerylonidae (cerylonid beetles)
　　Chrysomelidae (leaf beetles)
　　Ciidae (fungus beetles)
　　Cleridae (checkered beetles)
　　Coccinellidae (ladybird beetles)
　　Colydiidae (cylindrical bark
　　　beetles)
　　Corylophidae (minute fungus
　　　beetles)
　　Cryptophagidae (silken fungus
　　　beetles)
　　Cucujidae (flat bark beetles)
　　Curculionidae (weevils)
　　Dascillidae (soft-bodied plant
　　　beetles)
　　Dermestidae (dermistid beetles)
　　Dytiscidae (diving beetles)
　　Elateridae (click beetles)
　　Endomychidae (fungus beetles)
　　Erotylidae (fungus beetles)
　　Eucinetidae (plate-thigh beetles)
　　Eucnemidae (false click beetles)
　　Euglenidae (antlike leaf beetles)
　　Haliplidae (crawling water beetles)
　　Helodidae (marsh beetles)
　　Histeridae (hister beetles)
　　Hydrophilidae (water beetles)
　　Inopeplidae (flat rove beetles)
　　Lagriidae (long-jointed beetles)
　　Lampyridae (fireflies)
　　Languriidae (lizard beetles)
　　Lathridiidae (scavenger beetles)
　　Leiodidae (round fungus beetles)
　　Limnichidae (minute marsh
　　　beetles)
　　Limulodidae (horseshoe crab
　　　beetles)
　　Lucanidae (stag beetles)

Lyctidae (powderpost beetles)
Lymexylidae (ship timber beetles)
Melandryidae (false darkling beetles)
Melyridae (soft-winged flower beetles)
Micromalthidae
Mordellidae (flower beetles)
Mycetophagidae (hairy fungus beetles)
Mycteridae (mycterid beetles)
Nitidulidae (sap beetles)
Noteridae (burrowing water beetles)
Ostomatidae (bark-gnawing beetles)
Paussidae
Pedilidae
Phalacridae (shining flower beetles)
Platypodidae (flat-footed beetles)
Pselaphidae (short-winged mold beetles)
Ptiliidae (feather-winged beetles)
Ptilodactylidae (ptilodactylidid beetles)
Ptinidae (spider beetles)
Pyrochroidae (fire-colored beetles)
Rhipiphoridae (wedge-shaped beetles)
Rhizophagidae (root-eating beetles)
Rhysodidae (wrinkled bark beetles)
Scaphidiidae (shining fungus beetles)
Scarabaeidae (scarab beetles)
Scolytidae (bark beetles)
Scydmaenidae (antlike stone beetles)
Staphylinidae (rove beetles)
Tenebrionidae (darkling beetles)
Throscidae (throscid beetles)
Collembola (springtails)
Entomobryidae
Isotomidae
Sminthuridae
Diplura
Procampodeidae
Diptera (flies)
Acroceridae (small-headed flies)
Agromyzidae (leaf miners)
Anisopodidae (wood gnats)
Asilidae (robber flies)
Asteiidae (asteiid flies)

Aulacigastridae (aulacigastrid flies)
Bibionidae (March flies)
Bombyliidae (bee flies)
Carnidae (carnid flies)
Cecidomyiidae (gall midges)
Ceratopogonidae (biting midges)
Chamaemyiidae
Chaoboridae (phantom midges)
Chironomidae (midges)
Chloropidae (chloropid flies)
Clusiidae
Culicidae (mosquitoes)
Dixidae (dixid midges)
Dolichopodidae (long-legged flies)
Drosophilidae (small fruit flies)
Empididae (dance flies)
Ephydridae (shore flies)
Lauxaniidae (lauxaniid flies)
Micropezidae (stilt-legged flies)
Milichiidae (milichiid flies)
Muscidae (muscid flies)
Mycetophilidae (fungus gnats)
Odiniidae (odiniid flies)
Otitidae (picture-winged flies)
Pachyneuridae
Periscelidae (periscelid flies)
Phoridae (phorid flies)
Pipunculidae (big-headed flies)
Platypezidae (flat-footed flies)
Psychodidae (moth flies, sand flies)
Rhagionidae (snipe flies)
Richardiidae (richardiid flies)
Scatopsidae (scavenger flies)
Sciaridae (fungus gnats)
Simuliidae (blackflies)
Sphaeroceridae (sphaerocerid flies)
Stratiomyidae (soldier flies)
Syrphidae (syrphid flies)
Tabanidae (horseflies, deerflies)
Tachinidae (tachinid flies)
Tephritidae (fruit flies)
Tipulidae (crane flies)
Trixoscelididae (trixoscelidid flies)
Dermaptera (earwigs)
No family specified
Embioptera (web spinners)
Anisembiidae
Teratembiidae
Ephemeroptera (mayflies)
Baetidae
Leptophlebiidae
Hemiptera (bugs)
Anthocoridae (minute pirate bugs)
Aradidae (flat bugs)

Dipsocoridae (jumping ground bugs)
Enicocephalidae (gnat bugs)
Gerridae (water striders)
Leptopodidae (spiny shore bugs)
Lygaeidae (seed bugs)
Miridae (plant bugs)
Nabidae (damsel bugs)
Pentatomidae (stink bugs)
Reduviidae (ambush bugs)
Schizopteridae (jumping ground bugs)
Thaumastocoridae (royal palm bugs)
Tingidae (lace bugs)
Homoptera (clear-winged bugs)
Aetalionidae (treehoppers)
Aleyrodidae (whiteflies)
Aphididoidea (plant lice)
Cercopidae (spittlebugs)
Cicadellidae (leafhoppers)
Cixiidae (planthoppers)
Coccidae (scales)
Delphacidae (planthoppers)
Dictyopharidae (planthoppers)
Flatidae (planthoppers)
Fulgoroidea (planthoppers)
Issidae (planthoppers)
Membracidae (treehoppers)
Ortheziidae (coccids)
Psyllidae (psyllids)
Hymenoptera (ants, wasps, bees)
Apidae (bees)
Argidae (argid sawflies)
Bethylidae (bethylids)
Braconidae (braconids)
Chalcididae (chalcidids)
Colletidae
Cynipidae (gall wasps)
Dryinidae (dryinids)
Eulophidae (eulophids)
Eupelmidae (eupelmids)
Evaniidae (ensign wasps)
Formicidae (ants)
Ichneumonidae (ichneumonids)
Mutillidae (velvet ants)
Mymaridae (fairyflies)
Mymaridae (fairyflies)
Platygasteridae (platygasterids)
Proctotrupidae (proctotrupids)
Pteromalidae (pteromalids)
Scelionidae (scelionids)
Sphecidae (sphecids)
Tiphiidae (tiphiids)

Torymidae (torymids)
Vespidae (vespids)
Isoptera (termites)
Kalotermitidae
Mastotermidae
Rhinotermitidae
Termitidae
Lepidoptera (moths, butterflies)
Blastobasidae
Cosmopterygidae
Gelechiidae
Noctuidae
Tineidae
Tortricidae
Neuroptera (membraneous-winged flies)
Chrysopidae (green lacewings)
Coniopterygidae (dusty-wings)
Hemerobiidae (brown lacewings)
Odonata (dragonflies)
No family specified
Orthoptera (grasshoppers and allies)
Acrididae (short-horned grasshoppers)
Blattidae (cockroaches)
Gryllidae (crickets)
Mantidae (praying mantids)
Phasmatidae (walking sticks)
Tettigoniidae (long-horned grasshoppers)
Plecoptera
Perlidae
Psocoptera (psocids)
Amphientomidae
Archipsocidae
Caeciliidae
Cladiopsocidae
Dolabellopsocidae
Epipsocidae
Lepidopsocidae
Liposcelidae
Philotarsidae
Polypsocidae
Pseudocaeciliidae
Psocidae
Psoquillidae
Ptiloneuridae
Siphonaptera (fleas)
Rhopalopsyllidae
Strepsiptera (twisted-winged insects)
Elenchidae
Thysanoptera (thrips)
No family specified
Thysanura (bristletails)
Nicoletiidae

Trichoptera (caddis flies)
  Atopsychidae
  Glossosomatidae
  Helicopsychidae
  Hydropsychidae
  Hydroptilidae
Leptoceridae
Philopotamidae
Polycentropodidae
Zoraptera (zorapterans)
  Zorotypidae

Reference Matter

References Cited

Abdullah, M. 1964. New heteromerous beetles (Coleoptera) from the Baltic amber of eastern Prussia and gum copal of Zanzibar. *Trans. R. Entomol. Soc. London* 116: 329–46.

Acra, A., R. Milki, and F. Acra. 1972. The occurrence of amber in Lebanon. *Abst. Reunion Scientific Assoc. Advancement of Sciences, Beirut* 4: 76–77 (UNESCO).

Agricola, G. 1546. *De natura fossilium.* English edition by M. A. and J. A. Bandy, Geological Society of America, special paper 63, 1955.

Ahmadjian, V., and S. Paracer. 1986. *Symbiosis.* Hanover: Univ. Press New England. 212 pp.

Alekseyev, V. N., and A. P. Rasnitsyn. 1981. Late Cretaceous Megaspilidae (Hymenoptera) from amber of the Taymyr. *Paleontol. J.* 15: 124–28.

Alexander, C. P. 1931. Crane flies of the Baltic amber (Diptera). *Bernstein-Forschungen* 2: 1–135.

Ander, K. 1941. Die Insektenfauna des Baltischen Bernstein nebst damit verknüpften zoogeographischen Problemen. *Lunds Univ. Ärsskrift.* N.F. 38: 3–82.

Andersen, N. M. 1982. The semiaquatic bugs (Hemiptera, Gerromorpha). Phylogeny, adaptations, biogeography and classification. *Entomonograph* 3: 1–455.

Andersen, N. M., and G. O. Poinar, Jr. 1992. Phylogeny and classification of an extinct water strider genus (Hemiptera, Gerridae) from Dominican amber, with evidence of mate guarding in a fossil insect. *Zeitsch. f. zoologisches Systematik und Evolutionsforschung* (in press).

Anderson, W. F. 1977. Infra-roodspectrometrisch Onderzoek van Barnstein. *Grondboor and Hamer* August 1977 (No. 4): 98–107.

Andrée, K. 1937. *Der Bernstein und seine Bedeutung in Natur- und Geisterwissenschaften, Junst und Kunstgewerbe, Technik, Industrie und Handel.* Königsberg: Grfe und Unzer. 219 pp.

———. 1939. Über neue Funde neolithischer Bernsteinartefakte Schwarzorter Stils an der Ostseeküste Kurlands und ihre Bedeutung für die Frage der Entstehung der Schwarzorter Bernstein-Lagerstätte. *Bernstein-Forschungen* 4: 48–51.

————. 1951. *Der Bernstein: Das Bernsteinland und sein Leben.* Stuttgart: Kosmos. 96 pp.

Anonymous. 1987. American Museum of Natural History. Cover photo: 177th Annual Report (1985–86).

Antropov, A. V., and W. J. Pulawski, 1989. A new species of *Pison* Jurine from Baltic amber (Hymenoptera: Sphecidae). *Pan Pac. Entomol.* 65: 312–18.

Arnold, C. A. 1947. *An Introduction to Paleobotany.* New York: McGraw Hill.

Arnoldi, L. V., V. V. Zherichin, L. M. Nikritin, and A. G. Ponomarenko. 1977. Mezozojskie Zestkokrylye. *Trudy Paleontol. Instit. Akad. Nauk SSSR* 161: 1–204.

Aycke, J. C. 1835. *Fragmente zur Naturgeschichte des Bernsteins.* Danzig.

Bachmayer, F. 1962. Fossile Pilzhypen im Flyschharz des Steinbruches im Höbersbachtal bei Gablitz in Niederösterreich. *Ann. Naturhistor. Mus. Wien* 65: 47–49.

————. 1968. Ein bemerkenswerter Fund: *Myrica*-Früchte im Flyschlarz. *Ann. Naturhistor. Mus. Wien* 72: 639–43.

Bachmayer, F., and E. Schulz. 1978. Ein bemerkenswerter Insektenrest im fossilen Harz des Glaukonitsandsteines (Eggenburgien) der Aufschlussbohrung "Herzogbirbaum 1" (Niederösterreich). *Ann. Naturhistor. Mus. Wien* 81: 113–120.

Bachofen-Echt, A. 1942. Ueber die Myriapoden des Bernsteins. *Palaeobiologica* 7: 394–403.

————. 1944. Einschlüsse von Federn und Haaren im Bernstein. *Palaeobiologica* 8: 113–19.

————. 1949. *Der Bernstein und seine Einschlüsse.* Wien: Springer-Verlag. 204 pp.

Balaguer, J. 1987. Decree on the exportation of amber fossils from the Dominican Republic. Decree No. 288–87, effective June 5, 1987, published in the Santo Domingo newspaper *El Caribe* on June 5, 1987 (p. 8).

Bandel, K., and N. Vavra. 1981. Ein fossiles Harz aus der Unterkreide Jordaniens. *N. Jb. Geol. Palont. Mk.* 1: 19–33.

Baroni Urbani, C. 1980A. First description of fossil gardening ants (Amber collection Stuttgart and Natural History Museum Basel: Hymenoptera: Formicidae I: Attini). *Stuttgarter Beitr. Naturk.* (Serie B) 54: 13 pp.

————. 1980B. *Anochetus corayi* n.sp., the first fossil Odontomachiti ant (Amber collection Stuttgart: Hymenoptera, Formicidae II: Odontomachiti). *Stuttgarter Beitr. Naturk.* (Serie B) 55: 6 pp.

————. 1980C. The first fossil species of the Australian ant genus *Leptomyrmex* in amber from the Dominican Republic (Amber collection Stuttgart: Hymenoptera, Formicidae III: Leptomyrmicini). *Stuttgarter Beitr. Naturk.* (Serie B) 62: 10 pp.

————. 1980D. The ant genus *Gnamptogenys* in Dominican amber (Amber collection Stuttgart: Hymenoptera, Formicidae IV: Ectatommini). *Stuttgarter Beitr. Naturk.* (Serie B) 64: 10 pp.

————. 1988. Phylogeny and behavioral evolution in ants. Paper presented Int. Conf. Evolution and Ecology of Social Behavior, Florence 19–24 March, 1988.

Baroni Urbani, C., and S. Graeser. 1987. REM-Analysen an einer pyritisierten Ameise aus Baltischem Bernstein. *Stuttgarter Beitr. Naturk.* (Serie B) 133: 16 pp.

Baroni Urbani, C., and J. B. Saunders. 1980. The fauna of the Dominican Republic amber: the present status of knowledge. *Proc. Ninth Caribbean Geological Conference* (Santo Domingo; August 1980) pp. 213–23.

Baroni Urbani, C., and E. O. Wilson. 1987. The fossil members of the ant tribe Leptomyrmecini (Hymenoptera: Formicidae). *Psyche* 94: 1–8.

Barthel, M., and H. Hetzer. 1982. Bernstein-Inklusen aus dem Miozän des Bitterfelder Raumes. *Zeitschr. angewandte Geologie* 28: 314–36.

Beal, R. S. 1972. A new fossil *Cryptorhopalum* (Dermestidae: Coleoptera) from Tertiary amber of Chiapas, Mexico. *J. Paleontol.* 46: 317–18.

Beardsley, J. W. 1969. A new fossil scale insect (Homoptera: Coccoidea) from Canadian amber. *Psyche* 76: 270–79.

Beck, C. W. 1972. Aus der Bernstein-Forschung. *Naturwissenschaften* 59: 294–8.

———. 1986. Spectroscopic investigations of amber. *Appl. Spectroscopy Rev.* 22: 57–110.

Beck, C. W., J. B. Lambert, and J. S. Frye. 1986. Beckerite. *Phys. Chem. Minerals* 13: 411–3.

Beck, C. W., E. Wilbur, and S. Meret. 1964. Infra-red spectra and the origin of amber. *Nature* 201: 256–7.

Becker, E. C. 1963. Three new fossil elaterids from the amber of Chiapas, Mexico, including a new genus (Coleoptera). *J. Paleontol.* 37: 125–8.

Beier, M. 1937. Pseudoscorpione aus dem baltischen Bernstein. *Festschrift für Embrik Strand* (Riga) 2: 302–16.

———. 1947. Pseudoscorpione im baltischen Bernstein und die Untersuchung von Bernstein-Einschlüssen. *Mikroskopie* (Vienna) 1: 188–99.

———. 1948. Phoresie und Phagophilie bei Pseudoscorpionen. *Osterr. zool. Zeitsch., Vienna* 1: 441–97.

———. 1955. Pseudoscorpione im baltischen Bernstein aus dem Geologischen Staatsinstitut in Hamburg. *Mitteilungen Geol. Staatsinst. Hamburg* 24: 48–54.

Berendt, G. C. 1856. *Die im Bernstein befindlichen organischen Reste der Vorwelt.* I–II. Berlin. 374 pp.

Berggren, W. A., and J. A. H. Van Couvering. 1974. The late Neogene. *Palaeogeogr. Palaeoclimat. Palaeoecol.* 16: 1–216.

Berkeley, M. J. 1848. On three species of mold detected by Dr. Thomas in the amber of East Prussia. *Ann. Mag. Nat. Hist.* (Series 2) 2: 380–83.

Berner, R. A., and G. P. Landis. 1987. Chemical analysis of gaseous bubble inclusions in amber: the composition of ancient air? *Amer. J. Sci.* 287: 757–62.

———. 1988. The major gas composition of ancient air: analysis of gas bubble inclusions in fossil amber. *Science* 239: 1406–8.

Bischoff, H. 1916. Bernstein hymenopteren. *Schr. phys. Ökon. Ges. Königsberg* 56: 139–44.

Blackwell, M. 1984. Myxomycetes and their arthropod associates. In: *Fungus-Insect Relationships*, pp. 67–90. Q. Wheeler & M. Blackwell, eds. New York: Columbia Univ. Press.

Blom, F. 1959. Historical notes relating to the pre-Columbian amber trade from Chiapas. *Mitteilungen aus dem Museum Volkerkunde, Hamburg* 25: 24–27.

Blunck, G. 1929. Bakterieneinschlüsse im Bernstein. *Centralblatt für Mineralogie, Geologie und Paläontologie* (Abt. B, no. 11) 554–5.

Böhme, W. 1984. Erstfund eines fossilen Kugelfingergeckos (Sauria: Gekkonidae: Sphaerodactylinae) aus Dominikanischem Bernstein (Oligozän von Hispaniola, Antillen). *Salamandra* 20: 212–20.

Bollow, H. 1940. Die erste Helminide (Col. Dryop.) aus Bernstein. *Mitteilungen der Münchner entomol. Ges.* 30: 117–9.

Borgmeier, T. 1968. A catalogue of the Phoridae of the World (Diptera, Phoridae). *Studia entomol.* 11: 1–367.

Borror, D. J., D. M. DeLong, and C. A. Triplehorn. 1981. *An Introduction to the Study of Insects*. 5th ed. New York: Saunders College Publishing.

Borror, D. J., C. A. Triplehorn, and N. F. Johnson. 1989. *An Introduction to the Study of Insects*. 6th ed. Philadelphia: Saunders College Publishing.

Botosaneanu, L. 1981. On a false and genuine caddis-fly from Burmese amber (Insecta: Trichoptera, Homoptera). *Bull. Zool. Museum Amsterdam* 8: 73–78.

Botosaneanu, L., and W. Wichard. 1983. Upper-Cretaceous Siberian and Canadian amber caddisflies (Insecta: Trichoptera). *Bijdragen tot de Dierkunde* 53: 187–217.

Boucot, A. J. 1990. *Evolutionary Paleobiology of Behavior and Coevolution*. Amsterdam: Elsevier. 725 pp.

Brischke, D. 1886. Die Hymenopteren des Bernsteins. *Schr. naturf. Ges. Danzig* N.F. 6: 278–9.

Brown, B. V., and E. M. Pike. 1990. Three new fossil phorid flies (Diptera: Phoridae) from Canadian Late Cretaceous amber. *Can. J. Earth Sci.* 27: 845–8.

Brown, W. L., Jr. 1973. A comparison of the Hylean and Congo-West African Rain forest ant faunas. In: *Tropical Forest Ecosystems in Africa and South America: a comparative review*, pp. 161–85. B. J. Meggers, E. S. Ayensu, and W. D. Duckworth, eds. Washington, D.C.: Smithsonian Inst. Press.

Brues, C. T. 1923. Some new fossil parasitic Hymenoptera from Baltic amber. *Proc. Amer. Acad. Sci., Boston* 58: 327–46.

———. 1926. A species of *Urocerus* from Baltic amber. *Psyche* 33: 168–70.

———. 1933A. The parasitic Hymenoptera of the Baltic amber. I. *Bernstein-Forschungen* 3: 4–178.

———. 1933B. Progressive change in the insect population of forests since the Early Tertiary. *Amer. Nat.* 67: 385–406.

———. 1937. Ichneumonoidea, Serphoidea, and Chalcidoidea from Canadian amber. *Univ. Toronto Studies. Geological Series* 40: 27–44.

———. 1939. New Oligocene Braconidae and Bethylidae from Baltic amber (Hymenoptera). *Ann. Entomol. Soc. Amer.* 32: 251–63.

———. 1940. Fossil parasitic Hymenoptera of the family Scelionidae from Baltic amber. *Proc. Amer. Acad. Sci. Boston* 74: 69–70.

Brundin, L. 1976. A Neocomian chironomid and Podonominae-Aphroteniinae (Diptera) in the light of phylogenetics and biogeography. *Zoologica Scripta* 5: 139–60.

Bryant, D. D. 1983. A recently discovered amber source near Totolapa, Chiapas, Mexico. *American Antiquity* 48: 354−7.

Buddhue, J. D. 1935. Mexican amber. *Rocks and Minerals* 10: 170−71.

Buffum, W. A. 1898. *The Tears of the Heliades or Amber as a Gem.* 3d and rev. ed. London: Sampson Low, Marston & Co. 108 pp.

Burnham, L. 1978. Survey of social insects in the fossil record. *Psyche* 85: 85−133.

Burr, M. 1911. Dermaptera (earwigs) preserved in amber from Prussia. *Trans. Linn. Soc. London* (2d Series) 11: 145−50.

Buskirk, R. E. 1985. Zoogeographic patterns and tectonic history of Jamaica and the northern Caribbean. *J. Biogeography* 12: 447−61.

Campbell, J. M. 1963. A fossil beetle of the genus *Hymenorus* (Coleoptera: Alleculidae) found in amber from Chiapas, Mexico. *Univ. Calif. Publ. Entomol.* 31: 41−42.

Cano, R. J., H. Poinar, and G. O. Poinar, Jr. 1992. Isolation and partial characterization of DNA from *Proplebeia dominicana* (Hymenoptera: Apidae) in 25−40 million year old amber. *Medical Science Res.* 20: 249−51.

Carpenter, F. M. 1954. The Baltic amber Mecoptera. *Psyche* 61: 31−40.

————. 1956. The Baltic amber snake-flies (Neuroptera). *Psyche* 63: 77−81.

————. 1976. Note on *Bittacus validus* in Baltic amber. *Psyche* 82: 303.

Carpenter, F. M., and L. Burnham. 1985. The geological record of insects. *Ann. Rev. Earth Planet. Sci.* 13: 297−314.

Carpenter, F. M., J. W. Folsom, E. O. Essig, A. C. Kinsey, C. T. Brues, M. W. Boesel, and H. E. Ewing. 1937. Insects and arachnids from Canadian amber. *Univ. Toronto Studies. Geological Series* 40: 7−62.

Case, G. R. 1982. *A Pictorial Guide to Fossils.* New York: Van Nostrand Reinhold Co. 514 pp.

Caspary, R., and R. Klebs. 1907. Die Flora des Bernsteins u. anderer fossiler Harze des ostpreussichen Tertiärs. *Abh. König. Preuss. Geol. L.-A.N.F.H.* 4: 182 pp.

Champetier, Y., M. Madre, J. C. Samama, and I. Tavares. 1980. Localisation de l'ambre au sein des sequences a lignites en Republique Dominicaine. *Proc. Ninth Caribbean Geological Conference* (August, 1980) pp. 277−9.

Chaney, R. W. 1951. How old are the Manchurian *Lotus* seeds? *The Garden Journal* (N.Y. Botanical Garden) 1: 137−9.

Chhibber, H. L. 1934. *The Mineral Resources of Burma.* London: Macmillan & Co.

Chopard, L. 1936. Orthoptères fossiles et subfossiles de l'ambre et du copal. *Ann. Soc. entomol. France* 105: 375−86.

Christiansen, K. 1971. Notes on Miocene amber Collembola from Chiapas. *Univ. Calif. Publ. Entomol.* 63: 45−48.

Cobben, R. H. 1971. A fossil shore bug from the Tertiary amber of Chiapas, Mexico (Heteroptera, Saldidae). *Univ. Calif. Publ. Entomol.* 63: 49−56.

Cobos, A. 1963. Comentarios criticos sobre algunos *Sternoxia* fósiles del ámbar del Báltico recientemente descritos (Coleoptera). *Eos* 39: 345−55.

Cockerell, T. D. A. 1909. Some European fossil bees. *The Entomologist* 42: 313−7.

————. 1917A. Arthropods in Burmese amber. *Psyche* 24: 40−45.

————. 1917B. Fossil insects. Appendix. *Ann. Entomol. Soc. Amer.* 10: 19−22.

———. 1917C. Some American fossil insects. *Proc. U.S. National Museum* 51: 89–106.

———. 1917D. Insects in Burmese amber. *Ann. Entomol. Soc. Amer.* 10: 323–9.

———. 1917E. Descriptions of fossil insects. *Proc. Biol. Soc. Washington* 30: 79–81.

———. 1917F. Arthropods in Burmese amber. *Amer. J. Sci.* 44: 360–68.

———. 1919. Two interesting insects in Burmese amber. *The Entomologist* 52: 193–5.

———. 1920A. A therivid fly from Burmese amber. *The Entomologist* 53: 169–70.

———. 1920B. Fossil arthropods in the British Museum–IV. *Ann. Mag. Nat. Hist.* 6: 211–214.

———. 1920C. Fossil arthropods in the British Museum: I. *Ann. Mag. Nat. Hist.* 5: 273–79.

———. 1922. Fossils in Burmese amber. *Nature* 109: 713–4.

Cokendolpher, J. C. 1986. A new species of fossil *Pellobunus* from Dominican Republic amber (Arachnida: Opiliones: Phalangodidae). *Carib. J. Sci.* 22: 205–11.

Cokendolpher, J. C., and G. O. Poinar, Jr. 1992. Tertiary harvestmen from Dominican Republic amber (Arachnida: Opiliones: Phalangodidae). *Bull. British Arachnol. Soc.* (in press).

Coney, P. J. 1982. Plate tectonic constraints on the biogeography of middle America and the Caribbean Region. *Ann. Missouri Bot. Garden* 69: 432–43.

Conwentz, H. 1886A. Die Bernsteinfichte. *Berichte der Deutschen Botanischen Gesellschaft* 4: 375.

———. 1886B. *Die Flora des Bernsteins. 2. Die Angiospermen des Bernsteins.* Danzig. 140 pp.

———. 1890. *Monographie der baltischen Bernsteinbäume.* Danzig. 203 pp.

Cook, E. F. 1971. Fossil Scatopsidae in Mexican amber (Diptera: Insecta). *Univ. Calif. Publ. Entomol.* 63: 57–61.

Cooper, B. S., and D. G. Murchison. 1969. Organic geochemistry of coal. In: *Organic Geochemistry*, pp. 669–726. G. Eglinton and M. T. J. Murphy, eds. Berlin: Springer-Verlag.

Cooper, K. W. 1964. The first fossil tardigrade: *Beorn leggi* Cooper, from Cretaceous amber. *Psyche* 71: 41–48.

Cox, C. B., and P. D. Moore. 1985. *Biogeography.* Boston: Blackwell Scientific Publications. 244 pp.

Crosskey, R. W. 1990. *The Natural History of Blackflies.* Chichester, G.B.: J. Wiley & Sons. 711 pp.

Crowson, R. A. 1973. On a new superfamily Artematopoidea of polyphagan beetles, with the definition of two new fossil genera from the Baltic amber. *J. Nat. Hist.* 7: 225–38.

———. 1981. Evolutionary history of beetles. In: *The Biology of the Coleoptera*, pp. 658–88. R. A. Crowson, ed. London: Academic Press.

———. 1984. The associations of Coleoptera with Ascomycetes. In: *Fungus-Insect Relationships*, pp. 256–85. Q. Wheeler and M. Blackwell, eds., New York: Columbia Univ. Press.

Crowson, R. A., W. D. I. Rolfe, J. Smart, C. D. Waterston, E. C. Willey, and

R. J. Wootton. 1967. Arthropoda: Chelicerata, Pycnogonida, Palaeoisopus, Myriapoda and Insecta. In: *The Fossil Record*, Part II, pp. 499–534. W. B. Harland, C. H. Holland, M. R. House, N. F. Hughes, A. B. Reynolds, M. J. S. Rudwick, G. E. Satterthwiate, L. B. H. Tarlo, and E. C. Willey, eds. Geological Society of London.

Cunninghamn, A., I. D. Gay, A. C. Oehlschlager, and J. H. Langenheim. 1983. ¹³CNMR and IR analyses of structure, aging and botanical origin of Dominican and Mexican ambers. *Phytochemistry* 22: 965–8.

Currado, I., and M. Olmi. 1983. Primo reperto di Driinide fossile in ambra della Repubblica Dominicana (Hymenoptera, Dryinidae). *Boll. Museo Region. Scienze Naturali* (Torino) 1: 329–34.

Czeczott, H. 1961. The flora of the Baltic amber and its age. *Prace Muzeum Ziemi (Paleobotaniczne) Warszawa* 4: 119–45.

Dahl, C. 1971. Trichoceridae (Diptera) from the Baltic amber. *Entomol. Scandia* 2: 29–40.

Dahms, P. 1906. Über den Brechungsquotienten des Succinit und einige Erscheinungen, die sich bei der künstlichen Behandlung dieses Bernsteins zeigen. *Schr. naturf. Ges. Danzig* N.F. 11: 25–49.

Dall, W. H. 1870. *Alaska and Its Resources*. London: Sampson Lawson and Marston. 625 pp.

Demoulin, G. 1970. Troisième contribution á la connaissance des Éphéméroptères de l'ámbre oligocène de la Baltique. *Bull. Inst. R. Soc. Nature Belg.* 46: 1–11.

Disney, R. H. L. 1987. Four species of scuttle fly (Diptera: Phoridae) from Dominican amber. *Pan Pac. Entomol.* 63: 377–80.

Dlussky, G. M. 1975. Formicidae. In: Hymenoptera Apocrita of the Mesozoic. A. P. Rasnitsyn, ed. *Trudy Paleontol. Inst. Acad. Nauk. SSSR* 147: 115–21 (in Russian).

———. 1983. A new family of Upper Cretaceous Hymenoptera: an intermediate link between the ants and the scolioids. *Paleontol. J.* 17: 63–76.

Dombrowski, H. J. 1960. Palaeobiologische Untersuchungen der Nauheimer Quellen. *Zentr. Bakteriol. Parasitenk.*, Abt. 1, 178: 83–90.

———. 1961. *Bacillus circulans* aus Zechsteinsalzen. *Zentr. Bakteriol. Parasitenk.*, Abt. 1, 183: 173–9.

Domke, W. 1952. Der erste sichere Fund eines Myxomyceten im baltischen Bernstein. *Mitteilungen Geol. Staatsinst. Hamburg* 21: 152–62.

Donnelly, T. W. 1985. Mesozoic and Cenozoic plate evolution of the Caribbean regions. In: *The Great American Biotic Interchange*, pp. 89–121. F. G. Stehli and S. D. Webb, eds. New York: Plenum.

Dorf, E. 1964. The use of fossil plants in paleoclimatic interpretations. In: *Problems of Paleoclimatology*, pp. 13–31, 46–48. A. F. M. Nairn, ed. New York: Wiley-Interscience.

Doutt, R. L. 1973A. The genus *Polynemoidea* Girault (Hymenoptera: Mymaridae). *Pan Pac. Entomol.* 49: 215–20.

———. 1973B. The fossil Mymaridae (Hymenoptera: Chalcidoidea). *Pan Pac. Entomol.* 49: 221–8.

Downes, J. A., and W. W. Wirth. 1981. Ceratopogonidae. In: *Manual of Nearctic Diptera.* J. F. McAlpine et al., eds. Research Branch Agriculture Canada (Ottawa) Monograph 27, 1: 393–421.

Duby, G. 1957. On the amber trail in Chiapas. *Pacific Discovery* 10: 8–14.

Durham, J. W. 1956. Insect bearing amber in Indonesia and the Philippine Islands. *Pan Pac. Entomol.* 32: 51–53.

———. 1957. Amber through the Ages. *Pacific Discovery* 10: 3–5.

Eberle, W., W. Hirdes, R. Muff, and M. Pelaez. 1980. The geology of the Cordillera Septentrional. Proc. Ninth Caribbean Geological Conference (Santo Domingo; August 1980) pp. 619–32.

Eckstein, K. 1890. Thierische Haareinschlüsse im baltischen Bernstein. *Schr. naturf. Ges. Danzig* 7: 90–93.

Edwards, F. W. 1923. Oligocene mosquitoes in the British Museum, with a summary of our present knowledge concerning fossil Culicidae. *Quart. J. Geol. Soc. London* 79: 139–55.

Edwards, R., and A. E. Mill. 1986. *Termites in Buildings, Their Biology and Control.* London: Rentokil Ltd.

Eldredge, N. 1985. *Time Frames.* New York: Simon & Schuster. 240 pp.

Emeljanov, A. F. 1983. Dictyopharidae from the Cretaceous deposits of the Taimyr Peninsula (Insecta, Homoptera). *Paleontol. J.* 17: 77–82.

Emerson, A. E. 1969. A revision of the Tertiary fossil species of the Kalotermitidae (Isoptera). *Amer. Mus. Novitates* 2359: 1–57.

———. 1971. Tertiary fossil species of the Rhinotermitidae (Isoptera), phylogeny of genera, and reciprocal phylogeny of associated flagellata (Protozoa) and the Staphylinidae (Coleoptera). *Bull. Amer. Mus. Nat. Hist.* 146: 243–304.

Emery, C. 1890. Le formiche dell' ambra Siciliana nel museo mineralogico dell' Universita di Bologna. *Memorie della R. Accademia delle Scienze dell' Istituto di Bologna* 1: 567–91.

Erwin, T. L. 1971. Fossil tachyine beetle from Mexican and Baltic amber with notes on a new synonymy of an extant group (Coleoptera: Carabidae). *Entomol. Scandia* 2: 233–6.

Evans, H. E. 1962. The genus *Epipompilus* in Australia (Hymenoptera: Pompilidae). *Pacific Insects* 4: 775–82.

———. 1969. Three new aculeate Cretaceous wasps (Hymenoptera). *Psyche* 76: 251–61.

———. 1973. Cretaceous aculeate wasps from Taimyr, Siberia (Hymenoptera). *Psyche* 80: 166–78.

Evans, J. W. 1956. Paleozoic and Mesozoic Hemiptera (Insecta). *Australian J. Zool.* 4: 165–258.

Evans, W. P. 1934. Microstructure of New Zealand lignites. Part 3. Lignites apparently not altered by igneous action. *New Zealand J. Sci. Technol.* 15: 365–85.

Fairchild, G. B., and R. S. Lane. 1989. A second species of fossil *Stenotabanus* (Diptera: Tabanidae) in amber from the Dominican Republic. *Florida Entomol.* 72: 630–32.

Fennah, R. G. 1963. New fossil fulgoroid Homoptera from the amber of Chiapas, Mexico. *Univ. Calif. Publ. Ent.* 31: 43–48.

————. 1987. A new genus and species of Ciixiidae (Homoptera: Fulgoroidea) from Lower Cretaceous amber. *J. Nat. Hist.* 21: 1237–40.

Fillding, M. J. 1950. Observations on the length of dormancy in certain plant infecting nematodes. *Proc. Helm. Soc. Washington* 18: 110–12.

Flint, O. S., Jr. 1978. Probable origins of the West Indian Trichoptera and Odonata faunas. *Proc. Second Inter. Symposium Trichoptera* (1977). pp. 215–23. The Hague: Junk.

Folinsbee, R. E., H. Baadsgaard, G. L. Cumming, and J. Nascimbene. 1964. Radiometric dating of the Bearpaw Sea. *Bull. Amer. Assoc. Petrol. Geol.* 48: 525.

Folsom, J. W. 1938. Order Collembola from Canadian amber. *Univ. Toronto Studies. Geological Series* 40: 14–17.

Fraas, O. 1878. Geologisches aus dem Libanon. *Jh. Ver. Vaterol. Naturkde. Württemberg* 34: 257–391.

Frost, S. H., and R. L. Langenheim, Jr. 1974. *Cenozoic Reef Biofacies. Tertiary Larger Foraminifera and Scleractenian Corals from Chiapas, Mexico.* De Kalb: Northern Illinois Univ. Press. 388 pp.

Futuyma, D. J., and M. Slatkin. 1983. Introduction. In: *Coevolution*, pp. 1–13. D. J. Futuyma and M. Slatkin, eds. Dunderland, Mass.: Sinauer Assoc.

Gagné, R. J. 1973. Cecidomyiidae from Mexican Tertiary amber (Diptera). *Proc. Entomol. Soc. Washington* 75: 169–71.

————. 1977. Cecidomyiidae (Diptera) from Canadian amber. *Proc. Entomol. Soc. Washington* 79: 57–62.

————. 1980. Mycetophilidae and Sciaridae (Diptera) in Mexican amber. *Proc. Entomol. Soc. Washington* 82: 152.

Galippe, V. 1920. Recherches sur la résistance des microzymas à l'action du temps et sur leur survivance dans l'ambre. *Comp. Rendu Acad. Sci.* (Paris) 170: 856–8.

Ghiurca, V. 1988. New considerations on Romanian amber. Sixth Meeting on Amber and Amber-bearing Sediments. October 1988. *Polish Acad. Sci.*, pp. 15–16.

Giebel, C. G. 1862. Wirbeltier und Insektenreste im Bernstein. *Z. Gesell. Naturwissen.* 20: 311–21.

Golenberg, E. M., D. E. Giannasi, M. T. Clegg, C. J. Smiley, M. Durbin, D. Henderson, and G. Zurawski. 1990. Chloroplast DNA sequence from a Miocene *Magnolia* species. *Nature* 344: 656–8.

Gómez, L. D. 1982. *Grammatis succinea*, the first New World fern found in amber. *Amer. Fern J.* 72: 49–52.

Goodey, T. 1923. Quiescence and reviviscence in nematodes, with special reference to *Tylenchus tritici* and *Tylenchus dipsaci*. *J. Helminthol.* 1: 47–52.

Göppert, H. R. 1836. Fossile Pflanzenreste des Eisensandes von Aachen. *N. Acta Acad. C. L. C. Nat. Cur.*, pp. 19–150.

————. 1853. Ueber die Bernsteinflora. *Monatsber. Konigl. Preuss. Akad. Wiss. Berlin* 60: 1–28.

Göppert, H. R., and G. C. Berendt. 1845. *Der Bernstein und die in ihm befindlichen Pflanzenreste der Vorwelt*. Vol. 1. Berlin.

Göppert, H. R., and A. Menge. 1883. *Die Flora des Bernsteins und ihre Beziehungen zur Flora der Tertiarformation und der Oegenwart*. Vol. 1. Danzig.

———. 1886. *Die Flora des Bernsteins*. Vol. 2. *Angiosperms*. Danzig.

Gough, L. J., and J. S. Mills. 1972. The composition of succinite (Baltic amber). *Nature* 239: 527–8.

Grabowska, J. 1883. *Polish Amber*. Warsaw: Interpress Publishers. 151 pp.

Grande, L. 1984. Paleontology of the Green River Formation, with a review of the fish fauna. *Geological Survey of Wyoming, Bull.* 63: 333 pp.

Gray, J., and A. J. Boucot. 1975. Color changes in pollen and spores: A review. *Geol. Soc. Am. Bull.* 86: 1019–33.

Gresset, J. L. 1963. A fossil chrysomelid beetle from the amber of Chiapas, Mexico. *J. Paleontol.* 37: 108–9.

———. 1971. A second fossil chrysomelid beetle from the amber of Chiapas, Mexico. *Univ. Calif. Publ. Entomol.* 63: 63–64.

Grimaldi, D. A. 1987. Amber fossil Drosophilidae (Diptera), with particular reference to the Hispaniolan taxa. *Amer. Mus. Novitates* 2880: 23 pp.

———. 1988. Relicts in the Drosophilidae (Diptera). In: *Zoography of Caribbean Insects*, pp. 183–213. J. K. Liebherr, ed. Ithaca, NY: Cornell Univ. Press.

———. 1989. The genus *Metopina* (Diptera: Phoridae) from Cretaceous and Tertiary ambers. *J. New York Entomol. Soc.* 97: 65–72.

———. 1991. Mycetobiine woodgnats (Diptera: Anisopodidae) from the Oligo-Miocene amber of the Dominican Republic, and Old World affinities. *Amer. Mus. Novitates* 3014: 24 pp.

Grimaldi, D. A., C. W. Beck, and J. J. Boon. 1989. Occurrence, chemical characteristics and paleontology of the fossil resins from New Jersey. *Amer. Mus. Novitates* 2948: 28 pp.

Grimalt, J. O., B. R. T. Simoneit, P. G. Hatcher, and A. Nissenbaum. 1987. The molecular composition of ambers. *Abst. European Assoc. Organic Geochemists. Venice.* 2 pp.

Grissell, E. E. 1980. New Torymidae from Tertiary amber of the Dominican Republic and a world list of fossil torymids (Hymenoptera: Chalcidoidea). *Proc. Entomol. Soc. Washington* 82: 252–9.

Grogan, W. L., Jr., and R. Szadziewski. 1988. A new biting midge from Upper Cretaceous (Cenomanian) amber of New Jersey (Diptera: Ceratopogonidae). *J. Paleontol.* 62: 808–12.

Grolle, R. 1988. Bryophyte fossils in amber. *The Bryological Times* 47: 4–5.

Grüss, J. 1931. Die Urform des *Anthomyces Reukaufii* und andere Einschlüsse in den Bernstein durch Insekten verschleppt. *Wochenschrift für Brauerei.* 48: 63–68.

Guérin, F. E. 1838. (Letter and note from Maravigna on insects found in Sicilian amber). *Revue Zoologique par la Société cuvierienne* 1(1838–1840): 168–70.

Gusovius, P. 1966. *Der Landkreis Samland*. Würzburg: Holzner Verlag.

Haczewski, J. 1838. (Untitled.) *O bursztynie*. Lwow: Sylwan. 14: 191–251.

Hale, E. E. 1891. *The Life of Christopher Colombus*. Chicago: G. L. Howe & Co. 320 pp.

Hamilton, K. G. A. 1971. A remarkable fossil Homopteran from Canadian Cretaceous amber representing a new family. *Canadian Entomol.* 103: 943–6.

Handschin, E. 1926. Über Bernsteincollembolen. Ein Beitrag zur ökologischen Tiergeographie. *Rev. Suisse Zool.* 33: 375–8.

Hardy, D. E. 1971. A new *Plecia* (Diptera: Bibionidae) from Mexican amber. *Univ. Calif. Publ. Entomol.* 63: 65–67.

Harrington, B. J. 1891. On the so-called amber of Cedar Lake, North Saskatchewan, Canada. *Amer. J. Sci.* 42: 332–8.

Heer, O. 1859. *Flora Tertiaria Helvetiae: Die tertiäre Flora der Schweiz*, Allgemeiner Thiel, pp. 308–10. Winterthur.

Heie, O. E. 1967. Studies on fossil aphids (Homoptera: Aphidoidea) especially in the Copenhagen collection of fossils in Baltic amber. *Spolia Zool. Mus. Haun.* 26: 274 pp.

———. 1972. Some new fossil aphids from Baltic amber in the Copenhagen Collection (Insecta, Homoptera, Aphidoidea). *Steenstrupia* 2: 247–62.

———. 1987. Palaeontology and phylogeny. In: *Aphids, Their Biology, Natural Enemies and Control*, Vol. 2, pp. 367–91. P. Harrewijn and A. K. Minks, eds. Amsterdam: Elsevier.

Heie, O. E., and G. O. Poinar, Jr. 1988. *Mindazerius dominicanus* nov. gen., nov. sp., a fossil aphid (Homoptera, Aphidoidea, Drepanosiphidae) from Dominican amber. *Psyche* 95: 153–65.

Helm, O. 1891. Mittheilungen über Bernstein. XIV. Über Rumänit, ein in Rumänien vorkommendes fossiles Harz. *Schr. naturf. Ges. Danzig* 7: 186–9.

———. 1892. On a new fossil, amber-like resin occurring in Burma, from Upper Burma. *Rec. Geol. Survey India* 25: 180–81.

Hennig, W. 1965. Die Acalyptratae des Baltischen Bernsteins und ihre Bedeutung für die Erforschung der phylogenetischen Entwicklung dieser Dipteren-Gruppe. *Stuttgarter Beitr. Naturk.* (Serie A) 145: 1–215.

———. 1966. *Fannia scalaris* Fabricius, eine rezente Art in Baltischen Bernstein? (Diptera: Muscidae). *Stuttgarter Beitr. Naturk.* (Serie A) 150: 1–12.

———. 1967. Neue Acalyptratae aus dem Baltischen Bernstein (Diptera: Cyclorrhapha). *Stuttgarter Beitr. Naturk.* (Serie A) 175: 1–27.

———. 1968. Ein weiterer Vertreter der Familie Acroceridae im Baltischen Bernstein (Diptera: Brachycera). *Stuttgarter Beitr. Naturk.* (Serie A) 185: 1–6.

———. 1969. Kritische Betrachtungen über die phylogenetische Bedeutung von Bernsteinfossilien. *Mem. Soc. entomol. Italiana* 48: 57–67.

———. 1970. Insektenfamilien aus der unteren Kreide. II. Empedidae (Diptera: Brachycera). *Stuttgarter Beitr. Naturk.* (Serie A) 214: 1–12.

———. 1971. Insektenfossilien aus der unteren Kreide. III. Empidiformia ("Microphorinae") aus der unteren Kreide und aus dem Baltischen Bernstein: ein Vertreter des Cyclorrhapha aus der unteren Kreide. *Stuttgarter Beitr. Naturk.* (Serie A) 232: 1–28.

———. 1972. Insektenfossilien aus der unteren Kreide. IV. Psychodidae (Phlebotominae), mit einer kritischen Übersicht über das phylogenetische System der Familie und die bisher beschriebenen Fossilien. *Stuttgarter Beitr. Naturk.* (Serie A) 241: 1–69.

Hickey, L. J., and J. A. Doyle. 1977. Early Cretaceous fossil evidence for angiosperm evolution. *Botanical Review* 43: 3–104.

Hieke, F., and E. Pietrzeniuk. 1984. Die Bernstein-Käfer des Museums für

Naturkunde, Berlin (Insecta, Coleoptera). *Mitteilungen Zool. Mus. Berlin* 60: 297–326.

Higuchi, R. G., B. Bowman, M. Freiberger, O. A. Ryder, and A. C. Wilson. 1984. DNA sequences from the Quagga, an extinct member of the horse family. *Nature* 312: 282–4.

Higuchi, R. G., and A. C. Wilson. 1984. Recovery of DNA from extinct species. *Federation Proc.* 43: 1557.

Hills, E. S. 1957. Fossiliferous Tertiary resin from Allendale, Victoria. *Proc. R. Soc. Victoria* 69: 15–19.

Hirschmann, W. 1971. A fossil mite of the genus *Dendrolaelaps* (Acarina: Meso-stigmata: Digamasellidae) found in amber from Chiapas, Mexico. *Univ. Calif. Publ. Entomol.* 63: 69–73.

Holl, F. 1829. *Handbuch der Petrefactenkunde*. Dresden: Hilscher. 489 pp.

Hong, Y.-Ch. 1981. *Eocene Fossil Diptera (Insecta) in Amber of Fushun Coalfield.* Beijing: Geological Publishing House. 166 pp.

Hong, Y.-Ch., T.-C. Yang, S.-T. Wang, S.-E. Wang, Y.-K. Li, M.-R. Sun, H.-C. Sun, and N.-C. Tu. 1974. Stratigraphy and palaeontology of Fushun coal-field, Liaoning Province. *Acta Geologica Sinica* 2: 113–49.

Hopfenberg, H. B., L. C. Witchey, and G. O. Poinar, Jr. 1988. Is the air in amber ancient? *Science* 241: 717–24.

Horibe, Y., and H. Craig. 1987. Trapped gas in amber: a paleobotanical and geo-chemical inquiry. *Eos* 68: 1513.

Hueber, F. M., and J. Langenheim. 1986. Dominican amber tree had African ancestors. *Geotimes* 31: 8–10.

Hummel, D. O. 1958. *Kunststoff-, Lack-, und Gummi-Analyse*. Munich: Hanser. 654 pp.

Hunger, R. 1979. *The Magic of Amber*. Radnor, PA: Chilton Book Co. 131 pp.

Hurd, P. D., Jr., and R. F. Smith. 1957. The meaning of Mexico's amber. *Pacific Discovery* 10: 6–7.

Hurd, P. D., Jr., R. F. Smith, and J. W. Durham. 1962. The fossiliferous amber of Chiapas, Mexico. *Ciencia* 21: 107–18.

Hurd, P. D., Jr., R. F. Smith, and R. L. Usinger. 1958. Cretaceous and Tertiary insects in arctic and Mexican amber. *Proc. Tenth Inter. Cong. Entomology* (1956) 1: 851.

James, T. 1971. A stratomyid fly (Diptera) from the amber of Chiapas, Mexico. *Univ. Calif. Publ. Entomol.* 63: 71–73.

Jansson, H.-B., and G. O. Poinar, Jr. 1986. Some possible fossil nematophagous fungi. *Trans. British Mycol. Soc.* 87: 471–4.

Janzen, D. H., ed. 1983. *Costa Rican Natural History*. Chicago: Univ. of Chicago Press. 816 pp.

Jaques, H. E. 1918. A long-lifed wood-boring beetle. *Proc. Iowa Acad. Sci.* 25: 175.

Jarzembowski, E. A. 1981. An early Cretaceous termite from southern England (Isoptera: Hodotermitidae). *Syst. Entomol.* 6: 91–96.

Jell, P. A., and P. M. Duncan. 1986. Invertebrates, mainly insects, from the fresh-water, Lower Cretaceous, Koonwarra fossil bed (Korumburra Group). South Gippsland, Victoria. *Assoc. Australian Palaeontologist's Memoir* 3: 111–205.

Jordan, K. H. C. 1953. Eine weitere fossile Notonectidae (Hem. Het.) von Rott im Siebengebirge. *Zool. Anzeiger* 150: 245–9.

Just, J. 1974. On *Palaeogammarus* Zaddach, 1864, with a description of a new species from western Baltic amber (Crustacea, Amphipoda, Crangonycidae). *Steenstrupia* 3: 93–99.

Kalugina, N. S. 1976. Non-biting midges of the subfamily Diamesinae (Diptera, Chironomidae) from the Upper Cretaceous of the Taimyr. *Paleontol. J.* 10: 78–83.

———. 1980. Cretaceous Aphroteniinae from North Siberia (Diptera, Chironomidae). *Electrotenia brundini* gen. nov. sp. nov. *Acta Universitatis Carolinae Biologica* (Prague) 1978: 89–93.

Kaszab, Z., and W. Schawaller. 1984. Eine neue Schwarzkäfer-Gattung und-art aus Dominikanischem Bernstein (Coleoptera, Tenebrionidae). *Stuttgarter Beitr. Naturk.* (Serie B) 109: 6 pp.

Katinas, V. 1983. *Baltijos Gintaras*. Vilnius: Mokslas. III pp.

Keilbach, R. 1939. Neue Funde des Strepsipterons *Mengea tertiaria* Menge im baltischen Bernstein. *Bernstein-Forschungen* 4: 1–7.

———. 1982. Bibliographie und Liste der Arten tierischer Einschlüsse in fossilen Harzen sowie ihrer Aufbewahrungsorte. *Deut. Entomol. Zeit.* N.F. 29: 129–286, 301–491.

Kelner-Pillault, S. 1969. Les Abeilles fossiles. *Memorie Società Entomologica Italiana* 48: 519–34.

Kinsey, A. C. 1937. Order Hymenoptera, family Cynipidae. Contributions to Canadian mineralogy. *Univ. Toronto Studies. Geological Series* 40: 21–27.

Kinzelbach, N. R. 1979. Das erste neotropische Fossil der Fächerflügler (Stuttgarter Bernsteinsammlung: Insecta, Strepsiptera). *Stuttgarter Beitr. Naturk.* (Serie B) 52: 1–14.

Kirchner, G. 1950. Amber inclusions. *Endeavour* 9: 70–75.

Klebs, R. 1886. Gastropoden im Bernstein. *Jahrb. Preuss. geol. Landesanst.* 188: 366–94.

———. 1897. Cedarit, ein neues bernsteinähnliches fossiles Harz-Canadas und sein Vergleich mit anderen fossilen Harzen. *Jahrbuch der Königlich preussischen geologischen Landesanstalt und Bergakademie zu Berlin für das Jahr 1896*, pp. 199–230.

———. 1910. Über Bernsteinelinschlüsse im allgemeinen und die Coleopteren meiner Bernsteinssammlung. *Schr. phys. Ökon. Ges. Königsberg* 51: 217–42.

Knowlton, F. H. 1896. American amber-producing tree. *Science* 3: 582–4.

Kononova, E. L. 1975. A new aphid family (Homoptera: Aphidinea) from the Upper Cretaceous of the Taimyr. *Entomol. Obozrenie Moskva* 54: 795–807 (in Russian).

———. 1976. Extinct aphid families (Homoptera: Aphidinea) of the Late Cretaceous. *Paleontol. Zhur. Moskva* 3: 117–26 (in Russian).

———. 1977. New species of aphids (Homoptera, Aphidinea) from the Upper Cretaceous deposits of Taimyr. *Rev. Entomol. URSS* 56: 588–600.

Kornilovich, N. 1903. Has the structure of striated muscle of insects in amber been preserved? *Prot. obshchestva estestro pri Itper. Yurev Univ.* (*Yurev*) 13: 198–206 (in Russian).

Kosmowska-Ceranowicz, B. 1987. Mineralogical-petrographic characteristics of the Eocene amber-bearing sediments in the area of Chłapowo, and the Palaeogene sediments of Northern Poland. *Biuletyn Instytutu Geologicznego* 356: 29–50.

Kosmowska-Ceranowicz, B., and C. Müller. 1985. Lithology and calcareous nannoplankton in amber bearing Tertiary sediments from boreholes Chłapowo (northern Poland). *Bull. Polish Acad. Sci.* 33: 119–29.

Koteja, J. 1985. Coccids (Hom. Coccinea) of Baltic amber. *Wiad. Entomol.* (Warsaw) 6: 195–206 (in Polish).

———. 1987. Current state of coccid paleontology. *Bollettino del Lab. di Entomol. agraria Filippo Silvestri* 43 (Suppl.): 29–34.

Kovalev, V. G. 1974. A new genus of the family Empididae (Diptera) and its phylogenetic relationships. *Paleontol. J.* 2: 196–204.

———. 1978. A new fly genus (Empididae) from Late Cretaceous retinites of the Taymyr. *Paleontol. J.* 12: 351–6.

Kozlov, M. A. 1975. Family Stigmaphronidae. Family Trupochalicididae. *In*: Hymenoptera Apocrita of the Mesozoic. A. P. Rasnitsyn, ed. *Trudy Paleontol. Inst. Acad. Nauk SSSR* 147: 75–83, 88–91 (in Russian).

———. 1987. New moth-like Lepidoptera from Baltic amber. *Paleontol. J.* 21: 56–65.

Kozlov, M. A., and A. P. Rasnitsyn. 1989. On the limits of the family Serphitidae (Hymenoptera, Proctotrupoidea). *Entomol. Obozrenie Moskva* 58: 402–16 (In Russian).

Krishna, K., and A. E. Emerson. 1983. A new fossil species of termite from Mexican amber, *Mastotermes electromexicus* (Isoptera: Mastotermitidae). *Amer. Mus. Novitates* 2767: 1–8.

Krishna, K., and D. A. Grimaldi. 1991. A new fossil species from Dominican amber of the living Australian termite genus *Mastotermes* (Isoptera: Mastotermitidae). *Amer. Mus. Novitates* 3021: 10 pp.

Krombein, K. V. 1986. Three cuckoo wasps from Siberian and Baltic amber (Hymenoptera: Chrysididae: Amiseginae and Elampinae). *Proc. Entomol. Soc. Washington* 88: 740–747.

Krzeminski, W. 1985A. Limoniidae (Diptera, Nematocera) from Baltic amber (in the collection of the Museum of the Earth in Warsaw). Part I. Subfamily Limoniinae. *Prace Muzeum Ziemi* 37: 113–7.

———. 1985B. A representative of Trichoceridae (Diptera, Nematocera) from Baltic amber (in the collection of the Museum of the Earth in Warsaw). *Prace Muzeum Ziemi* 37: 119–21.

Krzeminski, W., and A. W. Skalski. 1983. *Pseudolimnophila siciliana* sp.n. from Sicilian amber (Diptera, Limoniidae). *Animalia* 10: 303–307.

Krzeminski, W., and H. J. Teskey. 1987. New taxa of Limoniidae (Diptera: Nematocera) from Canadian amber. *Canadian Entomol.* 119: 887–92.

Kucharska, M., and A. Kwiatkowski. 1978. Research methods for the chemical composition of amber and problems concerning the origin of amber. *Prace Muzeum Ziemi* 29: 149–56 (in Polish).

———. 1979. Thin-layer chromatography of amber samples. *J. Chromatography* 169: 482–4.

Kühne, W. G., L. Kubig, and T. Schlüter. 1973. Eine Micropterygide (Lepidoptera, Homoneura) aus mittelkretazischem Harz Westfrankreichs. *Mitt. Deutsch Entomol. Ges.* 32: 61–65.

Kulicka, R. 1978. *Mengea tertiaria* (Menge, Strepsiptera) from the Baltic amber. *Prace Muzeum Ziemi* 29: 141–5.

Kulicka, R., W. Krzeminski, and R. Szadziewski. 1985. A collection of Diptera Nematocera in Baltic amber at the Museum of the Earth in Warsaw. *Prace Muzeum Ziemi* 37: 105–11 (in Polish).

Kunz, G. F. 1889. Mineralogical notes, on fluorite, opal, amber and diamond. *Amer. J. Sci.* 38: 72–74.

Kuznetsov, N. J. 1941. *A Revision of the Amber Lepidoptera.* Acad. Sci., USSR (Moscow-Leningrad). 136 pp.

La Baume, W. 1935. Zur Naturkunde und Kulturgeschichte des Bernsteins. *Schr. naturf. Ges. Danzig* N.F. 20: 5–48.

Lambert, J. B., C. W. Beck, and J. S. Frye. 1988. Analysis of European amber by Carbon-13 Nuclear Magnetic Resonance Spectroscopy. *Archaeometry* 30: 248–63.

Lambert, J. B., and J. S. Frye. 1982. Carbon functionalities in amber. *Science* 217: 55–57.

Lambert, J. B., J. S. Frye, T. A. Lee, Jr., C. J. Welch, and G. O. Poinar, Jr. 1989. Analysis of Mexican amber by Carbon-13 NMR Spectroscopy. *Archaeological Chemistry* 4: 381–8.

Lambert, J. B., J. S. Frye, and G. O. Poinar, Jr. 1985. Amber from the Dominican Republic: analysis of nuclear magnetic resonance spectroscopy. *Archaeometry* 27: 43–51.

———. 1990. Analysis of North American amber by Carbon-13 NMR spectroscopy. *Geoarchaeology* 5: 43–52.

Lane, R. S., and G. O. Poinar, Jr. 1986. First fossil tick (Acari: Ixodidae) in New World amber. *Inter. J. Acarol.* 12: 75–78.

Lane, R. S., G. O. Poinar, Jr., and G. B. Fairchild. 1988. A fossil horsefly (Diptera: Tabanidae) in Dominican amber. *Florida Entomol.* 71: 593–6.

Langenheim, J. H. 1964. Present status of botanical studies of ambers. *Botanical Museum Leaflets, Harvard University* 20: 225–87.

———. 1966. Botanical source of amber from Chiapas, Mexico. *Ciencias* 24: 201–10.

———. 1967. Preliminary investigations of *Hymenaea courbaril* as a resin producer. *J. Arnold Arboretum* 48: 203–27.

———. 1969. Amber: A botanical inquiry. *Science* 163: 1157–69.

———. 1973. Leguminous resin-producing trees in Africa and South America. In: *Tropical Forest Ecosystems in Africa and South America: A Comparative Review*, pp. 89–104. Washington, D.C.: Smithsonian Institution Press.

Langenheim, J. H., and C. W. Beck. 1965. Infrared spectra as a means of determining botanical sources of amber. *Science* 149: 52–55.

———. 1968. Catalogue of infrared spectra of fossil resin (ambers) 1. North and South America. *Botanical Museum Leaflets, Harvard University* 22: 65–120.

Langenheim, J. H., and Y. T. Lee. 1974. Reinstatement of the genus *Hymenaea* (Leguminosae: Caesalpinioideae) in Africa. *Brittonia* 26: 3–21.

Langenheim, J. H., Y. T. Lee, and S. S. Martin. 1973. An evolutionary and ecological perspective of Amazonian *Hylaea* species of *Hymenaea* (Leguminosae: Caesalpinioideae). *Acta Amazonica* 3: 5–38.

Langenheim, R. L., Jr., J. D. Buddhue, and G. Jelinek. 1965. Age and occurrence of the fossil resins bacalite, kansasite and jelinite. *J. Paleontology* 39: 283–87.

Langenheim, R. L., Jr., C. J. Smiley, and J. Gray. 1960. Cretaceous amber from the arctic coastal plain of Alaska. *Bull. Geol. Soc. Amer.* 71: 1345–56.

Larsson, S. G. 1978. Baltic amber—A palaeobiological study. *Entomonograph* Vol. 1. Klampenborg, Denmark. 192 pp.

Lawrence, P. N. 1985. Ten species of Collembola from Baltic amber. *Prace Muzeum Ziemi* 37: 101–4.

Lazell, J. D., Jr. 1965. An *Anolis* (Sauria, Iguanidae) in amber. *J. Paleontol.* 39: 379–82.

Lee, Y.-T., and J. H. Langenheim. 1974. Additional new taxa and new combinations in *Hymenaea* (Leguminosae, Caesalpinioideae). *J. Arnold Arboretum* 55: 441–52.

———. 1975. *Systematics of the genus Hymenaea L. (Leguminosae, Caesalpinioideae, Detarieae)*. Univ. Calif. Publ. Botany 69: 109 pp.

Legg, W. M. 1942. *Collection, preparation, and statistical study of fossil insects from Chemawinite*. Senior thesis, Department of Biology, Princeton University, 66 pp.

Lehmann, U., and G. Hillmer. 1983. *Fossil Invertebrates*. Cambridge, England: Cambridge University Press. 350 pp.

Lengweiler, W. 1939. Minerals in the Dominican Republic. *Rocks and Minerals* 14: 212–3.

Ley, W. 1951. *Dragons in Amber*. New York: The Viking Press.

Libby, W. F., and J. R. Arnold. 1951. Radiocarbon dates. *Science* 113: 111–20.

Loew, H. 1850. Über den Bernstein und die Bernsteinfauna. *Programm der Königlichen Realschule zu Meseritz* 1: 3–44.

Loveridge, A. 1942. Scientific results of a fourth expedition to forested areas in East and Central Africa. IV: Reptiles. *Bull. Mus. Comp. Zool.* 91: 240–373.

———. 1957. Checklist of the reptiles and amphibians of East Africa (Uganda: Kenya: Tanganyika: Zanzibar). *Bull. Mus. Comp. Zool.* 117: 153–362.

Lühe, M. 1904. Säugetierhaare im Bernstein. *Schrift. Phys. Ökon. Ges. Königsberg* 45: 62–63.

MacKay, M. R. 1969. Microlepidopterous larvae in Baltic amber. *Canadian Entomol.* 101: 1173–80.

———. 1970. Lepidoptera in Cretaceous amber. *Science* 167: 379–80.

MacLeod, E. G. 1970. The Neuroptera of the Baltic amber. I. Ascalaphidae, Nymphidae, Psychopsidae. *Psyche* 77: 147–80.

Mägdefrau, K. 1957. Flechten und Moose im baltischen Bernstein. *Ber. Deutsch. Bot. Ges.* 70: 433–35.

Malfait, B. T., and M. G. Dinkelman. 1972. Circum-Caribbean tectonic and igneous activity and evolution of the Caribbean Plate. *Bull. Geol. Soc. Amer.* 83: 251–72.

Mamikunian, G., and M. H. Briggs. 1965. *Current Aspects of Exobiology*. New York: Pergamon Press. 420 pp.

Manley, D. G., and G. O. Poinar, Jr. 1991. A new species of fossil *Dasymutilla* (Hymenoptera: Mutellidae) from Dominican amber. *Pan Pac. Entomol.* 67: 200–205.

Mari Mutt, J. A. 1983. Collembola in amber from the Dominican Republic. *Proc. Entomol. Soc. Washington* 85: 575–87.

Martinez, M. 1979. *Catálogo de nombres vulgares y científicos de plantas.* Mexicanas Fondo de Cultura Económica, Mexico.

Martinez, R., and D. Schlee. 1984. Die Dominikonischen Bernsteinminen der Nordkordillere, speziell auch aus der Sicht der Werkstätten. In: *Bernstein-Neigkeiten. Stuttgarter Beitr. Naturk.* (Serie C) 18: 78–84.

Martini, E. 1971. Standard Tertiary and Quaternary calcareous nannoplankton zonation. *Proc. Second Plankton Conf.* (Roma 1970) 2: 731–85.

Masner, L. 1969A. A scelionid wasp surviving unchanged since Tertiary (Hymenoptera: Proctotrupoidea). *Proc. Entomol. Soc. Washington* 71: 397–400.

———. 1969B. The geographic distribution of recent and fossil Ambrositrinae (Hymenoptera: Proctotrupoidea: Diapriidae). *Wandervers. Deutsch. Entomol. Ber.* (Dresden) 1965 (10): 105–109.

Matile, L. 1981. Description d'un Keroplatidae du Crétacé Moyen et données morphologiques et taxonomiques sur les Mycetophiloidea (Diptera). *Ann. Société entomol. France* 17: 99–123.

Mayr, L. 1868. *Die Ameisen des baltischen Bernsteins.* Beitr. Naturk. Preussens 1: 1–102.

McAlpine, J. F. 1973. A fossil ironomyiid fly from Canadian amber (Diptera: Ironomyiidae). *Canadian Entomol.* 105: 105–11.

———. 1981. *Morgea freidbergi* new species, a living sister-species of the fossil species *M. mcalpinei*, and a key to world genera of Pallopteridae (Diptera). *Canadian Entomol.* 113: 81–91.

McAlpine, J. F., and J. E. H. Martin. 1966. Systematics of Sciadoceridae and relatives with descriptions of two genera and species from Canadian amber and erection of the family Ironomyiidae (Diptera: Phoroidea). *Canadian Entomol.* 98: 527–44.

———. 1969. Canadian amber—a paleontological treasure-chest. *Canadian Entomol.* 101: 819–38.

Meinander, M. 1975. Fossil Coniopterygidae. *Notulae Entomol.* (Helsinki) 55: 53–57.

Melchior, I., and F. Brandenburg. 1990. *Quest: Searching for Germany's Nazi past: A young man's story.* Novato, California: Presidio Press. 321 pp.

Menge, A. 1855. Über die Scheerenspinnen, Chernetidae. *Schr. naturf. Ges. Danzig* 5: 1–41.

———. 1856. Lebenszeichen vorweltlicher im Bernstein eingeschlossene Thiere. *Programm Petrischule, Danzig.* pp. 1–32.

———. 1858. Beitrag zur Bernsteinflora. *Schr. naturf. Ges. Danzig* 6: 1.

———. 1863. Über ein Rhipidopteron und einige andere im Bernstein eingeschlossene Tiere. *Schr. naturf. Ges. Danzig* N.F. 1: 1–8.

———. 1869. Über einen Scorpion und zwei Spinnen im Bernstein. *Schr. naturf. Ges. Danzig* N.F. 2: 2–9.

————. 1872. Über eine im Bernstein eingeschlossene *Mermis. Schr. naturf. Ges. Danzig* N.F. 3: 1–2.

Meunier, F. 1904. Monographie des Cecidmyidae, des Sciaridae, des Mycetophilidae et des Chironomidae de l'ambre de la Baltique. *Ann. Soc. Sci. Bruxelles* 28: 1–264.

————. 1916. Sur quelques diptères (Bombylidae, Leptidae, Dolichopodidae, Conopidae et Chironomidae) de l'ambre de la Baltique. *Tydschr. Entomol.* 59: 274–86.

Meyer, A. B. 1887. Notiz über in Ostsee-Bernstein eingeschlossene Vogelfedern. *Schr. naturf. Ges. Danzig* N.F. 6: 206–8.

Michener, C. D. 1982. A new interpretation of fossil social bees from the Dominican Republic. *Sociobiology* 7: 37–45.

————. 1990. Classification of the Apidae (Hymenoptera). *Univ. Kansas Sci. Bull.* 54: 75–164.

Michener, C. D., and D. A. Grimaldi. 1988. A *Trigona* from Late Cretaceous amber of New Jersey (Hymenoptera: Apidae: Meliponinae). *Amer. Mus. Novitates* 2917: 1–10.

Mierzejewski, P. 1976A. Scanning electron microscope studies on the fossilization of Baltic amber spiders (preliminary note). *Ann. Med. Sect. Polish Acad. Sci.* 21: 81–82.

————. 1976B. On application of scanning electron microscope to the study of organic inclusions from Baltic amber. *Ann. Soc. Geol. Pologne* 46: 291–295.

————. 1978. Electron microscopy study on the milky impurities covering arthropod inclusions in Baltic amber. *Prace Muzeum Ziemi* 28: 79–84.

Mills, J. S., R. White, and L. J. Gough. 1984/85. The chemical composition of Baltic amber. *Chem. Geol.* 47: 15–39.

Miranda, F. 1963. Two plants from the amber of the Simojovel, Chiapas, Mexico, area. *J. Paleontol.* 37: 611–4.

Miyashiro, A. 1972. Metamorphism and related magnetism in plate tectonics. *Am. J. Science* 272: 629–56.

Mockford, E. L. 1969. Fossil insects in the order Psocoptera from the Tertiary amber of Chiapas, Mexico. *J. Paleontol.* 43: 1267–73.

————. 1972. New species, records and synonymy of Florida *Belaphotroctes* (Psocoptera: Liposcelidae). *Florida Entomol.* 55: 153–63.

————. 1986. A preliminary survey of Psocoptera from Tertiary amber of the Dominican Republic. *Entomol. Soc. Amer. Conf.* (Reno, Nevada; December 7–11, 1986) p. 112.

Mosini, V., and R. Samperi. 1985. Correlations between Baltic amber and *Pinus* resins. *Phytochemistry* 24: 859–61.

Muchmore, W. B. 1971. Phoresy by North and Central American pseudoscorpions. *Proc. Rochester Acad. Sci.* 12: 79–97.

Muesebeck, C. F. E. 1960. A fossil braconid wasp of the genus *Ecphylus* (Hymenoptera). *J. Paleontol.* 34: 495–6.

————. 1963. A new ceraphronid from Cretaceous amber (Hymenoptera: Proctotrupidae). *J. Paleontol.* 37: 129–30.

Muller, J. 1970. Palynological evidence on early differentiation of angiosperms. *Biol. Rev. Cambridge Philos. Soc.* 45: 417–50.

————. 1984. Significance of fossil pollen for angiosperm history. *Ann. Missouri Bot. Gardens* 71: 419–43.

Munro, J. 1981. Amber forever. *Aramco World Magazine* 32: 32–36.

Mustoe, G. E. 1985. Eocene amber from the Pacific Coast of North America. *Bull. Geol. Soc. Amer.* 96: 1530–6.

Negrobov, O. P. 1976. A new genus of the family Dolichopodidae (Diptera) from Paleogene retinites of the Sea of Okhotak. *Paleontol. J.* 10: 495–7.

————. 1978. Flies of the superfamily Empididoidea (Diptera) from Cretaceous retinite in northern Siberia. *Paleontol. J.* 12: 221–8.

Newton, A. F. 1984. Mycophagy in Staphylinoidea (Coleoptera). In: *Fungus-Insect Relationships*, pp. 302–53. Q. Wheeler and M. Blackwell, eds. New York: Columbia Univ. Press.

Nicoletti, R. 1975. Analisi di ambre: un nuova approccio. *Studi e Ricerche sulla Problematica dell' Ambra, Rome* 1: 299–305 (see also pp. 177–180).

Nikitsky, N. B. 1977. Two new genera of the Melandryidae (Coleoptera) from the Upper Cretaceous. *Paleontol. J.* 11: 267–70 (in Russian).

Nissenbaum, A. 1975. Lower Cretaceous amber from Israel. *Naturwissenchaften* 62: 341–2.

Noetling, F. 1892. Preliminary report on the economic resources of the amber and jade mines area in Upper Burma. *Rec. Geol. Surv. India* 25: 130–35.

Odrzywolska-Bienkowa, E., et al. 1981. The Polish part of the NW-European Tertiary Basin: a generalization of its stratigraphic section. *Bull. de l'Academie polonaise des sciences, terre* 29: 3–17.

Oke, C. G. 1957. Fossil Insecta from Cainozoic resin at Allendale, Victoria. *Proc. R. Soc. Victoria* 69: 29–31.

Ono, H. 1981. Erstnachiveis einer Krabbenspinne (Thomisidae) in Dominikanischem Bernstein (Stuttgarter Bernsteinsammlung: Arachnida, Araneae). *Stuttgarter Beitr. Naturk.* (Ser. B) 73: 13 pp.

Pampaloni, L. 1902. I resti organici nel disodile de Melilli in Sicilia. *Palaeontographia Italica* 8: 121–30.

Pęczalska, A. 1981. *Złoto Północy. Opowieści o burnsztynie.* Katowice: Wydawnictwo "Śląsk." 146 pp.

Peterson, B. V. 1975. A new Cretaceous bibionid from Canadian amber (Diptera: Bibionidae). *Canadian Entomol.* 107: 711–5.

Petrunkevitch, A. 1942. A study of amber spiders. *Trans. Conn. Acad. Arts Sci.* 34: 119–464.

————. 1950. Baltic amber spiders in the Museum of Comparative Zoology. *Bull. Mus. Comp. Zool., Harvard* 103: 259–337.

————. 1958. Amber spiders in European collections. *Trans. Conn. Acad. Arts Sci.* 41: 97–400.

————. 1963. Chiapas amber spiders. In: Studies of Fossiliferous Amber Arthropods of Chiapas, Mexico. *Univ. Calif. Publ. Entomol.* 31: 1–40.

————. 1971. Chiapas amber spiders, II. In: Studies of Fossiliferous Amber Arthropods of Chiapas, Mexico. Part II. *Univ. Calif. Publ. Entomol.* 63: 1–44.

Pictet, F. J. 1854. *Traité de Paléontologie ou histoire naturelle des animaux fossiles considérés dans leur rapports zoologiques et géologiques.* Paris. 727 pp.

Ping, C. 1931. On a blattoid insect in the Fushun amber. *Bull. Geol. Soc. China* 11: 205.

Piton, L. 1938. *Succinotettix chopardi* Piton, Orthoptère (Tetricinae) inédit de l'ambre de la Baltique. *Bull. Soc. Entomol. France* 43: 226–7.

Pliny, 1st century A.D. *Natural History*. Book 37, Chapters 11–13.

Poinar, G. O., Jr. 1977. Fossil nematodes from Mexican amber. *Nematologica* 23: 232–8.

———. 1981. Fossil dauer rhabditoid nematodes. *Nematologica* 27: 466–7.

———. 1982A. Amber—true or false? *Gems and Minerals* (April) 534: 80–84.

———. 1982B. Sealed in amber. *Nat. Hist.* 91: 26–32.

———. 1983. *The Natural History of Nematodes*. Englewood Cliffs, NJ: Prentice-Hall, Inc. 323 pp.

———. 1984A. First fossil record of parasitism by insect parasitic Tylenchida (Allantonematidae: Nematoda). *J. Parasitol.* 70: 306–8.

———. 1984B. *Heydenius dominicus* n.sp. (Nematoda: Mermithidae), a fossil parasite from the Dominican Republic. *J. Nematol.* 16: 371–5.

———. 1984C. Fossil evidence of nematode parasitism. *Revue Nématol.* 7: 201–3.

———. 1985A. Fossil evidence of insect parasitism by mites. *Intern. J. Acarol.* 11: 37–38.

———. 1985B. Tertiary forest in amber. *Pacific Horticulture* 46: 39–41.

———. 1987. Fossil evidence of spider parasitism by Ichneumonidae. *J. Arachnol.* 14: 399–400.

———. 1988A. Hair in Dominican amber: evidence for Tertiary land mammals in the Antilles. *Experientia* 44: 88–89.

———. 1988B. *Zorotypus palaeus* n.sp., a fossil Zoraptera (Insecta) in Dominican amber. *Proc. New York Entomol. Soc.* 96: 253–9.

———. 1988C. The amber ark. *Nat. Hist.* 97: 42–47.

———. 1991A. *Praecoris dominicana* gen.n., sp.n. (Holoptilinae: Reduviidae: Hemiptera) from Dominican amber with an interpretation of past behavior based on functional morphology. *Entomologica Scandinavica* 22: 193–99.

———. 1991B. Resinites, with examples from New Zealand and Australia. *Fuel Processing Tech.* 28: 135–48.

———. 1991C. The mycetophagous and entomophagous stages of *Iotonchium californicum* n. sp. (Iotonchiidae: Tylenchida). *Rev. de Nematologie* 14: 565–80.

———. 1991D. *Hymenaea protera* sp.n. (Leguminosae, Caesalpinioideae) from Dominican amber has African affinities. *Experientia* 47: 1075–82.

Poinar, G. O., Jr., and J. Brodzinsky. 1986. Fossil evidence of nematode (Tylenchida) parasitism in Staphylinidae (Coleoptera). *Nematologica* 32: 353–5.

Poinar, G. O., Jr., and D. C. Cannatella. 1987. An Upper Eocene frog from the Dominican Republic and its implications for Caribbean biogeography. *Science* 237: 1215–6.

Poinar, G. O., Jr., and J. T. Doyen. 1992. A termite bug, *Termitaradus protera*, new species (Termitaphidae: Hemiptera), from Mexican amber. *Entomologica Scandinavica* (in press).

Poinar, G. O., Jr., and D. Grimaldi. 1990. Fossil and extant macrochelid mites

(Acari: Macrochelidae) phoretic on drosophilid flies (Diptera: Drosophilidae). *J. New York Entomol. Soc.* 98: 88–92.

Poinar, G. O., Jr., and J. Haverkamp. 1985. Use of pyrolysis mass spectrometry in the identification of amber samples. *J. Baltic Studies* 16: 210–21.

Poinar, G. O., Jr., and R. Hess. 1982. Ultrastructure of 40-million-year-old insect tissue. *Science* 215: 1241–2.

———. 1985. Preservative qualities of recent and fossil resins: electron micrograph studies on tissue preserved in Baltic amber. *J. Baltic Studies* 16: 222–30.

Poinar, G. O., Jr., and C. Ricci. 1992. Bdelloid rotifers in Dominican amber: Evidence for parthenogenetic continuity. *Experientia* (in press).

Poinar, G. O., Jr., and B. Roth. 1991. Terrestrial snails (Gastropoda) in Dominican amber. *The Veliger* 34: 253–58.

Poinar, G. O., Jr., and J. A. Santiago-Blay. 1989. A fossil solpugid, *Happlodontus proterus* new genus, new species (Arachnida: Solpugida) from Dominican amber. *J. New York Entomol. Soc.* 97: 125–32.

Poinar, G. O., Jr., and R. Singer. 1990. Upper Eocene gilled mushroom from the Dominican Republic. *Science* 248: 1099–1101.

Poinar, G. O., Jr., and G. M. Thomas. 1982. An entomophthoralean fungus from Dominican amber. *Mycologia* 74: 332–34.

———. 1984. A fossil entomogenous fungus from Dominican amber. *Experientia* 40: 578–9.

Poinar, G. O., Jr., A. E. Treat, and R. V. Southcott. 1991. Mite parasitism of moths: Examples of paleosymbiosis in Dominican amber. *Experientia* 47: 210–12.

Poinar, G. O., Jr., K. Warheit, and J. Brodzinsky. 1985. A fossil feather in Dominican amber. *IRCS Med. Sci.* 13: 927.

Popov, Y. A. 1978. New species of Aradidae (Hemiptera) from Baltic amber. *Prace Muzeum Ziemi* 29: 137–40.

———. 1987A. A new species of the bug genus *Empicoris* Wolff from Dominican copal, with the redescription of *E. nudus* McAtee and Malloch (Heteroptera: Reduviidae: Emesinae). *Stuttgarter Beitr. Naturk.* (Serie B) 134: 9 pp.

———. 1987B. Synopsis of the neotropical bug genus *Malacopus* Stål, with the description of a new fossil species from Dominican amber (Heteroptera: Reduviidae, Emesinae). *Stuttgarter Beitr. Naturk.* (Serie B) 130: 15 pp.

———. 1989. *Alumeda* n.gen., a new bug genus erected for three fossil species from Dominican amber (Heteroptera: Reduviidae, Emesinae). *Stuttgarter Beitr. Naturk.* (Serie B) 150: 14 pp.

Pregill, G. K. 1981. An appraisal of the vicariance hypothesis of Caribbean biogeography and its application to West Indian terrestrial vertebrates. *Syst. Zool.* 30: 147–55.

Protescu, O. 1937. Étude géologique et paléobiologique de l'ambre Roumain. *Buletinul Societatii Române de Geologie* 3: 1–46.

Quate, L. W. 1961. Fossil Psychodidae (Diptera: Insecta) in Mexican amber; Part I. *J. Paleontol.* 35: 949–51.

———. 1963. Fossil Psychodidae in Mexican amber (Diptera: Insecta); Part II. *J. Paleontol.* 37: 110–18.

Raicevich, I. 1788. *Osservazione storiche, naturali e Politeche intorno la Valachia e Moldavia*. Naples: G. Raimondo. 328 pp.

Rasnitsyn, A. P. 1975. Hymenoptera Apocrita of Mesozoic. *Trudy Paleontol. Inst. Acad. Nauk SSSR* 147: 1–132. Moscow (in Russian).

———. 1977A. A new family of sawflies (Hymenoptera; Tenthredinoidea, Electrotomidae) from Baltic amber. *Zool. Zhurn.* 56: 1304–8 (in Russian).

———. 1977B. New Hymenoptera from the Jurassic and Cretaceous of Asia. *Paleontol. J.* 11: 349–57.

Rawlins, J. E. 1984. Mycophagy in Lepidoptera. In: *Fungus-Insect Relationships*, pp. 382–423. Q. Wheeler and M. Blackwell, eds. New York: Columbia Univ. Press.

Rebel, H. 1934. Bernstein-Lepidoptera. *Palaeobiol.* 6: 1–16.

———. 1935. Bernstein-Lepidopteren. *Deutsch. Entomol. Zeit. "Iris"* 49: 162–86.

Reed, A. H. 1972. *The Gumdiggers*. Wellington, N. Z.: A. H. and A. W. Reed. 193 pp.

Remm, R. H. Y. 1976. Midges (Diptera, Ceratopogonidae) from the Upper Cretaceous amber of the Khatanga depression. *Paleontol. Zhurn.* 3: 107–16 (In Russian).

Rice, P. C. 1980. *Amber, the Golden Gem of the Ages*. New York: Van Nostrand Reinhold. 289 pp.

Richards, W. R. 1966. Systematics of fossil aphids from Canadian amber. *Canadian Entomol.* 98: 746–60.

Riek, E. F. 1976. A new collection of insects from the Upper Triassic of South Africa. *Ann. Natal Mus.* 22: 791–820.

Rieppel, O. 1980. Green anole in Dominican amber. *Nature* 286: 486–7.

Rodendorf, B. B. 1961. Description of the first insect from the Devonian beds of the Timan. *Entomol. Rev.* 40: 260–62.

———. 1964. The Historical development of Diptera. *Trudy Paleontol. Inst. Akad. Nauk SSSR* 100: 1–311 (in Russian). (English-language edition, 1974, Edmonton, Alberta: University of Alberta Press.)

Rodendorf, B. B., and V. V. Zherichin. 1974. Paleontology and the preservation of nature. *Priroda* 5: 82–91 (in Russian).

Rosen, D. E. 1975. A vicariance model of Caribbean biogeography. *Syst. Zool.* 34: 431–64.

Ross, E. S. 1956. A new genus of Embioptera from Baltic Amber. *Mitteilungen Geologischen Staatsinstitut Hamburg* 25: 76–81.

Roth, V. D. 1965. Genera erroneously placed in the spider families Agelenidae and Pisauridae (Araneida: Arachnida). *Ann. Entomol. Soc. Amer.* 58: 284–292.

Rottländer, R. C. A. 1970. On the formation of amber from *Pinus* resin. *Archaeometry* 12: 35–51.

———. 1974. Die Chemie des Bernsteins. *Chemie Unserer Zeit.* 8: 78–83.

———. 1980/81. Neue Beiträge zur Kenntnis des Bernsteins. *Acta Praehist. Archaeol.* 11/12: 21–34.

———. 1984/85. Noch einnmal: neue Beiträge zur Kenntnis des Bernsteins. *Acta Praehist. Archaeol.* 16/17: 223–36.

Rozen, J. G., Jr. 1971. *Micromalthus debilis* LeConte from amber of Chiapas, Mexico (Coleoptera: Micromalthidae). *Univ. Calif. Publ. Entomol.* 63: 75–76.

Sabrosky, C. W. 1963. A new acalypterate fly from the Tertiary amber of Mexico. *J. Paleontol.* 37: 119–20.

Salt, G. 1931. Three bees from Baltic amber. *Bernstein-Forschungen* 2: 136–47.

Sandberger, F. V. 1887. Bemerkungen über einige Heliceen im Bernstein der pruessischen Küste. *Schr. Naturf. Ges. Danzig* N.F. 6: 137–141.

Sanderson, M. W., and T. H. Farr. 1960. Amber with insect and plant inclusions from the Dominican Republic. *Science* 131: 1313.

Santamaria, F. J. 1978. *Diccionario de Mejicanismos*. Editorial Porrua, S.A. Mexico. 724 pp.

Santiago-Blay, J. A., and G. O. Poinar, Jr. 1988. A fossil scorpion *Tityus geratus* new species (Scorpiones: Buthidae) from Dominican amber. *Historical Biology* 1: 345–54.

———. 1992. Millipedes from Dominican amber, with the description of two new species of *Siphonophora* (Diplopoda: Siphonophoridae). *Ann. Entomol. Soc. Amer.* (in press).

Santiago-Blay, J. A., W. Schawaller, and G. O. Poinar, Jr. 1990. A new specimen of *Microtityus ambarensis* (Scorpiones: Buthidae), fossil from Hispaniola: Evidence of taxonomic status and biogeographic implications. *J. Arachnol.* 18: 115–7.

Saunders, W. B., R. H. Mapes, F. M. Carpenter, and W. C. Elsik. 1974. Fossiliferous amber from the Eocene (Claiborne) of the Gulf Coast plain. *Geol. Soc. Amer. Bull.* 85: 979–84.

Savkevich, S. S. 1974. State of investigation and prospects for amber in USSR. *Intern. Geol. Rev.* 17: 919–24.

Savkevich, S. S., and T. N. Popkova. 1978. Donnes nouvelles dans l'etude mineralogique de resines fossiles de France. *Bulletin de Mineralogie* 101: 442–7.

Savkevich, S. S., and I. A. Shakhs. 1964A. Infrared absorption spectra of Baltic amber. *J. Applied Chem. USSR* 37: 1117–9.

———. 1964B. Infrared absorption spectra of Baltic amber. *J. Applied Chem. USSR* 37: 2717–9.

Sawkiewicz, S. S., and I. Arua. 1988. Amekit—a new fossil resin from Nigeria. *Abstracts of the Sixth International Conference on Amber and Amber-Bearing Sediments* (Oct. 20–21, 1988). Warsaw: Polish Acad. Sci., p. 27.

Schawaller, W. 1978. Neue Pseudoskorpione aus dem Baltischen Bernstein der Stuttgarter Bernsteinsammlung (Arachnida: Pseudoscorpionidea). *Stuttgarter Beitr. Naturk.* (Serie B) 42: 22 pp.

———. 1979A. Erstnachweis der Ordnung Geisselspinnen in Dominikanischem Bernstein (Stuttgarter Bernsteinsammlung: Arachnida, Amblypygi). *Stuttgarter Beitr. Naturk.* (Serie B) 50: 12 pp.

———. 1979B. Erstnachweis eines Skorpions in Dominikanischem Bernstein (Stuttgarter Bernsteinsammlung: Arachnida, Scorpionida). *Stuttgarter Beitr. Naturk.* (Serie B) 45: 15 pp.

———. 1980A. Erstnachweis tertiärer Pseudoskorpione (Chernetidae) in Dominikanischem Bernstein (Stuttgarter Bernsteinsammlung: Arachnida, Pseudoscorpionidea). *Stuttgarter Beitr. Naturk.* (Serie B) 57: 20 pp.

———. 1980B. Fossile Chthoniidae in Dominikanischem Bernstein, mit phylogenetischen Anmerkungen (Stuttgarter Bernsteinsammlung: Arachnida, Pseudoscorpionidea). *Stuttgarter Beitr. Naturk.* (Serie B) 63: 19 pp.

———. 1981A. Cheiridiidae in Dominikanischem Bernstein, mit Anmerkungen zur morphologischen Variabilität (Stuttgarter Bernsteinsammlung: Arachnida, Pseudoscorpionidea). *Stuttgarter Beitr. Naturk.* (Serie B) 75: 14 pp.

———. 1981B. Pseudoskorpione (Cheliferidae) phoretisch auf Käfern (Platypodidae) in Dominikanischem Bernstein (Stuttgarter Bernsteinsammlung: Pseudoscorpionidea und Coleoptera). *Stuttgarter Beitr. Naturk.* (Serie B) 71: 17 pp.

———. 1981C. Übersicht über Spinnen-Familien im Dominikanischem Bernstein und anderen tertiaren Harzen (Stuttgarter Bernsteinsammlung: Arachnida, Araneae). *Stuttgarter Beitr. Naturk.* (Serie B) 77: 10 pp.

———. 1981D. Die Spinnenfamilie Hersiliidae in Dominikanischem Bernstein (Stuttgarter Bernsteinsammlung: Arachnida, Araneae). *Stuttgarter Beitr. Naturk.* (Serie B) 79: 10 pp.

———. 1982A. Zwei weitere Skorpione in Dominikanischem Bernstein (Stuttgarter Bernsteinsammlung: Arachnida, Scorpionida). *Stuttgarter Beitr. Naturk.* (Serie B) 82: 14 pp.

———. 1982B. Der erste Pseudoskorpion (Chernetidae) aus Mexikanischem Bernstein (Stuttgarter Bernsteinsammlung: Arachnida, Pseudoscorpionidea). *Stuttgarter Beitr. Naturk.* (Serie B) 85: 9 pp.

———. 1982C. Neue Befunde an Geisselspinnen in Dominikanischem Bernstein (Stuttgarter Bernsteinsammlung: Arachnida, Amblypygi). *Stuttgarter Beitr. Naturk.* (Serie B) 86: 12 pp.

———. 1982D. Spinnen der Familien Tetragnathidae, Uloboridae und Dipluridae in Dominikanischem Bernstein und allgemeine Gesichtspunkte (Arachnida, Araneae). *Stuttgarter Beitr. Naturk.* (Serie B) 89: 10 pp.

———. 1984. Die Familie Selenopidae in Dominikanischem Bernstein (Arachnida, Araneae). *Stuttgarter Beitr. Naturk.* (Serie B) 103: 8 pp.

Schedl, K. E. 1962. New Platypodidae from Mexican amber. *J. Paleontol.* 36: 1035–8.

Schimper. 1870–72. *Traité de paléontologie végétale*, Vol. 2, p. 377. Paris, J. B. Baillière et fils.

Schlee, D. 1970. Verwandtschaftsforschung an fossilen und rezenten Aleyrodina (Insecta, Hemiptera). *Stuttgarter Beitr. Naturk.* (Serie B) 213: 72 pp.

———. 1972. Bernstein aus dem Lebanon. *Kosmos* (Stuttgart) 68: 460–3.

———. 1973. Harzkonservierte Vogelfedern aus der untersten Kreide. *J. Ornithol.* 114: 207–19.

———. 1980. *Bernstein-Raritäten*. Stuttgart: Staatliches Museum für Naturkunde. 88 pp.

———. 1984. *Bernstein-Neuigkeiten*. Stuttgart Beitr. Naturk. (Serie C) 18: 100 pp.

———. 1986. *Der Bernsteinwald*. Stuttgart: Staatliches Museum für Naturkunde. 15 pp.

———. 1990. *Das Bernstein-Kabinett*. Stuttgart Beitr. Naturk. (Serie C) 28: 100 pp.

Schlee, D., and H.-G. Dietrich. 1970. Insektenführender Bernstein aus der Unterkreide des Lebanon. *Neues Jahrb. Geol. Paläont. Monatshefte* (Stuttgart) 1: 40–50.

Schlee, D., and W. Glöckner. 1978. *Bernstein.* Stuttgarter Beitr. Naturk. (Serie C) 8: 72 pp.

Schlüter, T. 1976. Die Wollschweber-Gattung *Glabellula* (Diptera: Bombyliidae) aus dem oligozänen Harz der Dominikanischen Republik. *Entomol. Germanica* 2: 355–63.

———. 1978A. Zur Systematik und Paläokologie harzkonservierter Arthropoda einer Taphozönose aus dem Cenomanium von NW-Frankreich. *Berliner Geowiss. Abh. A* 9: 150 pp.

———. 1978B. Die Schmetterlingsmücken-Gattung *Nemopalpus* (Diptera: Psychodidae) aus dem oligozänen Harz der Dominikanischen Republik. *Entomol. Germanica* 4: 242–9.

Schlüter, T., and W. Stürmer. 1984. Die Identifikation einer fossilen Rhachiberothinae-Art (Planipennia: Berothidae oder Mantispidae) aus mittelkretazischem Bernstein NW-Frankreichs mit Hilfe röntgenographischer Methoden. In: *Progress in World's Neuropterology*, pp. 49–55. J. Gepp, H. Aspöck, and H. Hölzel, eds. Graz, Austria.

Schmalfuss, H. 1979. *Proceroplatus hennigi* n. sp., die erste Pilmücke aus dem Dominikanischen Bernstein (Stuttgarter Bernsteinsammlung: Diptera, Mycetophiloidea, Keroplatidae). *Stuttgarter Beitr. Naturk.* (Serie B) 49: 9 pp.

———. 1980. Die ersten Landasseln aus Dominikanischem Bernstein mit einer systematisch-phylogenetischen Revision der Familie Sphaeroniscidae (Stuttgarter Bernsteinsammlung: Crustacea: Isopoda, Oniscoidea). *Stuttgarter Beitr. Naturk.* (Serie B) 61: 12 pp.

———. 1984. Two new species of the terrestrial isopod genus *Pseudarmadillo* from Dominican amber (Amber-collection Stuttgart: Crustacea, Isopoda, Pseudarmadillidae). *Stuttgarter Beitr. Naturk.* (Serie B) 102: 14 pp.

Schram, F. R. 1970. Isopod from the Pennsylvanian of Illinois. *Science* 169: 854–5.

Schubert, K. 1953. Mikroskopische Untersuchung Pflanzlicher Einschlüsse des Bernsteins. *Palaeontographica* 93: 103–19.

———. 1961. Neue Untersuchungen über Bau und Leiben der Bernsteinkiefern [*Pinus succinifera* (Conw.) emend]. *Beihefte zum Geologischen Jahrbuch* 45: 149 pp.

Schumann, H. 1984. Erstnachweis einer Raubfliege aus dem Sächsischen Bernstein. *Deutsche entomol. Zeitsch.* 31: 217–23.

Schumann, H., and H. Wendt. 1989. Zur Kenntnis der tierischen Inklusen des Sächsischen Bernsteins. *Deutsche entomol. Zeitsch.* 36: 33–44.

Schweigger, A. F. 1819. *Beobachtungen auf naturhistorischen Reisen, Anhang: Bemerkungen über den Bernstein.* Berlin: Reimer. 127 pp.

Seevers, C. H. 1971. Fossil Staphylinidae in Tertiary Mexican amber (Coleoptera). *Univ. Calif. Publ. Entomol.* 63: 77–86.

Sellnick, M. 1931. Milben im Bernstein. *Bernstein-Forschungen* 2: 148–80.

Selvig, W. A. 1945. Resins in coal. United States Department of the Interior Technical Paper 680: 24 pp.

Seward, A. C. 1931. *Plant Life through the Ages.* New York: The MacMillan Company. 601 pp.

Shear, W. A. 1981. Two fossil millipedes from the Dominican amber (Diplopoda: Chytodesmidae, Siphonophoridae). *Myriapodologica* 1: 51–54.

———. 1987. Myriapod fossils from the Dominican amber. *Myriapodologica* 7: 43.

Shear, W. A., P. M. Bonamo, J. D. Grierson, W. D. I. Rolfe, E. L. Smith, and R. A. Norton. 1984. Early land animals in North America: evidence from Devonian age arthropods from Gilboa, New York. *Science* 224: 492–4.

Shellford, R. 1910. On a collection of Blattidae preserved in amber from Prussia. *J. Linn. Soc. London* 30: 336–55.

Silvestri, F. 1913. Die Thysanuren des baltischen Bernsteins. *Schr. phys.-ökon. Ges. Königsberg* 53: 42–66.

Simoneit, B. R. T., J. O. Grimalt, T. G. Wang, R. E. Cox, P. G. Hatcher, and A. Nissenbaum. 1986. Cyclic terpenoids of contemporary resinous plant detritus and of fossil woods, amber and coals. *Org. Geochem.* 10: 877–89.

Skalski, A. W. 1975. Notes on present status of botanical and zoological studies of ambers. In: *Studi e Ricerche sulla Problematica dell'ambra*, Vol. 1, pp. 153–175. Rome.

———. 1976. Les lepidopteres fossiles de l'ambre etat actuel de nos connaissances. *Linneana Belgica* (Rev. Belge d'Entomol.) 6: 154–69; 195–208; 221–33.

———. 1977. Studies on the Lepidoptera from fossil resins. *Prace Muzeum Ziemi* 26: 22 pp.

———. 1985. Butterflies (Lepidoptera) in Baltic amber. *Wiad. Entomol.* 6: 207–10 (in Polish).

Skalski, A. W., and A. Veggiani. 1988. Fossil resins in Sicily and northern Apennines; geology and organic content. *Abstracts of the Sixth International Conference on Amber and Amber-Bearing Sediments* (October 20–21, 1988). Warsaw: Polish Acad. Sci., p. 29b.

Smith, J. 1896. On the discovery of fossil microscopic plants in the fossil amber of the Ayrshire coal-field. *Trans. Geol. Soc. Glasgow* 10: 318–22.

Snyder, T. E. 1960. Fossil termites from Tertiary amber of Chiapas, Mexico. *J. Paleontol.* 34: 493–4.

Soom, M. 1984. Bernstein vom Nordrand der Schweizer Alpen. In: D. Schlee, ed., *Bernstein-Neuigkeiten*. Stuttgarter Beitr. Naturk. (Serie C) 18: 15–20.

Spahr, U. 1981A. Systematischer Katalog der Bernstein—und Kopal—Käfer (Coleoptera). *Stuttgarter Beitr. Naturk.* (Serie B) 80: 107 pp.

———. 1981B. Bibliographie der Bernstein—und Kopal—Käfer (Coleoptera). *Stuttgarter Beitr. Naturk.* (Serie B) 72: 21 pp.

———. 1985. Ergänzungen und Berichtigungen zu R. Keilbachs Bibliographie und Liste der Bernsteinfossilien—Ordnung Diptera. *Stuttgarter Beitr. Naturk.* (Serie B) 111: 146 pp.

———. 1987. Ergänzungen und Berichtigungen zu R. Keilbachs Bibliographie und Liste der Bernsteinfossilien—Ordnung Hymenoptera. *Stuttgarter Beitr. Naturk.* (Serie B) 127: 121 pp.

———. 1988. Ergänzungen und Berichtigungen zu R. Keilbachs Bibliographie und Liste der Bernsteinfossilien—Überordnung Hemipteroidea. *Stuttgarter Beitr. Naturk.* (Serie B) 144: 60 pp.

———. 1989. Ergänzungen und Berichtigungen zu R. Keilbachs Bibliographie und Liste der Bernsteinfossilien—Überordnung Mecopteroidea. *Stuttgarter Beitr. Naturk.* (Serie B) 157: 87 pp.

———. 1990. Ergänzungen und Berichtigungen zu R. Keilbachs Bibliographie und Liste der Bernsteinfossilien—Apterygota. *Stuttgarter Beitr. Naturk.* (Serie B) 166: 23 pp.

Spilman, T. J. 1971. Fossil *Stichtoptychus* and *Cryptorama* in Mexican amber (Coleoptera: Anobiidae). *Univ. Calif. Publ. Entomol.* 63: 87–89.

Stach, E., G. H. Taylor, M. Th. Mackowsky, M. Teichmüller, R. Teichmüller, and D. Chandra. 1975. *Coal Petrology.* Berlin: Gebrüder Borntraeger. 485 pp.

Stark, B. P., and D. L. Lentz. 1992. *Dominiperla antiqua*, the first stonefly from Dominican amber (Plecoptera: Perlidae). *J. Kansas Entomol. Soc.* 65: 93–96.

Steiner, W. E., Jr. 1984. A review of the biology of phalacrid beetles (Coleoptera). In: *Fungus-Insect Relationships*, pp. 424–45. Q. Wheeler and M. Blackwell, eds. New York: Columbia University Press.

Stewart, W. N. 1983. *Paleobotany and the evolution of plants.* Cambridge, England: Cambridge Univ. Press. 405 pp.

Strassen, R. zur. 1973. Fossile Fransenflügler aus mesozoischem Bernstein des Lebanon. *Stuttgart Beitr. Naturk.* (Serie A) 267: 1–51.

Stuart, M. 1923. Dating the amber-bearing beds of Burma. *Records of the Geological Survey of India* 54: 1–12.

Stubblefield, S. P., C. E. Miller, T. N. Taylor, and G. T. Cole. 1985. *Geotrichites glaesarius*, a conidial fungus from Tertiary Dominican amber. *Mycologia* 77: 11–16.

Stuckenberg, B. R. 1974. A new genus and two new species of Athericidae (Diptera) in Baltic amber. *Ann. Natal Mus.* 22: 275–88.

———. 1975. New fossil species of *Phlebotomus* and *Haematopota* in Baltic amber (Diptera: Psychodidae, Tabanidae). *Ann. Natal Mus.* 22: 455–64.

Sturtevant, A. H. 1963. A fossil periscelid (Diptera) from the amber of Chiapas, Mexico. *J. Paleontol.* 37: 121–2.

Stys, P. 1969. Revision of fossil and pseudofossil Enicocephalidae (Heteroptera) *Acta entomol. bohemoslov.* 66: 352–65.

Sussman, A. S. 1966. *Spores, Their Dormancy and Germination.* New York: Harper & Row. 354 pp.

Szabó, J., and J. Oehkle. 1986. Neue Proctotrupoidea aus dem Baltischen Bernstein. *Beitr. Entomol.* 36: 99–106.

Szadziewski, R. 1985. Biting midges of the genus *Eohelea* Petrunkevitch (Insecta, Diptera, Ceratopogonidae) from Baltic amber (in the collection of the Museum of the Earth). *Prace Muzeum Ziemi* 37: 123–30.

———. 1988. Biting midges (Diptera, Ceratopogonidae) from Baltic amber. *Polskie Pismo Entomol.* 57: 283 pp.

Teskey, H. J. 1971. A new soldier fly from Canadian amber (Diptera: Stratiomyidae). *Canadian Entomol.* 103: 1659–61.

Thomas, B. R. 1969. Kauri resins—modern and fossil. In: *Organic Geochemistry*, pp. 599–618. G. Eglinton and M. Murphy, eds. Berlin: Springer-Verlag.

———. 1970. Modern and fossil plant resins. In: *Phytochemical Phylogeny*, pp. 59–79. J. B. Harborne, ed. London: Academic Press.

Thomas, D. B., Jr. 1988. Fossil Cydnidae (Heteroptera) from the Oligo-Miocene amber of Chiapas, Mexico. *J. New York Entomol. Soc.* 96: 26–29.

Thomas, G. M., and G. O. Poinar, Jr. 1988. A fossil *Aspergillus* from Eocene Dominican amber. *J. Paleontol.* 62: 141–43.

Ting, W. S., and A. Nissenbaum. 1986. *Fungi in Lower Cretaceous Amber from Israel.* Special publication, Exploration and Development Research Center, Chinese Petroleum Corporation, Miaoli, Taiwan. 27 pp.

Townes, H. 1973. Three tryphonine ichneumonids from Cretaceous amber (Hymenoptera). *Proc. Entomol. Soc. Washington* 75: 282–7.

Troost, G. 1821. Description of a variety of amber, and of a fossil substance supposed to be the nest of an insect discovered at Cape Sable, Magothy River, Ann-Arundel County, Maryland. *Amer. J. Sci.* 3: 8–15.

Tshernova, O. A. 1971. A mayfly (Ephemeroptera, Leptophlebiidae) from fossil resin of Cretaceous deposits in the polar regions of Siberia. *Entomol. Obozr.* 50: 612–8 (in Russian).

Türk, E. 1963. A new tyroglyphid deuteronymph in amber from Chiapas, Mexico. *Univ. Calif. Publ. Entomol.* 31: 49–51.

Tyrrell, J. B. 1893. Summary report on the operations of the Geological Survey, for the year 1890. *Ann. Rept. Geol. Surv. Canada* (New series) 5: 30A–31A.

Ulmer, G. 1912. Die Trichopteren des baltischen Bernsteins. *Beitr. Naturk. Preuss.* 10: 1–380.

Urbanski, T., T. Glinka, and E. Wesolowska. 1976. On the chemical composition of Baltic amber. *Bull. Acad. Polon. Sci., Ser. Sci. Chem.* 24: 625–9.

Usinger, R. L. 1941. Two new species of Aradidae from Baltic amber (Hemiptera). *Psyche* 48: 95–100.

———. 1942. An annectent genus of Cimicoidea from Baltic amber (Hemiptera). *Psyche* 49: 41–46.

———. 1972. Autobiography of an entomologist. *Mem. Pacific Coast Entomol. Soc.* 4: 153–65.

Usinger, R. L., and R. F. Smith. 1957. Arctic amber. *Pacific Discovery* 10: 15–19.

Vaillant, L. 1873. Sur un geckotien de l'ambre jaune. *Bull. Soc. Philom.* (Paris) 10: 65–67.

Vávra, N. 1984. "Reich an armen Fundstellen": Übersicht über die fossilen Harze Osterreichs. In: D. Schlee, ed., *Bernstein-Neuigkeiten.* Stuttgarter Beitr. Naturk. (Serie C) 18: 9–14.

Vercammen-Grandjian, P. H. 1972. Of techniques and ortho-iconography. *Folia Parasitologica* (Praha) 19: 289–304.

Vishniakova, V. N. 1975. Psocoptera in Late Cretaceous insect bearing resins from the Taimyr. *Entomol. Rev. Washington* 54: 63–75.

Voigt, E. 1952. Ein Haareinschluss mit Phthirapteren-Eiern im Bernstein. *Mitteilungen Geol. Staatsinst.* 21: 59–74.

Walker, T. L. 1934. Chemawinite or Canadian amber. *Univ. Toronto Studies. Geological Series.* 36: 5–10.

Wasmann, E. 1929. Die Paussiden des baltischen Bernsteins und die Stammesgeschichte der Paussiden. *Bernstein-Forschungen* 1: 1–138.

———. 1933. Eine ameisenmordende Gastwanze (*Proptilocerus dolosus* n.g. n.sp.) im Baltischen Bernstein. *Bernstein-Forschungen* 3: 1–3.

Weidner, H. 1958. Einige interessante Insektenlarven aus der Bernsteininklusen-Sammlung des Geologischen Staatsinstituts Hamburg (Odonata, Coleoptera, Megaloptera, Planipinnia). *Mitteilungen Geol. Staatinst. Hamburg* 27: 50–68.

———. 1964. Eine Zecke, *Ixodes succineus*, sp.n., im baltischen Bernstein. *Veröff. Uberseemus. Bremen* 3: 143–51.

Weitschat, W. 1980. *Leben in Bernstein*. Geol.-Palaö. Inst. Univ. Hamburg. 48 pp.

Wells, A., and W. Wichard. 1989. Caddisflies of Dominican amber. VI. Hydroptilidae (Trichoptera). *Studies on Neotropical Fauna and Environment* 24: 41–51.

Wenzel, R. L. 1953. Research aided by amber collector. *Chicago Nat. Hist. Mus. Bull.* 24: 6–7.

Wert, C. A., and M. Weller. 1988. The polymeric nature of amber. *Bull. Amer. Phys. Soc.* 33: 497.

Whalley, P. 1977. Lower Cretaceous Lepidoptera. *Nature* 266: 526.

———. 1978. New taxa of fossil and recent Micropterigidae with a discussion of their evolution and a comment on the evolution of Lepidoptera (Insecta). *Ann. Transvaal Mus.* 31: 65–81.

———. 1980. Neuroptera (Insecta) in amber from the Lower Cretaceous of Lebanon. *Bull. British Mus. Nat. Hist. (Geol.)* 33: 157–64.

———. 1981. *Insects from Lebanese amber*. Unpublished report, British Mus. Nat. Hist. (Geol.). 11 pp.

———. 1983. *Fera venatrix* gen. and sp.n. (Neuroptera, Mantispidae) from amber in Britain. *Neuroptera Intern.* 2: 229–33.

Wheeler, M. R. 1963. A note on some fossil Drosophilidae (Diptera) from the amber of Chiapas, Mexico. *J. Paleontol.* 37: 123–4.

Wheeler, W. M. 1915. The ants of the Baltic amber. *Schr. phys.-ökon., Ges. Königsberg* 55: 1–142.

Wichard, W. 1981. Köcherfliegen des Dominikanischen Bernsteins. I. *Ochrotrichia doehleri* sp.nov. (Trichoptera, Hydroptilidae). *Mitteilungen Münch. Entomol. Ges.* 71: 161–2.

———. 1983A. Köcherfliegen des Dominikanischen Bernsteins. II. Fossile Arten der Gattung *Chimarra* (Trichoptera, Philopotamidae). *Mitteilungen Münch. Entomol. Ges.* 72: 137–45.

———. 1983B. Köcherfliegen des Dominikanischen Bernsteins. III. *Chimarra succini* n.sp. (Stuttgarter Bernsteinsammlung: Trichoptera, Philopotamidae). *Stuttgarter Beitr. Naturk.* (Serie B) 95: 1–8.

———. 1985. Köcherfliegen des Dominikanischen Bernsteins. IV. *Antillopsyche oliveri* spec. nov. (Trichoptera, Polycentropodidae). *Studies on Neotropical Fauna and Environment* 20: 117–24.

———. 1986A. Köcherfliegen des Baltischen Bernsteins I. *Marilla altrocki* sp.n. (Trichoptera, Odontoceridae) der Bernsteinsammlung Bachofen-Echt. *Mitteilungen Boyer Staatsslg. Paläont. Hist. Geol.* 26: 33–40.

———. 1986B. Köcherfliegen des Dominikanischen Bernsteins. V. *Palaehydropsyche fossilis* gen.n. sp.n. (Trichoptera, Hydropsychidae). *Studies on Neotropical Fauna and Environment* 21: 189–95.

———. 1987. Caribbean amber caddisflies—biogeographical aspects. In: *Proceed-*

ings of the Fifth International Symposium on Trichoptera (Lyon, France), pp. 67–69. M. Bournaud and H. Tachet, eds. Bordrecht: Junk Publishers.

———. 1989. Köcherfliegen des Dominikanischen Bernsteins. VII. Fossile Arten der Gattung *Cubanoptila* Sykora, 1973. *Mitteilungen Münch. Entomol. Ges.* 79: 91–100.

Wichterman, R. 1953. *The biology of Paramecium.* New York: The Blakiston Company. 527 pp. (see p. 94).

Wille, A. 1959. A new fossil stingless bee (Meliponini) from the amber of Chiapas, Mexico. *J. Paleontol.* 33: 849–52.

———. 1977. A general review of the fossil stingless bees. *Revista Biologia Tropical* (Costa Rica) 25: 43–46.

Wille, A., and L. C. Chandler. 1964. A new stingless bee from the Tertiary amber of the Dominican Republic. *Revista Biologia Tropical* 12: 187–95.

Williamson, G. C. 1932. *The Book of Amber.* London: Ernest Benn. 268 pp.

Wilson, E. O. 1985A. Ants of the Dominican amber (Hymenoptera: Formicidae). 1. Two new myrmicine genera and an aberrant *Pheidole. Psyche* 92: 1–9.

———. 1985B. Ants of the Dominican amber (Hymenoptera: Formicidae). 2. The first fossil army ants. *Psyche* 92: 11–16.

———. 1985C. Ants of the Dominican amber (Hymenoptera: Formicidae). 3. The subfamily Dolichoderinae. *Psyche* 92: 17–37.

———. 1985D. Ants from the Cretaceous and Eocene amber of North America. *Psyche* 92: 205–16.

———. 1985E. Invasion and extinction in the West Indian ant fauna: evidence from the Dominican amber. *Science* 229: 265–7.

Wilson, E. O., F. M. Carpenter, and W. L. Brown. 1967. The first Mesozoic ants, with the description of a new subfamily. *Psyche* 74: 1–19.

Wilson, G. S., and H. L. Shipp. 1938. Part 4—Bacteriological investigations. Chemistry and Industry. *Chem. Ind.* 57: 834–6.

Wittmer, W. 1963. A new Cantharid from the Chiapas amber of Mexico. *Univ. Calif. Publ. Entomol.* 31: 53.

Wolfe, J. A. 1971. Tertiary climatic fluctuations and methods of analysis of Tertiary floras. *Palaeogeogr. Palaeoclimatol. Palaeoecol.* 9: 27–57.

Womersley, H. 1957. A fossil mite (*Acronothrus ramus* n.sp) from Cainozoic resin at Allendale, Victoria. *Proc. R. Soc. Victoria* 69: 21–23.

Woodley, N. E. 1986. Parhadrestiinae, a new subfamily for *Parhadrestia* James and *Cretaceogaster* Teskey (Diptera: Stratiomyidae). *Syst. Entomol.* 11: 377–87.

Woolley, T. A. 1971. Fossil oribatid mites in amber from Chiapas, Mexico (Acarina: Oribatei = Cryptostigmata). *Univ. Calif. Publ. Entomol.* 63: 91–99.

Wunderlich, J. 1982. Sex in Bernstein: Ein fossiles Spinnenpaar. *Neue Entomol. Nachr.* 2: 9–112.

———. 1986. *Spinnenfauna gestern und heute. Fossil Spinnen in Bernstein und ihre heute lebenden Verwandten.* Wiesbaden: Erich Bauer Verlag. 283 pp.

———. 1988. Die fossilen Spinnen im Dominikanischen Bernstein. *Beitr. Araneologie* 2: 378 pp.

———. 1988. Die fossilen Spinnen (Araneae) im Baltischen Bernstein. *Beitr. Araneologie* 3: 280 pp.

Wygodzinsky, P. 1959. A new hemipteran (Dipsocoridae) from the Miocene amber of Chiapas, Mexico. *J. Paleontol.* 33: 853–4.

———. 1966. A monograph of the Emesinae (Reduviidae, Hemiptera). *Bull. Amer. Mus. Nat. Hist.* 133: 614 pp.

———. 1971. A note on a fossil machilid (Microcoryphia) from the amber of Chiapas, Mexico. *Univ. Calif. Publ. Entomol.* 63: 101–2.

Yablokov-Khnzorian, S. M. 1961. Circaeidae—a new family of Coleoptera from amber. *Dokl. Akad. Nauk. SSSR* 136: 209–10 (in Russian).

———. 1962. Some *Stenoxia* (Coleoptera) from Baltic amber. Family Throscidae Bach, 1849. *Paleontol. Zhurn.* 1962/3: 81–89 (in Russian).

Yoshimoto, C. M. 1975. Cretaceous chalcicoid fossils from Canadian amber. *Canadian Entomol.* 107: 499–529.

Zaddach, G. 1864. Eine Amphipode im Bernstein. *Schr. phys.-ökon., Ges. Königsberg* 5: 1–12.

Zaitzev, V. F. 1981. Vergleich der tertiären (Oligozän und Miozän) und rezenten Bombyliidenfauna Europas (Diptera, Bombyliidae). *Acta Entomol. Jugoslavica* 17: 103–6.

———. 1987. New species of Cretaceous fossil bee flies and a review of paleontological data on the Bombyliidae (Diptera). *Entomol. Review* 66: 150–60.

Zherichin, V. V. 1978. Various ways of examining past faunistic complexes. *Trudy Paleontol. Inst. Akad. Nauk SSSR (Moscow)* 165: 1–198 (in Russian).

Zherichin, V. V., and I. D. Sukacheva. 1973. On Cretaceous insect bearing ambers (retinites) of northern Siberia. *Reports of the 24th Annual Readings in Memory of N. A. Kholodkovsky* (Leningrad), pp. 3–48 (In Russian).

Zimmerman, E. C. 1948. *Insects of Hawaii.* Vol. 1. *Introduction.* Honolulu: University of Hawaii Press. 206 pp.

———. 1971. Mexican Miocene amber weevils (Insecta: Coleoptera: Curculionidae). *Univ. Calif. Publ. Entomol.* 63: 103–6.

Zoppis, R. 1949. *Informe sobre los Yacimientos del ambar y lignito de Santiago.* Archivo Centro de Documentacion, Santo Domingo (unpublished report). 7 pp.

Index

In this index an "f" after a number indicates a separate reference on the next page, and an "ff" indicates separate references on the next two pages. A continuous discussion over two or more pages is indicated by a span of page numbers, e.g., "57–59." *Passim* is used for a cluster of references in close but not consecutive sequence.

Index

Library of Congress Cataloging-in-Publication Data

Poinar, George O.
    Life in amber / George O. Poinar, Jr.
       p. cm.
    Includes bibliographical references and index.
    ISBN 0−8047−2001−0
    1. Fossils. 2. Amber. I. Title.
QE742.P65 1992
560—dc20   91−5045
                                         CIP

  ∞ This book is printed on acid-free paper

Printed in the USA
CPSIA information can be obtained
at www.ICGtesting.com
JSHW011510221024
72173JS00014B/1641/J